PERGAMON INTERNATIONAL LIBRARY
of Science, Technology, Engineering and Soci[al Studies]
The 1000-volume original paperback library in aid o[f]
industrial training and the enjoyment of leisure
Publisher: Robert Maxwell, M.C.

NEW TERTIARY MATHEMATICS

Volume 2 Part 2
Further Applied Mathematics

MATHS DEPARTMENT
K. G. V.

THE PERGAMON TEXTBOOK
INSPECTION COPY SERVICE

An inspection copy of any book published in the Pergamon International Library will gladly be sent to academic staff without obligation for their consideration for course adoption or recommendation. Copies may be retained for a period of 60 days from receipt and returned if not suitable. When a particular title is adopted or recommended for adoption for class use and the recommendation results in a sale of 12 or more copies, the inspection copy may be retained with our compliments. The Publishers will be pleased to receive suggestions for revised editions and new titles to be published in this important International Library.

Some other Pergamon titles of interest

C. PLUMPTON & W. A. TOMKYS
Sixth Form Pure Mathematics
Volumes 1 & 2

Theoretical Mechanics for Sixth Forms
2nd SI Edition Volumes 1 & 2

D. T. E. MARJORAM
Exercises in Modern Mathematics
Further Exercises in Modern Mathematics
Modern Mathematics in Secondary Schools

D. G. H. B. LLOYD
Modern Syllabus Algebra

E. Œ. WOLSTENHOLME
Elementary Vectors
3rd SI Edition

NEW TERTIARY MATHEMATICS

C. Plumpton
Queen Mary College, London

P. S. W. MacIlwaine

Volume 2 Part 2
Further Applied Mathematics

PERGAMON PRESS
Oxford · New York · Toronto · Sydney · Paris · Frankfurt

UK	Pergamon Press Ltd., Headington Hill Hall, Oxford OX3 0BW, England
USA	Pergamon Press Inc., Maxwell House, Fairview Park, Elmsford, New York 10523, U.S.A.
CANADA	Pergamon of Canada, Suite 104, 150 Consumers Road, Willowdale, Ontario, M2J 1P9, Canada
AUSTRALIA	Pergamon Press (Aust.) Pty. Ltd., P.O. Box 544, Potts Point, N.S.W. 2011, Australia
FRANCE	Pergamon Press SARL, 24 rue des Ecoles, 75240 Paris, Cedex 05, France
FEDERAL REPUBLIC OF GERMANY	Pergamon Press GmbH, 6242 Kronberg-Taunus, Hammerweg 6, Federal Republic of Germany

Copyright © 1981 C. Plumpton and P. S. W. MacIlwaine

All Rights Reserved. No part of this publication may be reproduced, stored in a retrieval system or transmitted in any form or by any means: electronic, electrostatic, magnetic tape, mechanical, photocopying, recording or otherwise, without permission in writing from the publishers

First edition 1981

British Library Cataloguing in Publication Data

Plumpton, Charles
New tertiary mathematics. – (Pergamon international library).
Vol. 2, Part 2: Further Applied mathematics
1. Mathematics – 1961 –
I. Title   II. MacIlwaine, Patrick Sydney Wilson
510   QA36   80–49934

ISBN 0–08–025037–8 Hardcover
ISBN 0–08–025026–2 Flexicover
ISBN 0–08–025036–X Flexicover (non-net)
ISBN 0–08–021646–3 (4 volume set, hard only)

Printed in Great Britain by A. Wheaton & Co. Ltd. Exeter

Preface

In this volume, the basic "Applied Mathematics" of Volume 1 Part 2 is extended and developed to cover almost all the topics of Theoretical Mechanics and Probability included in Advanced Level Applied Mathematics and Further Mathematics syllabuses of the seven Examining Boards. The aim of the book is to provide an adequate course and source book for mathematical pupils in Sixth Forms of Schools, Sixth Form Colleges, Colleges of Further Education and Technical Colleges, and students engaged in private study.

Chapters 7–10 on Theoretical Mechanics are strongly "vector-based", wherever this is appropriate. Chapter 7 extends the statics of Volume 1 to the analysis of systems of forces in three dimensions, considers more complicated equilibrium problems, and deals with virtual work, stability and the catenary. In Chapters 8 and 9, the dynamics of a particle in one and two dimensions is discussed. In particular, the implications of the Galilean transformation and the general theorems of motion for a system of particles are examined. Chapter 10 deals with the applications of these theorems to simple cases of the motion of a rigid body. Chapter 11 extends the work on Probability of Volume 1 to the Normal and Poisson distributions, Markov chains and miscellaneous problems.

In the belief that worked examples are of great value, many examples of varying difficulty are worked in the text and exercises are added after each major topic is covered. The miscellaneous exercises at the end of each chapter are intended to help revision of earlier work.

The sections, exercises, equations and figures are numbered to correspond to the chapters, e.g. §7:4 is the fourth section of Chapter 7; Ex. 9.5 is the exercise at the end of §9:5; equation (10.7) is the seventh (numbered) equation of Chapter 10. Only those equations to which subsequent reference is made are numbered. Certain sections are starred, not so much to imply excessive difficulty, as to indicate that they can be omitted without affecting the logical sequence of the particular chapter.

We are grateful to the following G.C.E. examination boards for permission to reproduce questions for their papers:

The University of London (L.);
The Northern Universities Joint Matriculation Board (N.);
The Oxford and Cambridge Schools Examination Board,
The Schools Mathematics Project, Mathematics in Education and
    Industry (O.C.) (O.C.S.M.P.) (O.C.M.E.I.);
The Cambridge Local Examinations Syndicate (C.);
The Oxford Delegacy (O.);
The Associated Examination Board (A.E.B.).

The questions are theirs, the answers are ours!

C. Plumpton
P. S. W. MacIlwaine

Contents

Glossary of Symbols and Abbreviations   x

Chapter 7   Further Statics   231

7:1	Forces in three dimensions	231
	Exercise 7.1	234
7:2	Conditions of equilibrium for a rigid body acted upon by a system of coplanar forces	236
	Exercise 7.2	238
7:3	Hinged bodies—equilibrium problems involving more than one body	239
	1. Smooth hinges	239
	2. Rough hinges	240
	3. The action of one part of a body on another	241
	4. Problems of equilibrium involving more than one body	241
	5. The equilibrium of the hinge	244
	Exercise 7.3	247
*7:4	The principle of virtual work	248
	The equilibrium of a particle	248
	Applications of the principle of virtual work	249
	*Exercise 7.4	251
7:5	Determination of equilibrium positions	252
7:6	Types of equilibrium—stability	255
	Exercise 7.6	260
7:7	A uniform flexible inelastic string hanging under gravity	261
	The tightly stretched wire	266
	Exercise 7.7	267
7:8	A rope in contact with a rough surface	268
	Exercise 7.8	269
	Miscellaneous Exercise 7	270

Chapter 8   Further Dynamics of a Particle with one Degree of Freedom   274

8:1	Damped harmonic oscillations	274
	Forced oscillations	277
	Exercise 8.1	279

8:2	Motion in a straight line under variable forces	280
	Exercise 8.2(a)	282
	Miscellaneous Problems	283
	Motion under gravity in a resisting medium	284
	1. Resistance proportional to speed	284
	2. Resistance proportional to the square of the speed	286
	Exercise 8.2(b)	290
*8:3	The rectilinear motion of bodies with variable mass	292
	*Exercise 8.3	297
	Miscellaneous Exercise 8	298

Chapter 9 Dynamics of a Particle with two Degrees of Freedom   303

9:1	Further differentiation of vectors—applications to kinematics	303
	1. Polar coordinates	304
	2. Intrinsic coordinates	305
	Exercise 9.1	306
9:2	Motion referred to cartesian axes	307
	Exercise 9.2	308
9:3	Motion of a particle on a smooth curve	309
	Exercise 9.3	312
9:4	Motion under a central force—polar coordinates	313
	1. A central orbit is a plane curve	314
	2. The angular momentum integral	314
	3. The theorem of areas	315
	Exercise 9.4	317
9:5	The motion of projectiles	319
	1. The range on an inclined plane	319
	Exercise 9.5(a)	321
	2. The bounding parabola	322
	Exercise 9.5(b)	327
9:6	Oblique impact of elastic bodies	328
	Exercise 9.6	332
9:7	The Galilean transformation	334
	1. Frames with uniform relative velocity	335
	2. Frames with uniform relative acceleration	335
9:8	The motion of a system of particles—general theorems	336
	1. Notation	336
	2. Linear momentum	337
	3. Kinetic energy	337
	4. Moment of momentum	338
	5. The motion of the centre of mass	338
	6. Motion about the centre of mass	338
	7. Motions generated by simultaneously applied impulses	339
9:9	The motion of connected particles	339
	Exercise 9.9	346

CONTENTS   ix

9:10 The two-dimensional motion of a projectile in a resisting medium   347
      Exercise 9.10   350
      Miscellaneous Exercise 9   352

Chapter 10 An Introduction to the Dynamics of a Rigid Body   358

10:1 Rotation of a lamina about a fixed axis   358
10:2 Momentum and energy equations for angular motion of a lamina   361
      Exercise 10.2   364
10:3 The compound pendulum   366
      Exercise 10.3   370
10:4 The force exerted on the axis of rotation   372
      Exercise 10.4   374
10:5 Impulse and angular momentum   375
      Exercise 10.5   379
10:6 Note on the relationship between the equations of angular motion of a rigid body and the equations of motion of a particle moving in a straight line   380
10:7 The motion of a lamina in its own plane—instantaneous centre of rotation   381
      Exercise 10.7   384
10:8 The general motion of a rigid lamina in its own plane   384
10:9 Application to miscellaneous problems   388
      Exercise 10.9   397
      Miscellaneous Exercise 10   399

Chapter 11 Further Probability   405

11:1 The binomial and geometric probability distributions   405
      Exercise 11.1   406
11:2 Continuous probability distributions   409
      Exercise 11.2   411
11:3 The Poisson distribution   413
      Exercise 11.3   417
11:4 The normal distribution   419
      Exercise 11.4   425
11:5 Transition matrices for Markov chains   428
11:6 Steady-state vector and matrix   429
11:7 An equivalence relation on transition states   433
      Exercise 11.7   434
11:8 Further probability problems   436
      Exercise 11.8   438
      Miscellaneous Exercise 11   441

Answers to the Exercises   446

Index   xi

Glossary of Symbols and Abbreviations

Mechanics

L, M, T	(dimensions) length, mass, time
m, kg, s, N	(units) metre, kilogram, second, newton
$m\,s^{-1}, m\,s^{-2}$	metres per second, metres per second per second
k	(as a prefix) kilo, $\times 1000$, e.g. kilometre $= 10^3$ metres
$\mathbf{g}, g$	the acceleration due to gravity, its magnitude
J, W	joules, watts

Probability and statistics

μ	the mean of a population (a theoretical value)
σ, σ^2	the standard deviation, variance of a population (theoretical values)
$\bar{x}$	the mean of a sample (measured value)
s, s^2	the standard deviation, variance of a sample (measured values)
$P(E_1)$	the probability that the event E_1 takes place
$E_i', \overline{E}_i, \sim E_i$	the complement of the event E_i (the set of all elements in the sample space not in E_i)
$P(A\mid B)$	the conditional probability of A given B
$E(X)$	the expected value or expectation of X
$^nP_r, {}_nP_r$	the number of permutations or ways of arranging r objects from n distinct objects $= (n!)/(r!)$
$^nC_r, {}_nC_r$	the number of combinations or ways of choosing r objects from n distinct objects $= (n!)/[(n-r)!r!]$
$F(x)$	the normal probability integral $= \dfrac{1}{\sqrt{2\pi}} \displaystyle\int_{-\infty}^{x} e^{-t^2/2}\,dt$

7 Further Statics

7:1 FORCES IN THREE DIMENSIONS

The moment of a force about a point and about a line

In elementary two-dimensional mechanics the moment of a force about a point (or, more precisely, the turning-moment of the force about an axis through that point perpendicular to the plane containing the force and the point) is defined as the product pF, where F is the magnitude of the force and p is the perpendicular distance of its line of action from the point. This moment is represented by twice the area of the triangle OAB, which has O as a vertex and its base AB, of length F, on the line of action of the force. We now extend this definition to three dimensions. Suppose that (Fig. 7.1) O represents the origin of position vectors and $\mathbf{F}$ is a force acting along a given line. If $\mathbf{r}_A$ is the position vector of any point A on the line of action then the vector $\mathbf{M} = \mathbf{r}_A \times \mathbf{F}$ is

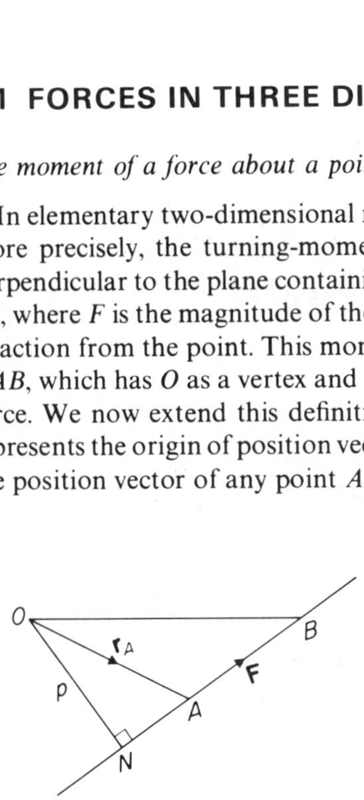

FIG. 7.1. 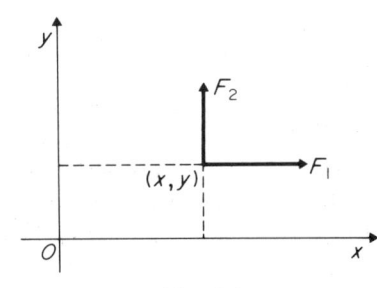　　　　　　　　　　FIG. 7.2.

twice the vector area of the triangle OAB. The magnitude of $\mathbf{r}_A \times \mathbf{F}$ is the turning-moment of $\mathbf{F}$ about an axis through O perpendicular to OAB. We notice that $\mathbf{r}_A$ may be replaced by the position vector $\mathbf{r}$ of any other point on the line of action, for

$$\mathbf{r} = \mathbf{r}_A + t\mathbf{F},$$

and

$$\mathbf{r} \times \mathbf{F} = (\mathbf{r}_A + t\mathbf{F}) \times \mathbf{F} = \mathbf{r}_A \times \mathbf{F} = \mathbf{M}. \tag{7.1}$$

Let us consider the turning-moment of $\mathbf{F}$ about the axis Oz. The element, F_3, parallel to this axis will have no turning-moment. The remaining two elements will give a turning-moment (see Fig. 7.2)

$$xF_2 - yF_1$$

about Oz, where $(x, y, 0)$ are the coordinates of the point of intersection of $\mathbf{F}$ with Oxy.

This turning-moment is the component M_3 of the vector $\mathbf{M}$ in equation (7.1), which defines the moment of the force $\mathbf{F}$ about the point O.

Thus, by using a vector product we introduce the *vector moment of a force about a point*, which has been defined in such a way that its component in any direction is the turning-moment of the force about a line through the point drawn in the given direction. In general, the turning-moment L of $\mathbf{F}$ about a line through $\mathbf{a}$ drawn in the direction of the unit vector $\hat{\mathbf{l}}$ is given (defined) by

$$L = \{(\mathbf{r} - \mathbf{a}) \times \mathbf{F}\} . \hat{\mathbf{l}}, \tag{7.2}$$

where $\mathbf{r}$ is any point on the line of action of $\mathbf{F}$. As is usual with vector products the order of the factors in (7.1) or (7.2) is important; the position vector of a point on the line of action of $\mathbf{F}$ comes before $\mathbf{F}$ in equation (7.1).

Throughout this section we take the units of length and force to be the metre and newton respectively.

Example. Find the moment about the line

$$\mathbf{r} = \mathbf{i} + 2\mathbf{j} + 4\mathbf{k} + t(2\mathbf{i} + 3\mathbf{j} - 2\mathbf{k}) \text{ of the force } 7\mathbf{i} + 2\mathbf{j} - 3\mathbf{k}$$

acting through the point P with position vector $4\mathbf{i} + 2\mathbf{j} + 6\mathbf{k}$.

If A is the point $\mathbf{i} + 2\mathbf{j} + 4\mathbf{k}$, the moment of $\mathbf{F}$ about A is

$$\mathbf{M} = \overrightarrow{AP} \times \mathbf{F} = (3\mathbf{i} + 2\mathbf{k}) \times (7\mathbf{i} + 2\mathbf{j} - 3\mathbf{k})$$
$$= -4\mathbf{i} + 23\mathbf{j} + 6\mathbf{k}.$$
$$\hat{\mathbf{l}} = (2\mathbf{i} + 3\mathbf{j} - 2\mathbf{k})/\sqrt{17}$$
$$\Rightarrow \mathbf{M} . \hat{\mathbf{l}} = 49/\sqrt{17}$$

(the result being measured in newton metres).

Varignon's theorem

If $\mathbf{P}$ and $\mathbf{Q}$ are two forces acting through the point A and $\mathbf{R}$ is their resultant, and if $\mathbf{r}$ is the position vector of A with respect to O, then the moment of $\mathbf{R}$ about O is

$$\mathbf{r} \times \mathbf{R} = \mathbf{r} \times (\mathbf{P} + \mathbf{Q}) = \mathbf{r} \times \mathbf{P} + \mathbf{r} \times \mathbf{Q}.$$

This result implies that *the moment of the vector sum of two forces about a point is equal to the sum of the moments of the forces about the point*. The result can be extended by induction to relate to the vector sum of any number of forces, and it is the three-dimensional extension of the result for coplanar forces in §4:3.

The moment of a force about a point can therefore be represented by a free vector in a direction at right angles to the plane containing the force and the point, and the sum of such moments can be obtained by the laws of vector addition.

Couples as vectors

It follows that a couple can be represented by a free vector perpendicular to the plane of the couple, and, with this method of representation, we deduce at once the following property of couples:

Couples of the same moment in parallel planes are equivalent.

We can compound (add) couples by adding their components.

Equilibrium of a system of forces in three dimensions

If forces $P_1, P_2, P_3, \ldots$ of a set act at the respective points with position vectors $r_1, r_2, r_3, \ldots$, we *define equilibrium of this set of forces* to be determined by the conditions

$$R = P_1 + P_2 + P_3 + \ldots = 0, \qquad (7.3)$$

$$N(A) = N_1(A) + N_2(A) + N_3(A) + \ldots = 0. \qquad (7.4)$$

Here $N_1(A), N_2(A), \ldots$ are the moments of the respective forces about an arbitrary point A, which can conveniently be chosen as the origin. We adopt the conditions (7.3) and (7.4) to *define* equilibrium. The reasons for choosing these conditions are that equation (7.3) shows that the forces do not tend to cause translation and that equation (7.4), which is a generalisation of the principle of the lever, shows that the forces do not tend to cause rotation about any point. [Note that $N_1(A) = r_1 \times P_1$.]

Example 1. A system of three forces consists of a force $F_1 = (i - j + k)$ acting through the point whose position vector is $(2i + 2k)$, a force $F_2 = (j + 2k)$ acting through the point whose position vector is $(i - k)$, and a force F_3.
(a) If F_3 acts through the origin and the system reduces to a couple, find F_3 and the magnitude of the couple.
(b) If the system is in equilibrium and F_3 acts through the point $(ai + j + 4k)$, find the value of a. (L.)

(a) The system reduces to a couple

$$\Rightarrow F_1 + F_2 + F_3 = 0$$
$$\Rightarrow F_3 = -(i - j + k) - (j + 2k) = (-i - 3k).$$

The total moment of the system about O is

$$(2i + 2k) \times (i - j + k) + (i - k) \times (j + 2k) = (3i - 2j - k)$$
$$\Rightarrow \text{magnitude of couple is } \sqrt{14} \text{ N m}.$$

(b) The moment of F_3 about O is $-(3i - 2j - k)$

$$\Rightarrow (ai + j + 4k) \times (-i - 3k) = (-3i + 2j + k)$$
$$\Rightarrow [-3i + (3a - 4)j + k] = (-3i + 2j + k)$$
$$\Leftrightarrow 3a - 4 = 2 \Leftrightarrow a = 2.$$

Example 2. A particle with position vector $2i$ is kept in equilibrium by the following five forces:

$$P = (7i - 15j + k), \quad Q = (5i - 8j - 7k),$$

R_1 acting along the line $r = 2i + \lambda(3i + 4k)$,
R_2 acting along the line $r = 2i + \mu(2i - j + 2k)$,
R_3 acting along the line $r = 2i + \gamma(-3i + 4j)$.
Find the magnitudes of R_1, R_2, and R_3. (L.)

Let $R_1 = a(3i + 4k)$, $R_2 = b(2i - j + 2k)$, $R_3 = c(-3i + 4j)$.

The resultant of the five forces acting at $2i$ must be zero

$$\Rightarrow (7i - 15j + k) + (5i - 8j - 7k) + a(3i + 4k) + b(2i - j + 2k) + c(-3i + 4j) = 0$$
$$\Rightarrow (12 + 3a + 2b - 3c)i + (-23 - b + 4c)j + (-6 + 4a + 2b)k = 0$$
$$\Rightarrow 3a + 2b - 3c = -12,$$
$$-b + 4c = 23,$$
$$4a + 2b = 6$$
$$\Rightarrow a = 3, b = -3, c = 5.$$

Hence the magnitudes of R_1, R_2, R_3 are 15, 9, 25 newtons respectively.

Example 3. A force **F** of magnitude 10 N acts along a diagonal of a cube whose edges have length 2m. Calculate the moment of the force about a diagonal of a face of the cube. The line of action of **F** and the diagonal do not intersect.

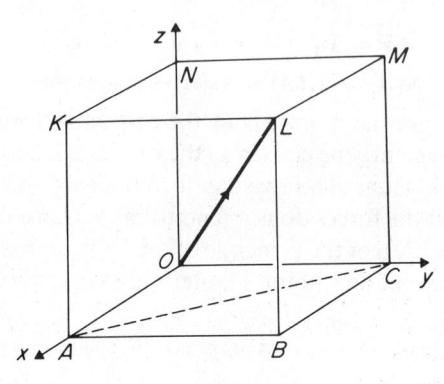

FIG. 7.3.

Let the cube $OABCKLMN$ be placed so that OA, OB, OC lie along the coordinate axes Ox, Oy, Oz respectively (Fig. 7.3). We let the force **F** act along OL so that $\mathbf{F} = (10/\sqrt{3})(\mathbf{i}+\mathbf{j}+\mathbf{k})$ and obtain the magnitude of its moment about the line AC. By definition the required moment M is

$$(\vec{AO} \times \mathbf{F}).\vec{AC}/|\vec{AC}|.$$

Since $\vec{AO} = -\mathbf{i}$, $\vec{AC} = (-\mathbf{i}+\mathbf{j})$,

$$M = (-\mathbf{i}) \times (10/\sqrt{3})(\mathbf{i}+\mathbf{j}+\mathbf{k}).(-\mathbf{i}+\mathbf{j})/\sqrt{2}$$
$$\Rightarrow M = 10/\sqrt{6}.$$

Example 4. Forces $\mathbf{F}_1 = (2\mathbf{i}+2\mathbf{j})$, $\mathbf{F}_2 = (\mathbf{i}-\mathbf{j}-2\mathbf{k})$ and $\mathbf{F}_3 = (-3\mathbf{i}-\mathbf{j}+2\mathbf{k})$ act through the points A_1, A_2 and A_3 with position vectors $(\mathbf{i}+4\mathbf{j})$, $(3\mathbf{j}+\mathbf{k})$ and $(-\mathbf{i}+\mathbf{k})$ respectively. Show that this system is equivalent to a couple and find the magnitude of the moment of this couple. (L.)

$$\mathbf{F}_1 + \mathbf{F}_2 + \mathbf{F}_3 = (2\mathbf{i}+2\mathbf{j}) + (\mathbf{i}-\mathbf{j}-2\mathbf{k}) + (-3\mathbf{i}-\mathbf{j}+2\mathbf{k}) = 0.$$

Therefore the system reduces to a couple or is in equilibrium (if the moment of the couple is zero). The moment of the system of forces about the origin is

$$\mathbf{M} = (\mathbf{i}+4\mathbf{j}) \times (2\mathbf{i}+2\mathbf{j}) + (3\mathbf{j}+\mathbf{k}) \times (\mathbf{i}-\mathbf{j}-2\mathbf{k}) + (-\mathbf{i}+\mathbf{k}) \times (-3\mathbf{i}-\mathbf{j}+2\mathbf{k})$$
$$\Rightarrow \mathbf{M} = (-4\mathbf{i}-8\mathbf{k}).$$

Since $\mathbf{M} \neq 0$, the system reduces to a couple of moment **M**. (This is the moment of the system about any (arbitrary) point.) The magnitude of the couple is $|\mathbf{M}|\,\mathrm{N\,m} = 4\sqrt{5}\,\mathrm{N\,m}$.

Exercise 7.1

1. A system of forces consists of three forces, $\mathbf{F}_1 = (\mathbf{i}-5\mathbf{j}+\mathbf{k})$ and $\mathbf{F}_2 = (\mathbf{j}+\mathbf{k})$ acting through points whose position vectors are $(5\mathbf{i}-10\mathbf{j}+10\mathbf{k})$ and $(4\mathbf{i}-7\mathbf{j}+7\mathbf{k})$ respectively, and a third force $\mathbf{F}_3$.
   (a) If the system is in equilibrium, find the magnitude of $\mathbf{F}_3$ and a vector equation of its line of action.
   (b) If $\mathbf{F}_3$ acts through the origin and the system reduces to a couple, find the magnitude of this couple.
   (L.)

2. Forces $\mathbf{F}_1 = (\mathbf{i}+\mathbf{j}+\mathbf{k})$, $\mathbf{F}_2 = -2\mathbf{j}$ and $\mathbf{F}_3 = (-\mathbf{i}+\mathbf{j}-\mathbf{k})$ act through the points P_1, P_2 and P_3 with position vectors $(2\mathbf{i}-\mathbf{k})$, $3\mathbf{j}$ and $4\mathbf{i}$ respectively.
Show that this system is equivalent to a couple and find the magnitude of the moment of this couple.

3. A force of magnitude $\sqrt{6}$ newtons acts along the line

$$\mathbf{r} = (s-4)\mathbf{i} + (s+1)\mathbf{j} + (2s+3)\mathbf{k}$$

and a second force of magnitude $\sqrt{14}$ newtons acts along the line

$$\mathbf{r} = (t-7)\mathbf{i} + (2t-3)\mathbf{j} + (2-3t)\mathbf{k}.$$

Show that these two lines meet and find the position vector of their point of intersection P.
Find the magnitude of the resultant of these two forces and a vector equation of its line of action. If these two forces act on a particle of mass 2 kg which is initially at rest at P, find the velocity vector and the position vector of the particle after 4 seconds.

4. A force $\mathbf{F}_1 = 2\mathbf{i} + 6\mathbf{j} - 9\mathbf{k}$ acts at a point A with position vector $5\mathbf{i} + 10\mathbf{j} - 15\mathbf{k}$ and a second force $\mathbf{F}_2 = 9\mathbf{i} - 12\mathbf{k}$ acts at a point B with position vector $4\mathbf{i} - 2\mathbf{j} - \mathbf{k}$, the units of force and distance being the newton and metre respectively. Show that the lines of action of these forces intersect and find the vector equation of the line of action of their resultant. A third force acts at a point C with position vector $2\mathbf{j} + 3\mathbf{k}$ and combines with the other two forces to form a couple. Show that the magnitude of the moment of this couple is nearly 100 N m. (L.)

5. A rigid body is acted on by a force $\mathbf{F}_1 = 4\mathbf{i} - 3\mathbf{j} + 2\mathbf{k}$ at the point A with position vector $9\mathbf{i}$, a force $\mathbf{F}_2 = 3\mathbf{i} + 2\mathbf{j} - 4\mathbf{k}$ at the point B with position vector $-\mathbf{i} - \mathbf{j} + 6\mathbf{k}$, and a force $\mathbf{F}_3$. If the body is in equilibrium, find $\mathbf{F}_3$ and show that the vector equation of the line of action of $\mathbf{F}_3$ can be written in the form

$$\mathbf{r} = (5\mathbf{i} + 3\mathbf{j} - 2\mathbf{k}) + s(-7\mathbf{i} + \mathbf{j} + 2\mathbf{k}),$$

where s is a parameter. (L.)

6. A force $2\sqrt{14}$ newtons acts in the direction $\mathbf{i} + 2\mathbf{j} + 3\mathbf{k}$ through the point $A(1, 0, 3)$ and a force $\sqrt{41}$ newtons acts in the direction $6\mathbf{i} + 2\mathbf{j} + \mathbf{k}$ through the point $B(6, 0, 1)$. Verify that the lines of action of these forces intersect at $C(0, -2, 0)$ and find the magnitude of the resultant force. (L.)

7. The vertices of a tetrahedron $ABCD$ have position vectors $\mathbf{a}$, $\mathbf{b}$, $\mathbf{c}$, $\mathbf{d}$ respectively, where

$$\mathbf{a} = 3\mathbf{i} - 4\mathbf{j} + \mathbf{k}, \quad \mathbf{b} = 4\mathbf{i} + 4\mathbf{j} - 2\mathbf{k},$$
$$\mathbf{c} = 4\mathbf{i} + \mathbf{k}, \quad \mathbf{d} = \mathbf{i} - 2\mathbf{j} + \mathbf{k}.$$

Forces of magnitude 30 and $3\sqrt{13}$ N act along CB and CD respectively. A third force acts at A. If the system reduces to a couple, find the magnitude of this couple and the force at A. Find also a unit vector along the axis of the couple. (L.)

8. The position vectors of the vertices B and C of a triangle ABC are, respectively, $(8\mathbf{i} + 3\mathbf{j} + 5\mathbf{k})$ and $(6\mathbf{i} + 4\mathbf{j} + 9\mathbf{k})$. Two forces $(3\mathbf{i} + 2\mathbf{j} + \mathbf{k})$ and $(4\mathbf{i} + 5\mathbf{j} + 6\mathbf{k})$ act along AB and AC respectively. A third force $\mathbf{F}$ acts through A. The system of forces is in equilibrium. Find
(a) the magnitude of the force $\mathbf{F}$,
(b) the position vector of A,
(c) an equation of the line of action of $\mathbf{F}$ in vector form. (L.)

9. A uniform cube of mass $2m$ and side $2a$ has its centre at the point (a, a, a), one vertex at the origin O and the three edges through O along the coordinate axes. Particles of mass m are attached to the cube at each of the points $(0, 0, a)$, $(a, 0, a)$ and $(a, 2a, a)$. Find the position of the centre of mass of the body. The body is acted on by a force $F(2\mathbf{i} - 2\mathbf{j} - \mathbf{k})$ at the point $(2a, 2a, 0)$ amd a force $F(-4\mathbf{i} + \mathbf{j} - 4\mathbf{k})$ at the point $(0, 2a, 2a)$. Find the moment of this set of forces about the line joining O to the point $(2a, 2a, 2a)$.

10. The position vectors of the vertices A, B, C of a triangle are respectively

$$4\mathbf{i} + 2\mathbf{j} + 2\mathbf{k}, \quad 2\mathbf{i} + \mathbf{j} + 4\mathbf{k}, \quad -3\mathbf{i} + 6\mathbf{j} - 6\mathbf{k}.$$

Find the position vector of G, the centroid of the triangle.
Forces $3\overrightarrow{AB}$ and $2\overrightarrow{AC}$ act along AB and AC. Find the magnitude of their resultant and the position vector of the point P in which the line of action of this resultant meets BC. Show that the area of the triangle GAP is $\tfrac{1}{2}\sqrt{5}$ and hence find the moment of the resultant about an axis through G perpendicular to the plane ABC.

7:2 CONDITIONS OF EQUILIBRIUM FOR A RIGID BODY ACTED UPON BY A SYSTEM OF COPLANAR FORCES

In § 4:5 we discussed the conditions of equilibrium for a body acted upon by three forces only and in § 4:6 we discussed the equilibrium of a rigid body acted upon by a system of coplanar forces. We now classify sets of conditions, each set being necessary and sufficient to ensure the equilibrium of a rigid body, as follows.

1. For a rigid body acted upon by two forces only

The forces must be equal in magnitude and opposite in direction, and they must act in the same straight line.

2. For a rigid body acted upon by three non-parallel forces only

(a) The forces must act at a point, and
(b) The vector sum of the forces must be zero.

Condition (b) may be tested by any one of: the triangle of vectors, Lami's theorem, the law of the resolved parts.

3. For a rigid body acted upon by any system of parallel forces

EITHER   (a) the vector sum of the forces must be zero, and
          (b) the sum of the moments of the forces about an arbitrary point in the plane must be zero,

OR   the sum of the moments of the forces about any two points of the plane, which do not lie in a line parallel to the forces, separately vanish.

4. For a rigid body acted upon by any system of forces

(This is the general case which includes the three special cases discussed above.)

EITHER   (a) the vector sum of the forces must be zero, and
          (b) the sum of the moments of the forces about an arbitrary point in the plane must be zero,
OR   the sum of the moments of the forces about any three noncollinear points in the plane must separately vanish,
OR   any other set of conditions, which ensure that the vector sum of the forces is zero and that the moment-sum of the forces about *every point* in the plane is zero, must be satisfied.

The examples and exercises which follow involve these principles and also, in some cases, the principles of friction enunciated in Chapter 4.

Example 1. A uniform straight rod has one end resting on a fixed smooth horizontal cylinder and the other end in contact with a smooth vertical wall whose plane is parallel to the axis of the cylinder. In equilibrium the rod makes an angle ϕ with the horizontal and the radius of the cylinder through the point where the rod rests on the cylinder makes an angle θ with the horizontal. Prove that $\tan \theta = 2 \tan \phi$.

CONDITIONS OF EQUILIBRIUM FOR A RIGID BODY

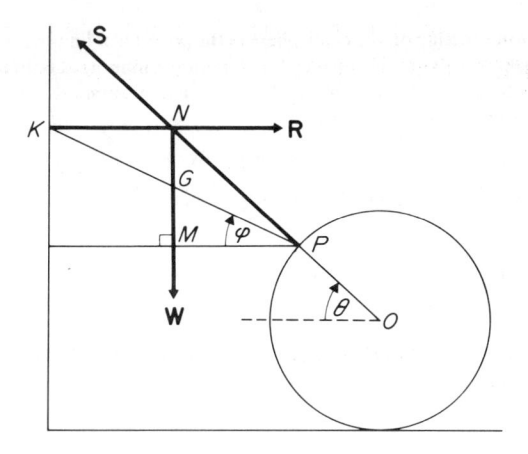

FIG. 7.4.

The rod PK is in equilibrium under the action of the following three forces, as illustrated in Fig. 7.4,
(a) its weight **W** acting through its mid-point G,
(b) the normal reaction **R** of the wall on the rod at K,
(c) the normal reaction **S** of the cylinder on the rod at P.

The lines of action of these three forces are therefore concurrent; hence the rod must lie in the vertical plane through O perpendicular to the wall.

If these lines meet at N and NG produced meets the horizontal through P at M,

$$\tan \theta = NM/MP, \qquad \tan \phi = GM/MP,$$

and, since triangles KNG and PMG are congruent, $NG = GM$.

Therefore, $NM = 2GM$ and hence $\tan \theta = 2 \tan \phi$.

Example 2. A uniform solid hemisphere rests with its curved surface against a rough horizontal plane and against a rough vertical plane, the radius through the mass-centre being in a vertical plane perpendicular to each of the other planes. If the coefficient of friction at the horizontal plane is μ, and that at the vertical plane μ', show that in the position of limiting equilibrium the radius through the mass-centre will be inclined to the vertical at an angle θ given by

$$\sin \theta = \frac{8\mu(1+\mu')}{3(1+\mu\mu')}.$$

Show also that, when $\mu = \mu'$, the hemisphere will rest at any inclination if $\mu \geqslant \tfrac{1}{5}(\sqrt{31}-4)$. (N.)

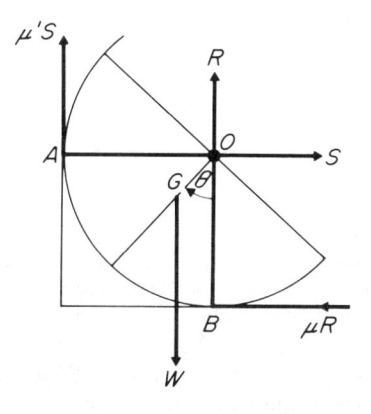

FIG. 7.5.

Figure 7.5 shows the forces acting on the hemisphere in the position of limiting equilibrium. Since slipping at one contact involves slipping also at the other contact, friction is limiting at both contacts in the position of limiting equilibrium. Necessary and sufficient conditions of equilibrium are:

1. The sum of the vertical components vanishes

$$\Rightarrow R + \mu' S = W.$$

2. The sum of the horizontal components vanishes

$$\Rightarrow S = \mu R.$$

3. The sum of the moments about O vanishes

$$\Rightarrow \mu' S r + \mu R r = W \cdot \tfrac{3}{8} r \sin \theta,$$

where r is the radius of the hemisphere. (The weight of the hemisphere acts through the centre of gravity G on the axis of symmetry distant $\tfrac{3}{8} r$ from the centre of the hemisphere.)

From (1) and (2),

$$R = \frac{W}{1 + \mu \mu'}, \qquad S = \frac{\mu W}{1 + \mu \mu'}.$$

Therefore from (3),

$$\frac{\mu \mu' W}{1 + \mu \mu'} + \frac{\mu W}{1 + \mu \mu'} = \tfrac{3}{8} W \sin \theta$$

$$\Leftrightarrow \sin \theta = \frac{8\mu(1 + \mu')}{3(1 + \mu \mu')}.$$

If $\mu = \mu'$, the position of limiting equilibrium is given by

$$\sin \theta = \frac{8\mu(1 + \mu)}{3(1 + \mu^2)}.$$

Since $\sin \theta \leqslant 1$ this position can only exist provided that

$$\frac{8\mu(1 + \mu)}{3(1 + \mu^2)} \leqslant 1,$$

i.e. provided that $5\mu^2 + 8\mu - 3 \leqslant 0$.

This inequality obtains only for values of $\mu \leqslant$ the positive root of

$$5\mu^2 + 8\mu - 3 = 0, \qquad \text{i.e. for} \qquad \mu \leqslant \frac{-4 + \sqrt{31}}{5}.$$

It follows that if

$$\mu > \frac{-4 + \sqrt{31}}{5}$$

friction is not limiting at any inclination and therefore the hemisphere rests at all inclinations. Note that when μ is equal to this critical value, equilibrium is limiting at $\theta = \tfrac{1}{2}\pi$, i.e. when the base is in a vertical plane.

Exercise 7.2

1. A uniform circular hoop of mass M rests on a rough rail which is inclined at an angle of 30° to the horizontal. The hoop and the rail are in the same vertical plane. The hoop is held in equilibrium by means of a string which leaves the hoop tangentially and is inclined at 60° to the horizontal, this angle being measured in the same sense as the angle of inclination of the rail. Show that the coefficient of friction between the rail and the hoop is not less than $\tfrac{1}{2}(\sqrt{3} - 1)$, and find the tension in the string. (N.)

2. A uniform lamina in the form of an equilateral triangle of side l rests in a vertical plane perpendicular to a smooth vertical wall, one vertex being in contact with the wall and another attached to a point in the wall by

a string of length l. Show that in the position of equilibrium the angle between the string and the vertical is $\cot^{-1}(3\sqrt{3})$. (L.)

3. A uniform rod, of mass M, rests partly within and partly without a thin smooth hemispherical bowl, of mass $3M\sqrt{3}$ and radius a, which itself rests on a horizontal plane. The centre of gravity of the bowl is at the middle point of its radius of symmetry. If the rod is inclined at an angle of $30°$ to the horizontal, prove that the plane of the rim of the bowl is inclined at an angle θ to the horizontal, where $\cot\theta = 4\sqrt{3}$. Show also that the length of the rod is $12a/7$. (N.)

4. A uniform rod PQ, of weight w, rests in contact with a smooth fixed right circular cylinder whose axis is horizontal, the vertical plane through the rod being perpendicular to the axis. The ends P, Q of the rod are fastened by strings to the cylinder and, when the rod is in equilibrium, the strings are *horizontal* and tangential to the cylinder. The string with the shorter horizontal length is uppermost. If the rod makes an angle $\theta(<\pi/2)$ with the horizontal, show that the lower string will not be taut unless

$$\cos^2\theta + \cos\theta < 1.$$ (L.)

5. A uniform rod of mass M and length l rests on a rough horizontal table, the coefficient of friction between the rod and the table being μ. A gradually increasing horizontal force is applied perpendicularly at one end of the rod. Assuming that the vertical reaction is distributed uniformly along the rod, show that the rod begins to turn about a point distant $l/\sqrt{2}$ from the end at which the force is applied, and find the corresponding magnitude of the applied force. (N.)

6. A thin uniform heavy ring hangs in equilibrium over a small rough peg. To the lowest point of the ring a string is attached and pulled in a horizontal direction with steadily increasing force, the ring and string being in one vertical plane. The angle of friction at the peg being $\lambda(<\sin^{-1}\tfrac{1}{3})$, show that slipping is about to commence there when the diameter through the peg is inclined at an angle $\tfrac{1}{2}\{\sin^{-1}(3\sin\lambda)+\lambda\}$ to the vertical, and that if $\lambda > \sin^{-1}\tfrac{1}{3}$, slipping will never occur whatever the tension in the string. (N.)

7. A uniform circular cylinder with its plane ends normal to its axis rests horizontally upon two equally rough inclined planes which are fixed in opposite senses at an angle θ to the horizontal, the axis of the cylinder being parallel to the line of intersection of the planes. A couple G is applied to the cylinder tending to turn it about its axis. If w is the weight of the cylinder, a its radius and λ the angle of friction, show that when the cylinder is about to slip

$$G = \tfrac{1}{2}wa\sin 2\lambda\sec\theta.$$ (N.)

8. An elastic string of natural length $3a$ and modulus of elasticity kW has its ends fixed at the same horizontal level at a distance $3a$ apart, and equal weights W are suspended from points distant a and $2a$ from one end. Show that in the position of equilibrium the two end portions of the string are inclined at an angle θ to the horizontal given by $3\cot\theta = 2k(1-\cos\theta)$, and that the length of the middle portion is $a(5-2\cos\theta)/3$.

Show that if k is large θ is approximately $(3/k)^{1/3}$. (L.)

7:3 HINGED BODIES—EQUILIBRIUM PROBLEMS INVOLVING MORE THAN ONE BODY

1. Smooth hinges

In Example 1 of § 4:5 we stated that a body is smoothly hinged when the action of the hinge on the body is a single force through the centre of the hinge. Figure 7.6 (i) shows two rods AB and AC smoothly hinged together at A. The hinge acts on the rod AB at A with a force, the components X_1, Y_1 of which are shown in Fig. 7.6 (ii), and the rod AB reacts on the hinge with an equal and opposite force. Similarly the hinge acts on the rod AC at A with a force of components X_2, Y_2, Fig. 7.6 (iii), and the rod AC reacts on the hinge with an equal and opposite force. It follows that if the hinge is in equilibrium, Fig. 7.6 (iv), $X_1 = X_2$ and $Y_1 = Y_2$ *provided that no other force is acting on the hinge.* If this latter condition is fulfilled, the forces can be represented as an action and an equal and opposite reaction between the rods themselves, Fig. 7.6 (v).

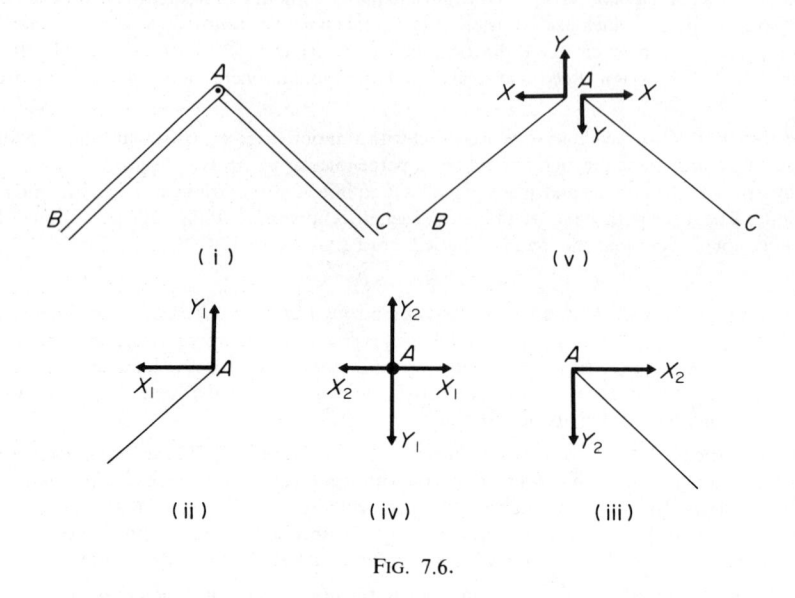

FIG. 7.6.

2. Rough hinges

If one body is joined to another by a hinge which is not smooth, the forces on the hinge at the surface of contact between the hinge and the body do not pass through the centre of the hinge and the resultant of these forces does not, in general, pass through the centre of the hinge. This resultant can, however, be replaced *by a single force acting through the centre of the hinge and a couple*. In cases where no other forces act on the hinge, the action of the hinge on the two bodies can then be represented by equal and opposite forces and couples as shown in Fig. 7.7.

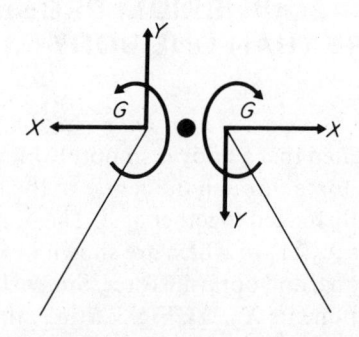

FIG. 7.7.

3. The action of one part of a body on another

Figure 7.8 shows a uniform horizontal rod AB fixed to a vertical wall at A. The part CB of the rod is in equilibrium under the action of

(i) its weight acting through its mid-point, and
(ii) the action of the part AC.

Equilibrium is only possible in this case if (ii) consists of an upward vertical force w at C and a couple of moment $\frac{1}{2}w \cdot CB$ counterclockwise, where w is the weight of the part CB. The part of the rod AC will act with an equal and opposite force and an equal and opposite couple on the part of the rod CB.

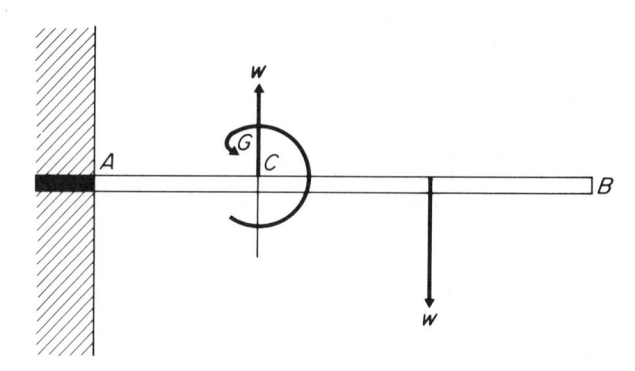

Fig. 7.8.

4. Problems of equilibrium involving more than one body

When there are two or more bodies concerned in a problem of equilibrium, there are, in general, three conditions of equilibrium for each of the bodies. There are also three conditions of equilibrium for the system as a whole, and there are three conditions of equilibrium for any part of the system considered as a whole. Clearly, all these sets of conditions are not independent sets, since, for example, the equilibrium of each body of the system, separately, implies the equilibrium of the system as a whole. In consequence, the solution of problems involving two or more bodies involves the careful selection from the equations of equilibrium of those equations which are independent and which constitute the minimum number necessary to determine whatever factors of the system are required.

Example 1. Two uniform rods AB, BC, of lengths $2a$, $2b$, and of weights $2W$, $3W$ respectively, are smoothly jointed at B. They are in equilibrium in a vertical plane with A resting on a rough horizontal plane and C resting against a smooth vertical guide. The point A is further from the vertical guide than B and α is the acute angle between AB and the horizontal plane. The angle between CB and the downward vertical at C is β. Show that

$$3 \tan \alpha \tan \beta = 8.$$

If μ is the coefficient of friction between the rod AB and the horizontal plane, show that $\beta \leqslant \tan^{-1}(10\mu/3)$.

(N.)

Figure 7.9 shows the forces acting on each of the rods.

For the equilibrium of the whole system,
(1) the sum of the vertical components is zero $\Rightarrow T = 5W$,
(2) the sum of the horizontal components is zero $\Rightarrow S = F$,
(3) the sum of the moments about A is zero
$$\Rightarrow 2Wa\cos\alpha + 3W(2a\cos\alpha + b\sin\beta) = S(2a\sin\alpha + 2b\cos\beta).$$

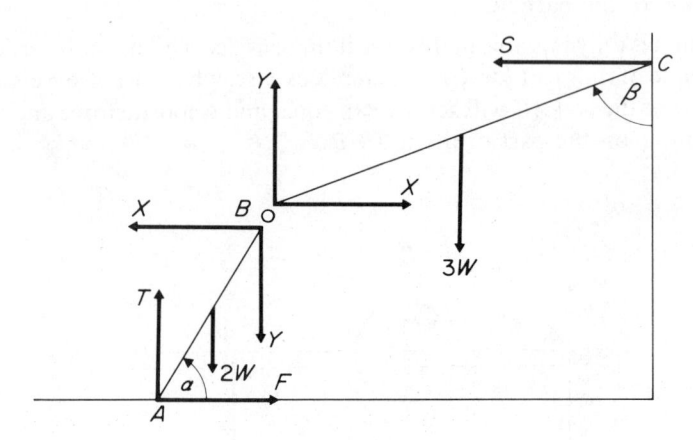

FIG. 7.9.

For the equilibrium of the rod BC,
(4) the sum of the moments about B is zero
$$\Rightarrow S2b\cos\beta = 3Wb\sin\beta$$
$$\Leftrightarrow S = \frac{3W}{2}\tan\beta.$$

From equations (3) and (4),
$$2a\cos\alpha + 6a\cos\alpha + 3b\sin\beta = \tfrac{3}{2}(2a\sin\alpha + 2b\cos\beta)\tan\beta$$
and this simplifies to the result $3\tan\alpha\tan\beta = 8$.

Also, for the equilibrium of BC, $F/T \leq \mu$.
$$\Rightarrow \left(\frac{3W\tan\beta}{2}\right)\bigg/(5W) \leq \mu$$
$$\Leftrightarrow \tan\beta \leq \frac{10\mu}{3},$$
and since, for $0 \leq \beta < \tfrac{1}{2}\pi$, $\tan\beta$ increases with β
$$\Rightarrow \beta \leq \tan^{-1}(10\mu/3).$$

Example 2. Two uniform rods AB and BC, each of length $2a$ and weight W, are maintained at right angles to each other by a stiff hinge at B. The friction at the hinge is such that the greatest couple which either rod can exert on the other has moment K. The end A is smoothly pivoted to a fixed point from which the rods hang. A vertical force F acting at C holds the rods in equilibrium with AB inclined at an angle ϕ to the downward vertical at A, and with C higher than B. Find F, and show that the moment of the reaction couple at B is $2Wa\sin\phi/(1 + \tan\phi)$.

By considering the maximum value of this function of ϕ, show that equilibrium is possible for any value of ϕ between 0 and $\pi/2$, provided that $K \geq Wa/\sqrt{2}$.

(N.)

Ch. 7 §7:3   HINGED BODIES—EQUILIBRIUM PROBLEMS   243

Figure 7.10 shows the forces and couple acting on each of the rods. The rod BC makes an angle $\frac{1}{2}\pi - \phi$ with the upward vertical at B.

For the equilibrium of the whole system, the sum of the moments about A is zero

$$\Rightarrow Wa \sin \phi + W(2a \sin \phi + a \cos \phi) - F(2a \sin \phi + 2a \cos \phi) = 0$$
$$\Rightarrow F = W(3 \tan \phi + 1)/(2 + 2\tan \phi).$$

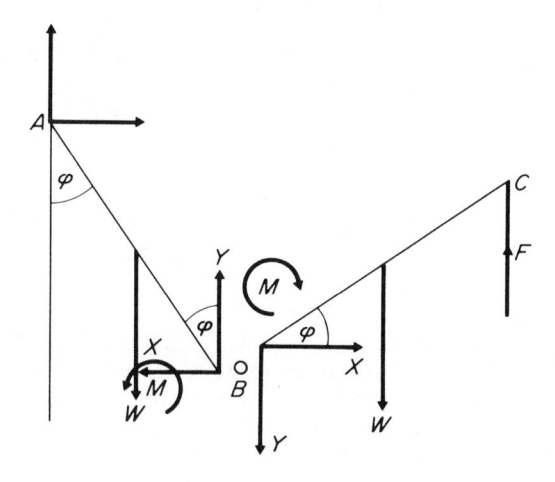

FIG. 7.10.

For the equilibrium of the rod BC, the sum of the moments about B is zero

$$\Rightarrow M = F2a \cos \phi - Wa \cos \phi = Wa \left\{ \frac{3 \sin \phi + \cos \phi}{1 + \tan \phi} - \frac{\cos \phi + \sin \phi}{1 + \tan \phi} \right\}$$

$$= 2Wa \sin \phi/(1 + \tan \phi)$$

$$\Rightarrow \frac{dM}{d\phi} = 2Wa \left\{ \frac{\cos \phi + \sin \phi - \sin \phi \sec^2 \phi}{(1 + \tan \phi)^2} \right\}$$

$$= 2Wa \left\{ \frac{\cos^3 \phi - \sin^3 \phi}{(1 + \tan \phi)^2} \right\} \sec^2 \phi.$$

M therefore has a stationary value when $\tan^3 \phi = 1$, i.e. when $\phi = \frac{1}{4}\pi$ for $0 \leq \phi \leq \frac{1}{2}\pi$, and this stationary value is a maximum since $dM/d\phi$ is positive when $\phi < \frac{1}{4}\pi$ and negative when $\phi > \frac{1}{4}\pi$ for values of ϕ in this range. This maximum value of M is $(2Wa \sin \frac{1}{4}\pi)/(1 + \tan \frac{1}{4}\pi) = Wa/\sqrt{2}$. It follows that provided $K \geq Wa/\sqrt{2}$ equilibrium is possible for all values of ϕ between 0 and $\frac{1}{2}\pi$.

Example 3. Two equal uniform solid rough cylinders, each of weight W, lie on a rough horizontal plane in contact along a generator of each. A horizontal force P is applied to one of the cylinders in a line through and at right angles to the axes of both cylinders. This line passes through the centres of gravity of the cylinders. The coefficient of friction between the cylinders is equal to μ and that between either cylinder and the plane is μ'. Show that while equilibrium is not broken the force of friction between each cylinder and the plane is equal to the force of friction between the two cylinders. Find the forces exerted by the cylinders on one another and the force exerted on each cylinder by the plane. Deduce that for equilibrium to be possible for any value of P, $\mu \geq 1$, and find how equilibrium will be broken as P is gradually increased.

Figure 7.11 shows the vertical section through the centres of gravity A and B of the cylinders and also shows the forces acting on each of the cylinders.

244 **FURTHER STATICS** Ch. 7 §7:3

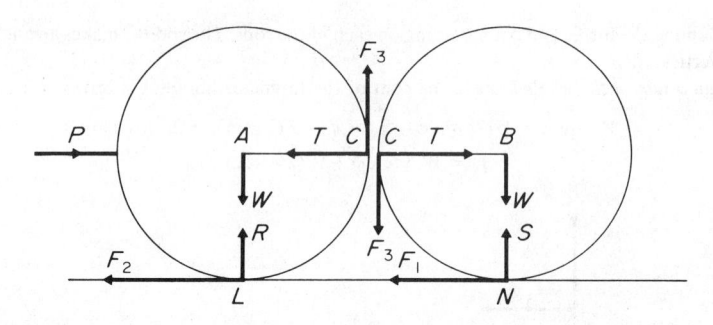

Fig. 7.11.

For the equilibrium of each cylinder separately, the sum of the moments about the axis is zero

$$\Rightarrow F_1 = F_3 \quad \text{and} \quad F_2 = F_3.$$

Therefore the forces of friction at all contacts are equal to F(say). Consideration of the horizontal components of the forces acting on the whole system shows that the frictional forces at the plane must act in the direction shown in order to balance P. Also, the sum of the horizontal components for each cylinder is zero and therefore

$$T = F = \tfrac{1}{2}P.$$

It follows that, whatever the value of P, for equilibrium at C, we must have $\mu \geqslant F/T$, i.e. $\mu \geqslant 1$. Further, if this condition obtains, there will not be slipping between the cylinders at C whatever the value of P. Equilibrium is therefore broken first by slipping at either L or N, the cylinder A rolling on the cylinder B at C.

The sum of the vertical components of the forces acting on each of the cylinders is zero, and therefore

$$R = W - F = W - \tfrac{1}{2}P,$$
$$S = W + F = W + \tfrac{1}{2}P$$
$$\Rightarrow \frac{F}{R} = \frac{x}{2-x} = \frac{2}{2-x} - 1, \quad \frac{F}{S} = \frac{x}{2+x} = 1 - \frac{2}{2+x},$$

where $x = P/W$. Therefore as x increases from 0 to 2, F/R increases from 0 without limit and F/S increases from 0 to $\tfrac{1}{2}$. Also

$$\frac{F}{R} - \frac{F}{S} = \frac{2x^2}{4 - x^2}$$

and this expression is positive for $0 < x < 2$. It follows that, as P increases, friction becomes limiting and slipping takes place at L. In fact, as P increases, F/R approaches the critical value μ' at L and equilibrium is broken when

$$\frac{x}{2-x} > \mu', \quad \text{i.e. when} \quad P > \frac{2\mu' W}{1 + \mu'}.$$

When equilibrium is broken, the cylinder A rotates counter-clockwise, slipping at L, while the cylinder B rolls clockwise on the plane and the spheres roll on one another at C.

---

5. *The equilibrium of the hinge*

On p. 239 we discussed the actions on the hinge of each of two rods connected by the hinge in the case in which no other forces acted on it. In this case Newton's third law operated between the two rods through the agency of the hinge. When, however, there are other forces acting on the hinge, the actions of the hinge on the two rods connected

HINGED BODIES—EQUILIBRIUM PROBLEMS

by it are not necessarily equal and opposite, and separate equations expressing the necessary conditions for the equilibrium of the hinge are required to complete the analysis of the total equilibrium. The examples which follow illustrate the use of this method.

In § 7:4, where the method of Virtual Work is discussed, we consider an alternative method for such problems in which the actions of the hinges on the rods need not be considered unless they are specifically required.

Example 1. Four equal uniform rods, each of weight W, are smoothly jointed together at their ends to form a rhombus $ABCD$. The rhombus is suspended from A and is maintained in equilibrium, with C below A and with $D\hat{A}B = 2\theta$, by a light string connecting the joints at A and C. Find the horizontal and vertical components of the force exerted by AB on BC. Hence, or otherwise, find the tension in the string. (N.)

The system is symmetrical both geometrically and mechanically about AC. The forces exerted by the hinge at C on the rods CB and CD are therefore as shown in Fig. 7.12 (i) which also shows the weights of the rods AB and BC and the forces exerted by the hinge B on those rods. Figure 7.12 (ii) shows the forces acting on the hinge C. We denote by T the tension in the string AC.

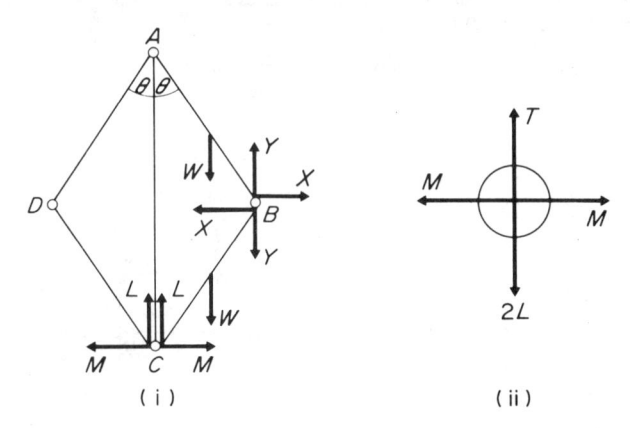

Fig. 7.12.

For the equilibrium of the rod AB, the sum of the moments about A is zero

$$\Rightarrow 2X \cos \theta + 2Y \sin \theta = W \sin \theta,$$

and, for the equilibrium of the rod BC, the sum of the moments about C is zero

$$\Rightarrow 2X \cos \theta - 2Y \sin \theta = W \sin \theta$$
$$\Rightarrow X = \tfrac{1}{2} W \tan \theta, \qquad Y = 0.$$

For the equilibrium of the rod BC, the sum of the vertical components is zero

$$\Rightarrow L = W + Y = W.$$

Finally, for the equilibrium of the hinge C, the sum of the vertical components is zero

$$\Rightarrow T = 2L = 2W.$$

The tension in the string is $2W$.

Note that this result follows also by resolving vertically for the equilibrium of the whole of the framework below the line BD.

Example 2. We solve the same problem as in Example 1 above except that the string AC is removed and equilibrium is maintained by a light string joining the mid-points E, F of AD, CD respectively.

As before C is vertically below A and the components of reaction at B are the same as in Example 1. However, the system is not symmetrical about AC. The tension T' in EF can be found by taking moments about D for the portion BCD of the framework. We find

$$T' = 4W.$$

Example 3. Six equal uniform rods, each of weight W, are smoothly jointed so as to form the hexagon ABCDEF. They are suspended from the vertex A and kept in the shape of a regular hexagon by means of a light horizontal rod PQ, of length $AB\sqrt{3}$, the ends P and Q being smoothly hinged to points of BC and EF respectively. If the action at the hinge D is X, and the thrust in PQ is T, express the horizontal and vertical components of the actions at C and B in terms of X, T, W, and show that $T = 6X = 3W\sqrt{3}$.
Show also that $BC = 6BP$. (L.)

Since the system is symmetrical about AD, the action at D is horizontal, X say, Fig 7.13. Then by considering the equilibrium of the hinges at C and B, and resolving horizontally and vertically for the rods DC, CB (starting at the bottom of the framework) the forces are as shown.

Taking moments about C for CD gives

$$X \cdot CD \cos 60° = W \cdot \tfrac{1}{2}CD \sin 60°$$
$$\Rightarrow X = \tfrac{1}{2}W\sqrt{3}.$$

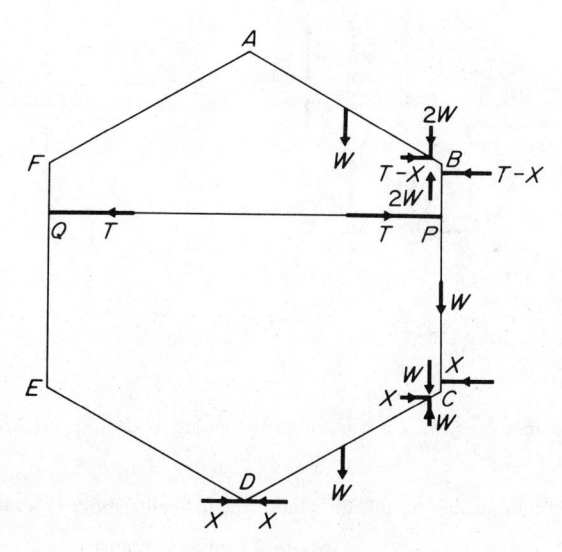

Fig. 7.13.

Taking moments about A for AB

$$\Rightarrow W \cdot \tfrac{1}{2}AB \sin 60° + 2W \cdot AB \sin 60° = (T - X) \cdot AB \cos 60°$$
$$\Leftrightarrow T - X = 5\sqrt{3}W/2$$
$$\Rightarrow T = 3\sqrt{3}W.$$

Taking moments about B for BC

$$\Rightarrow X \cdot BC = T \cdot BP$$
$$\Rightarrow BC = 6BP.$$

Exercise 7.3

1. A smooth uniform circular cylinder of weight W is kept in equilibrium with its curved surface in contact with a smooth inclined plane and its axis horizontal by means of a smooth uniform rod AB of weight $2W$ freely hinged to the inclined plane below the cylinder at A, the rod being tangential to the cylinder at its upper end B. The vertical plane through AB intersects the inclined plane in a line of greatest slope and passes through the centre of gravity of the cylinder. If α is the inclination of the plane to the horizontal and θ the acute angle which the rod makes with the inclined plane, show that

$$\tan \alpha = \frac{\sin 2\theta}{3 - \cos 2\theta}.$$ (N.)

2. Two planes each making an angle ϕ with the horizontal form a V-groove. Three equal uniform cylinders, each of radius a and weight W, are laid symmetrically in the groove with their axes horizontal and parallel to the line of intersection of the planes. The two lower cylinders touch one another and each touches one of the planes and the third cylinder rests on the other two. There is no friction at any point of contact. Show that equilibrium is not possible unless $\tan \phi \geq \sqrt{3}/9$. (N.)

3. Two equal uniform circular cylinders B and C, each of weight W, rest side by side on a rough horizontal plane; their axes are horizontal and parallel and the cylinders are almost in contact. A third equal cylinder A is gently placed so as to rest in equilibrium upon them and in contact with each of them along a generator. Show, by considering the equilibrium of C, that the force of friction between C and A and that between C and the plane are equal in magnitude. Show also that their common magnitude is $\frac{1}{2}W(2 - \sqrt{3})$.
If μ is the coefficient of friction at all contacts, show that $\mu > 2 - \sqrt{3}$. (N.)

4. A uniform ladder AB, of weight W and length $2a$, is placed at an inclination θ to the horizontal with A on a rough horizontal floor and B against a rough vertical wall. The ladder is in a vertical plane perpendicular to the wall and the coefficients of friction at A and B are μ and μ' respectively. Show that the ladder will be in equilibrium if

$$\tan \theta \geq (1 - \mu\mu')/2\mu.$$

If this condition is not satisfied and a couple of moment K is applied in the vertical plane so that the ladder is just prevented from slipping down, show that

$$K = Wa \cos \theta (1 - \mu\mu' - 2\mu \tan \theta)/(1 + \mu\mu').$$ (N.)

5. Two equal uniform rods AB, AC, each of length $2a$, are freely jointed at A. They rest symmetrically over two small smooth pegs at the same level and at a distance $2c$ apart. Prove that, if the rods make an angle α with the horizontal, $c = a \cos^3 \alpha$. (O.C.)

6. Two uniform rods AB, BC of equal lengths, but of masses 4 kg and 3 kg respectively, are freely hinged at B and rest with A and C on a rough horizontal plane with B vertically above AC. The ends A and C are originally near together and the angle 2θ between AB and BC is gradually increased. Assuming the coefficient of friction between each rod and the plane to have the same value μ, show that slipping will first occur at the point C and when θ reaches the value given by $7 \tan \theta = 13\mu$. (L.)

7. Four equal uniform rods, each of weight W, are smoothly jointed to form a square framework $ABCD$. They hang freely from the corner A and the shape of the square is maintained by a light string joining the midpoints of the rods AB and BC. Prove that the reaction at the joint D is horizontal and of magnitude $\frac{1}{2}W$, and that the tension in the string is $4W$.

8. A rhombus $ABCD$ of side $2a$ is composed of four freely jointed rods, and lies on a smooth horizontal plane. The rod AB is fixed and the mid-points of the rods BC and CD are joined by a string which is kept taut by a couple of moment M applied to the rod DA. If the angle ABC is 2θ, find the tension in the string and show that it is twice the reaction at the joint C. (L.)

9. Three uniform rods AB, BC, CD, each of length $2a$ and of weights $W, 4W, W$ respectively, are smoothly jointed at B and C. The rods are in equilibrium in a vertical plane with A and D resting on rough horizontal ground and AB, CD each inclined at $60°$ to the horizontal so that $AD = 4a$.
  (i) Calculate the vertical components and the frictional components of the forces at A and D.
  (ii) Calculate the horizontal and vertical components of the force exerted by BC on AB.
  (iii) If μ is the coefficient of friction at A and D, find the inequality satisfied by μ.
(The directions of the forces acting on each rod must be clearly indicated in a diagram.) (N.)

*7:4 THE PRINCIPLE OF VIRTUAL WORK

The concept of work forms an essential part of an alternative method of investigating conditions of equilibrium or the equivalence of sets of forces. The adjective "virtual" is used because the work is done, or would be done, by the forces in displacements which the system on which they act is *imagined* to undergo.

The equilibrium of a particle

A particle is subject to the action of a number of forces of which the vector sum is **F**, see Fig. 7.14. We can *imagine* this particle given a small displacement $\delta\mathbf{s}$; if it actually made this displacement, the forces would do an amount of work $\mathbf{F}\cdot\delta\mathbf{s}$; this work is the *virtual work* of the forces.

If the particle is in equilibrium, $\mathbf{F} = \mathbf{0}$ and the virtual work is zero. If the virtual work is zero *for all possible virtual displacements*, then **F** must be zero. (The scalar product $\mathbf{F}\cdot\delta\mathbf{s}$ cannot vanish for all possible displacements unless $\mathbf{F} = \mathbf{0}$.) Hence a necessary and sufficient condition for the equilibrium of the particle is that the virtual work of the applied forces is zero for all possible displacements.

Expressed in terms of components referred to rectangular axes, if the forces have resultant sums (X, Y, Z) and the displacement has components $(\delta x, \delta y, \delta z)$, then the virtual work is

$$\delta W = X\delta x + Y\delta y + Z\delta z.$$

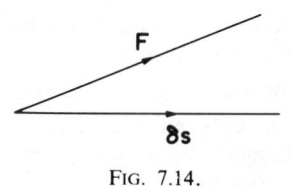

FIG. 7.14.

If $X = 0$, $Y = 0$, $Z = 0$, then $\delta W = 0$; if $\delta W = 0$ for arbitrary values of δx, δy, δz, we deduce that $X = 0$, $Y = 0$, $Z = 0$.

This result can be extended to the cases of a rigid body or system of bodies acted upon by a number of forces. In effect we take *as an axiom determining the equilibrium of a set of forces* the principle of virtual work or principle of virtual displacements in the following form.

If a set of forces acts upon a mechanical system, a necessary and sufficient condition for equilibrium is that the virtual work of the forces is zero, correct to the first order, when the system is given an arbitrary virtual displacement consistent with the constraints of the system.

The phrase "consistent with the constraints" needs some explanation. We have proved the principle for a particle. Most idealised mechanical systems consist of particles and rigid bodies which are connected by hinges, strings, etc. Conditions may also be imposed on the system such as one body remaining in contact with a given surface, a particle sliding on a smooth wire, and so on. When a system is given a virtual displacement "consistent with the constraints", the displacement is such that these conditions (constraints) are not violated and that any forces acting on the system because of these constraints (e.g. the reactions between surfaces, the tensions in inextensible strings, the interactions between particles of a rigid body) *do no work* in the displacement. If work is done in the displacement (perhaps a surface may be rough) it is usual to remove the constraint and add appropriate (initially unknown) forces to the given set acting on the system.

The principle of virtual work can be used to compare two sets of forces. If the two sets do the same work in an arbitrary virtual displacement, they are equivalent sets of forces. The principle is capable of greater generalisation than the methods of resolution and is used extensively in more advanced mechanics to discuss very general systems and obtain results of wide applicability.

Applications of the principle of virtual work

The following examples illustrate the use of the principle of virtual work. Using the principle frequently avoids the necessity for the introduction of unknown internal actions and reactions unless a particular reaction or "force of constraint" is required. Such a force can often be determined by violating the corresponding constraint in the virtual displacement. (The force of constraint becomes one of the set of applied forces when the constraint is ignored.)

Example. A uniform rod AB, of length $2a$ and weight W, rests in equilibrium with the end A against a smooth vertical wall and the end B on rough horizontal ground. The vertical plane through the rod is perpendicular to the plane of the wall and AB makes the angle θ with the downward vertical. Calculate the force of friction at B.

This example can easily be solved by the methods of §4:9. However, we give here a solution by using the principle of virtual work to illustrate the method.

The only external forces acting on the rod AB are the normal reactions R, S, at A and B, the force of friction F at B and the weight W at G, as shown in Fig. 7.15. In a virtual displacement consistent with the geometrical constraints of the system, A, B remain in contact with the wall and ground respectively and so the displacements of A and B are perpendicular to the forces R and S, which, therefore, do no work in the displacement.

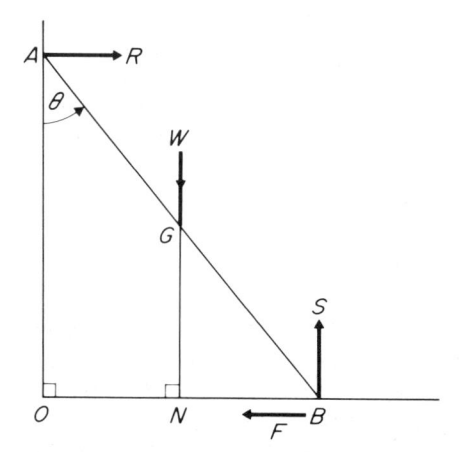

FIG. 7.15.

250   FURTHER STATICS   Ch. 7 §7:4

The increase in the height of G above the (fixed) horizontal plane is $\delta(NG)$ and so the virtual work of W is $-W\delta(NG)$. The negative sign arises because an increase in NG moves G upwards against the direction of W which therefore does work $(-W) \times \delta(NG)$. Similarly the work done by the force of friction F, corresponding to an increase $\delta(OB)$ in the distance of its point of application from O, is $-F\delta(OB)$. Therefore the equation of virtual work is

$$-W\delta(NG) - F\delta(OB) = 0.$$

But $NG = a\cos\theta$, $OB = 2a\sin\theta$

$$\Rightarrow -W\delta(a\cos\theta) - F\delta(2a\sin\theta) = 0.$$

To the first order in $\delta\theta$ this gives

$$Wa\delta\theta.\sin\theta - 2Fa\delta\theta.\cos\theta = 0$$

$$\Leftrightarrow F = \tfrac{1}{2}W\tan\theta.$$

Note: In the above example it is easy to see that, if θ is *increased* by $\delta\theta$, then OB increases and GN decreases so that the virtual works of F and W are $-2Fa\delta\theta.\cos\theta$ and $Wa\delta\theta.\sin\theta$ respectively and have correct signs. However, in more complicated cases this is not always apparent. In general it is best to use the fact that the increase of any length l is δl and to give the sign to the virtual work of a force, the point of application of which moves a distance δl, on the supposition that δl is positive. Correct use of the calculus will then ensure correct subsequent work.

Example 1. Four uniform rods AB, BC, CD, DA, each of weight W and length $2a$, are smoothly jointed to form a rhombus $ABCD$. The system is pivoted to a fixed point at A and prevented from collapsing by a light inextensible string AC. Calculate the tension in AC.

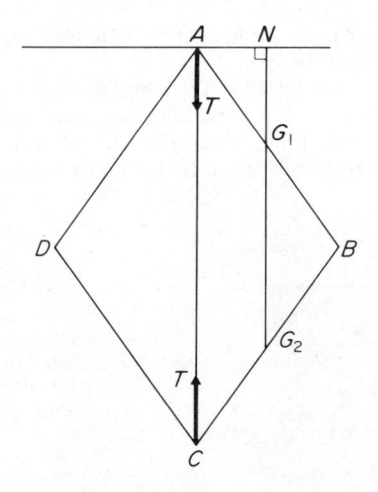

Fig. 7.16.

The tension T in AC, Fig. 7.16, is an internal force of the system, so that, in order to find this force, we remove the "constraint" AC, replace it by the forces T acting at A, C, and use the principle of virtual work. In a virtual displacement of the system in which A remains fixed and C remains vertically below A, the centre G_1 of AB falls a distance $\delta(G_1 N)$, since N is at the fixed level of A. Similarly G_2 falls a distance $\delta(G_2 N)$ and so the virtual work of the weights of all the rods is

$$2\{W\delta(G_1 N) + W\delta(G_2 N)\} = 2W\delta(AC). \tag{1}$$

To find the virtual work of T we note that we have removed the "constraint" that AC is an inextensible string, i.e. is of fixed length, and have inserted the forces T acting at A and C in place of the constraint. The forces T each act in the line AC, so that contributions to the total work done will only arise from components of displacement which take place in the line AC. The force T at A does no work, and for the force T at C the virtual work is

$$-T\delta(AC).$$

[If T were a thrust, the virtual work would be $T\delta(AC)$ because then the displacement $\delta(AC)$ is in the direction of the forces T.]

The force at A and the reaction at the joints do no net work in the displacement. (This will be assumed without explicit statement in subsequent examples.) The equation of virtual work is, therefore,

$$2W\delta(AC) - T\delta(AC) = 0$$
$$\Rightarrow T = 2W$$

and is independent of the angle BAD.

Example 2. We consider the same problem as in Example 1 above except that the string AC is removed and the system maintained in the shape of a square by a light rigid rod joining BD.

We calculate the thrust in this special case after writing down the equation of virtual work of the system under consideration in the *general case*. We do not insert the value of θ corresponding to the specified configuration until the equation of virtual work has been written down and simplified.

Consider the general case in which angle $BAD = 2\theta$. We remove the constraint BD, and replace it by forces P acting at B and D in the directions OB and OD respectively. Then, if BD meets AC at O, the virtual work of these forces is

$$P\delta(OB) + P\delta(OD) = P\delta(BD).$$

As before the virtual work of the weights of the rods is $2W\delta(AC)$ and so the equation of virtual work is

$$P\delta(BD) + 2W\delta(AC) = 0$$
$$\Rightarrow P\delta(4a\sin\theta) + 2W\delta(4a\cos\theta) = 0$$

and so, to the first order in $\delta\theta$,

$$P\delta\theta.\cos\theta - 2W\delta\theta.\sin\theta = 0$$
$$\Leftrightarrow P = 2W\tan\theta.$$

Therefore, for a square, $(\theta = \tfrac{1}{4}\pi)$, $P = 2W$.

---

*Exercise 7.4

*1. Four uniform straight rods AB, BC, CD, DE, each of length $5l$ and weight W, are freely hinged together at B, C and D. The ends A and E are held at the same level at a distance $14l$ apart; B and D are joined by a light strut of length $8l$; and the system hangs in equilibrium in a vertical plane. Use the principle of virtual work to find the stress in the strut BD.

*2. Four smoothly jointed, weightless rods, each of length a, form a rhombus $ABCD$. It rests in a vertical plane with A vertically above C, the lower rods BC, CD supported on fixed smooth pegs which are on the same level and distant b apart. A light rod BD maintains an angle 2θ at A when a load W hangs from A. Show by the method of virtual work, or otherwise, that when BD acts as a strut, the thrust it exerts on the framework is

$$\frac{W}{4a\cos\theta}(b\operatorname{cosec}^2\theta - 4a\sin\theta).$$

Show also that if $ABCD$ is a square, BD can only act as a tie and find the tension in BD. (L.)

*3. Four uniform rods AB, BC, CD, DE, each of length a and weight W, are smoothly jointed together at their ends B, C, and D, and the ends A, E are smoothly jointed to fixed points at a distance $2a$ apart in the same horizontal line. If AB, BC make angles θ, ϕ respectively with the horizontal when the system hangs in equilibrium, show, by the principle of virtual work, that $3 \cot \theta = \cot \phi$.

Show also that if B and D are now connected by an inextensible string of length a, the tension in this string is $W/\sqrt{3}$. (L.)

*4. Four equal uniform rods each of weight W are freely jointed together at their ends and are kept in the form of a square $ABCD$ by means of a rod BD of weight $2W$. The framework is suspended freely from A and particles of weight $2nW$ are attached to each of the joints B, C, and D. Prove that the horizontal stress in the rod BD is $(4n+3)W$. (N.)

*5. Three uniform rods AB, BC, CD each of weight W and length $2a$ are freely jointed at B and C. They rest in equilibrium with BC horizontal and uppermost, with AB, CD making equal angles of $\theta = 30°$ with the vertical and resting over two smooth pegs E and F respectively at the same horizontal level. Prove by the *principle of virtual work* that $EF = 13a/6$.

If θ is decreased to $\sin^{-1} \frac{5}{13}$ by means of a light elastic string of natural length $2a$ joining A and D, show that the modulus of the string is approximately $0.32 W$. (L.)

*6. Four uniform rods, each of length a and weight w, are freely jointed to form a rhombus $ABCD$ which is freely suspended from A. A smooth uniform disc, of radius r and weight w, rests between the rods CB and CD and in the same vertical plane. The joints B and D are connected by a light elastic string of natural length $3a/4$ and, in equilibrium, $BD = a$. Use the principle of virtual work to show that the modulus of elasticity of the string is $3(2r - \sqrt{3}a)w/a$ if $r > \sqrt{3}a/2$. (L.)

7:5 DETERMINATION OF EQUILIBRIUM POSITIONS

Suppose that a mechanical system acted upon by a conservative set of forces has one degree of freedom, i.e. its configuration is specified by one coordinate x, say. [Such a system is illustrated in Examples 1 and 2 following, where, in Example 2, the parameter is the angle θ.] Let F be the resultant "force" tending to increase the coordinate x and let δV, δW respectively be the change in potential energy and the work done by F when x increases by δx. Then, since the forces are conservative,

$$\delta V + \delta W = 0.$$

But $\delta W = F \delta x$ and therefore

$$F = -\frac{dV}{dx}. \tag{7.5}$$

This relation is very important. The minus sign implies that the "force" acts in a direction which tends to decrease the potential energy.

Note that, if x is an angle, then F as defined by (7.5) is the couple which tends to increase x.

Example 1. The potential energy of a stretched string is $V = \lambda x^2/2a$. The force tending to *increase* the coordinate x is

$$X = -\frac{dV}{dx} = -\frac{\lambda x}{a}.$$

This is the tension in the string.

Example 2. A uniform rod of weight W and length $2a$ can turn freely about one extremity A, which is fixed. A light inextensible string attached to the other extremity passes through a small smooth ring fixed at a point

C, distant $2a$ from A and at the same level as A, and carries at its other end a weight w. When the rod is at an inclination θ to the horizontal and below it, determine the magnitude of the couple which tends to increase θ.

(L.)

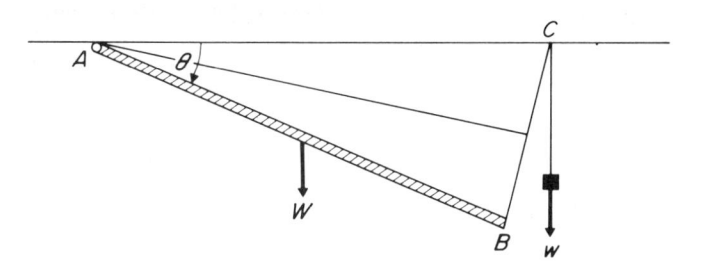

FIG. 7.17.

In Fig. 7.17 the moment of the forces about A in the clockwise direction, i.e. in the sense of θ increasing, is

$$G(A) = Wa \cos \theta - w.2a \cos \tfrac{1}{2}\theta.$$

If we take the standard position to be one in which AB coincides with AC, the potential energy in the position shown is

$$V = -Wa \sin \theta + w.4a \sin \tfrac{1}{2}\theta$$

(The centre of mass, where the weight W acts, is at a vertical distance $a \sin \theta$ below its standard position, and the weight w is distance BC above its standard position.)

$$\Rightarrow \frac{dV}{d\theta} = -Wa \cos \theta + w.2a \cos \tfrac{1}{2}\theta = -G(A).$$

This shows that the derivative of V gives the negative value of the couple (torque) tending to increase θ.

When a system, which has one degree of freedom specified by the coordinate x, is acted upon by a conservative set of forces in equilibrium, the force F tending to increase x must vanish. Therefore, the possible positions of equilibrium are given by

$$\frac{dV}{dx} = 0. \tag{7.6}$$

Application of this result enables equilibrium positions to be determined *without the introduction of internal reactions* as illustrated in the following examples. It must be noted, however, that these examples can be solved by other methods (e.g. those of § 7:3, § 7:4).

Example 1. A uniform rod AB, of length $2a$ and weight W, rests with its ends A, B on two smooth planes inclined at angles α, β to the horizontal ($\alpha > \beta$). The vertical plane through the rod contains a line of greatest slope of each of the planes. Show that the inclination of the rod to the horizontal is θ where

$$2 \sin \alpha \sin \beta \tan \theta = \sin (\alpha - \beta).$$

The system is shown in Fig. 7.18; here O is the point where the line of intersection of the planes meets the vertical plane through the rod. In this case, since the planes are smooth, the reactions at A and B are normal to the planes and do no work in a displacement in which the ends A, B remain in contact with the planes. Hence the potential energy V of the rod, *considered as a function of* θ, must be stationary. But, taking O as the level of zero potential energy,

$$V = \tfrac{1}{2} W(OA \sin \alpha + OB \sin \beta)$$
$$= Wa \{ \sin(\theta + \beta) \sin \alpha + \sin(\alpha - \theta) \sin \beta \} / \sin(\alpha + \beta)$$
$$\Rightarrow \frac{dV}{d\theta} = Wa \{ \cos(\theta + \beta) \sin \alpha - \cos(\alpha - \theta) \sin \beta \} / \sin(\alpha + \beta).$$

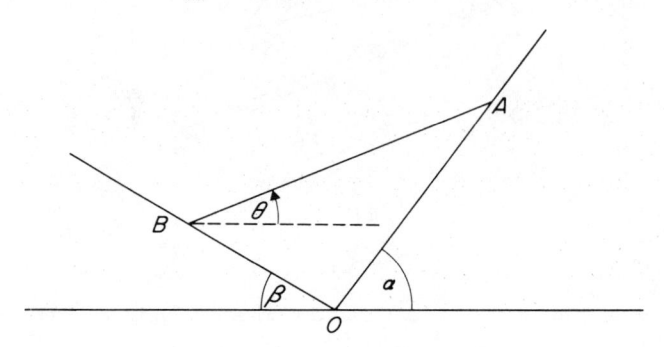

FIG. 7.18.

The equilibrium position is given by $dV/d\theta = 0$

$$\Rightarrow \cos(\theta + \beta) \sin \alpha - \cos(\alpha - \theta) \sin \beta = 0,$$

and this condition reduces to

$$2 \sin \alpha \sin \beta \tan \theta = \sin(\alpha - \beta).$$

Example 2. Two uniform rods AB, AC, each of weight W and length $2c$, are smoothly jointed at A. The rods rest on a smooth circular cylinder of radius a with their plane perpendicular to the axis of the cylinder which is horizontal. When the rods are in equilibrium in a vertical plane with A vertically above the axis of the cylinder, $B\hat{A}C = 2\theta$. Show that $c \sin^3 \theta = a \cos \theta$ and deduce that there is only one such position of equilibrium.

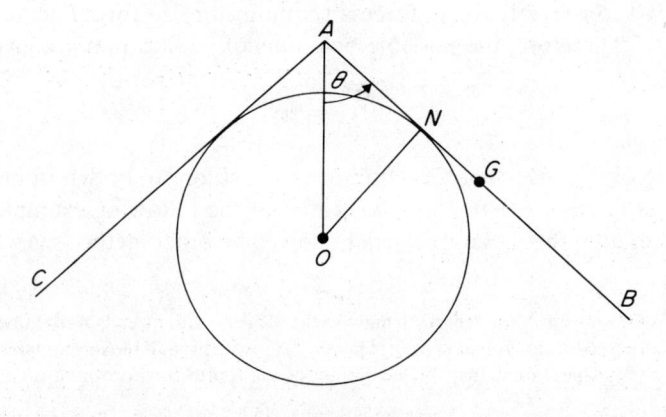

FIG. 7.19.

Let O, Fig. 7.19, be the point on the axis of the cylinder in the plane of the rods and N the point of contact of

AB with the cylinder. Then the height of G, the centre of mass of AB, above the level of O is

$$h = ON \sin\theta - NG \cos\theta$$
$$= a \sin\theta - (c - a \cot\theta) \cos\theta$$
$$= a \operatorname{cosec} \theta - c \cos\theta.$$

Therefore the potential energy of the system, referred to O as the level of zero energy, is

$$V = 2W(a \operatorname{cosec}\theta - c \cos\theta)$$
$$\Rightarrow \frac{dV}{d\theta} = 2W(c \sin\theta - a \operatorname{cosec}\theta \cot\theta).$$

The equilibrium positions occur where $dV/d\theta = 0$

$$\Rightarrow c \sin^3\theta = a \cos\theta. \tag{1}$$

As θ increases from 0 to $\tfrac{1}{2}\pi$ (within which range the equilibrium positions must lie) the l.h. side of (1) must increase steadily from 0 to c whereas the r.h. side decreases steadily from a to 0. Hence the expressions $c \sin^3\theta$ and $a \cos\theta$ must be equal for one and only one value of θ in the range $0 < \theta < \tfrac{1}{2}\pi$. Therefore there is just one position of equilibrium.

7:6 TYPES OF EQUILIBRIUM—STABILITY

We have shown that a stationary value of the potential energy function for a configuration depending on one variable corresponds to an equilibrium position of the system. In fact two types of stationary point—maximum and minimum—correspond to two types of equilibrium, which can therefore be distinguished by the usual calculus procedure. We shall consider simple examples and not give formal proofs of the results.

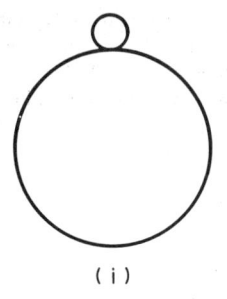

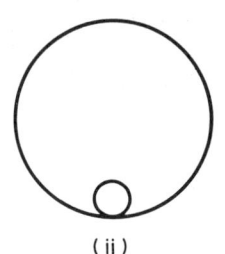

FIG. 7.20.

Consider first a small sphere in equilibrium on top of a large fixed smooth sphere [Fig. 7.20(i)]. In this position the potential energy of the small sphere is greater than in any other position of contact with the large sphere. Hence, if the small sphere is slightly displaced, its potential energy is reduced; but P.E. + K.E. = constant, hence the sphere acquires some kinetic energy, which will cause a further reduction in potential energy. The process is cumulative, so that the sphere cannot return to its equilibrium position. We say that this type of equilibrium, corresponding to a maximum of the P.E. function, is *unstable*.

Consider now the small sphere in equilibrium inside the large (hollow) sphere

[Fig. 7.20(ii)]. Here its potential energy is a minimum; if displaced its potential energy is increased, and hence when released it must return towards its equilibrium position. We say that this situation represents *stable* equilibrium. A position of *neutral equilibrium* is such that on receiving a small disturbance the system neither tends to return to the equilibrium position nor to move further away but is in equilibrium in the disturbed position. Cases of strictly neutral equilibrium are rare, being typified by our uniform sphere resting with its curved surface on a rough horizontal plane. (However, if the sphere is given a small velocity it will continue to move indefinitely, see Chapter 10; it can then hardly be said to "deviate slightly" from its original position, and so such a position could be called unstable.)

Since the determination of stability of equilibrium of a system with one degree of freedom, specified by the coordinate x, is effectively the same as the investigation of the nature of the stationary points of the potential energy $V(x)$, we may sum up as follows:
  (i) Positions of equilibrium occur where $V'(x) = 0$.
  (ii) If $V(x)$ is stationary when $x = x_0$, i.e. $V'(x_0) = 0$, then
     (a) the position where $x = x_0$ is **stable** if V has a **minimum** there.
     (b) the position $x = x_0$ is **unstable** if V has a **maximum** there.

Although we have not proved that positions of minimum and maximum potential energy are positions of stable and unstable equilibrium respectively, we shall assume that this is so (and apply the "change in sign of the derivative" test or consider the sign of higher derivatives as convenient).

Example 1. A uniform rod AB, of weight $12w$ and length $2a$, can turn freely about one extremity A, which is fixed. A light inextensible string attached to the other extremity, B, passes through a small smooth ring fixed at a point C, distant $2a$ from A and at the same level as A, and carries at its other end a particle P of weight w. Show that the system is in stable equilibrium when $8 \cos \theta = 1$, where θ is the inclination of AB to the horizontal (see Fig. 7.21).

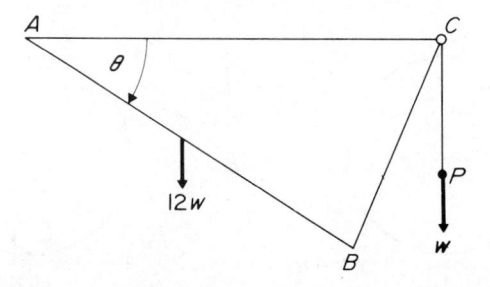

FIG. 7.21.

The potential energy is

$$V = -12wa \sin \theta + 4wa \sin \tfrac{1}{2}\theta$$

measured from the configuration of the system when AB is horizontal as origin. We determine the stationary values of V and the nature of these stationary values.

$$\frac{dV}{d\theta} = -12wa \cos \theta + 2wa \cos \frac{1}{2}\theta = -2wa(12c^2 - c - 6)$$

$$\Rightarrow \frac{dV}{d\theta} = -2wa(4c - 3)(3c + 2),$$

where $c = \cos \tfrac{1}{2}\theta$. But $dV/d\theta$ vanishes where $c = \tfrac{3}{4}$ and $c = -\tfrac{2}{3}$. As θ increases through the value

$2\cos^{-1}(\tfrac{3}{4}) = \cos^{-1}(\tfrac{1}{8})$, so that as c decreases through the value $\tfrac{3}{4}$, $dV/d\theta$ changes from negative to positive; hence V has a minimum and $c = \tfrac{3}{4}$ gives a position of stable equilibrium.

Example 2. A uniform square lamina of side $2a$ rests in a vertical plane with two of its sides in contact with horizontal smooth pegs distant b apart, and in the same horizontal line. Find the positions of equilibrium and discuss their stability. Show in particular that, if $a/\sqrt{2} < b < a$, a non-symmetrical position of unstable equilibrium is possible in which

$$b(\sin\theta + \cos\theta) = a,$$

where θ is the inclination of a side of the square to the horizontal. (L.)

Figure 7.22 shows that the height of G above the fixed level PQ is

$$a\sqrt{2}\sin(\theta + \tfrac{1}{4}\pi) - x\cos\theta, \quad \text{where} \quad x = AQ = b\sin\theta$$

$$\Rightarrow \frac{V}{W} = a\sqrt{2}\sin(\theta + \tfrac{1}{4}\pi) - b\sin\theta\cos\theta$$

$$\Rightarrow \frac{1}{W}\frac{dV}{d\theta} = a\sqrt{2}\cos(\theta + \tfrac{1}{4}\pi) - b\cos 2\theta$$

$$= (\cos\theta - \sin\theta)\{a - b(\cos\theta + \sin\theta)\}$$

$$\Leftrightarrow \frac{1}{W}\frac{dV}{d\theta} = (\cos\theta - \sin\theta)\{a - b\sqrt{2}\sin(\theta + \tfrac{1}{4}\pi)\}$$

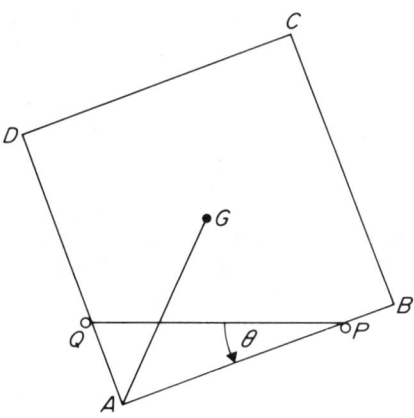

FIG. 7.22.

Equilibrium positions occur when $dV/d\theta = 0$, i.e. when $\cos\theta = \sin\theta$ which gives the symmetrical position, $\theta = \tfrac{1}{4}\pi$, and when

$$\sin(\theta + \tfrac{1}{4}\pi) = \frac{a}{b\sqrt{2}}$$

$$\Leftrightarrow b(\sin\theta + \cos\theta) = a. \tag{1}$$

This latter value gives an unsymmetrical position. The corresponding value of $\theta(< \tfrac{1}{4}\pi)$ is real if $b > a/\sqrt{2}$. Moreover, $\theta > 0$ for equilibrium to be possible and hence for the existence of unsymmetrical positions

$$\frac{a}{b\sqrt{2}} > \frac{1}{\sqrt{2}} \Leftrightarrow b < a.$$

Therefore unsymmetrical positions occur when

$$\frac{a}{\sqrt{2}} < b < a.$$

As θ is made to increase through $\tfrac{1}{4}\pi$, $dV/d\theta$ changes from positive to negative if $a \geqslant b\sqrt{2}$. Hence if $b \leqslant a/\sqrt{2}$ there is only one position of equilibrium and this position is unstable.

On the other hand, if $b > a/\sqrt{2}$, $dV/d\theta$ changes from negative to positive as θ increases through $\tfrac{1}{4}\pi$ and the symmetrical position is stable. In this case, too, the unsymmetrical positions exist (provided $b < a$) and, as θ increases through the values given by equation (1), $dV/d\theta$ changes from positive to negative showing that these positions are unstable.

If $a = b\sqrt{2}$, all three positions of equilibrium coincide in the position $\theta = \tfrac{1}{4}\pi$. Since

$$\frac{1}{W}\frac{d^2V}{d\theta^2} = -a\sqrt{2}\sin\left(\theta + \frac{1}{4}\pi\right) + 2b\sin 2\theta,$$

we see that, in this case, $d^2V/d\theta^2 = 0$ in the equilibrium position. We have shown above that this is an unstable position.

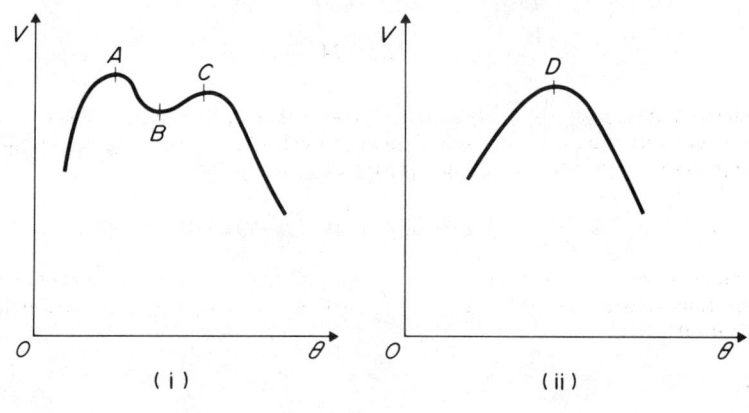

Fig. 7.23.

This example illustrates two features which occur frequently when V is a function of a single parameter. Fig. 7.23(i) shows that, in the graph of V, maxima and minima alternate, so that two positions of unstable equilibrium are separated by a position of stable equilibrium, and vice versa. This must be so for any function V which is continuous. The situation shown in Fig. 7.23(ii) is the limiting case in which the three points A, B, C, of Fig. 7.23(i) coincide in one stationary point D. Although $d^2V/d\theta^2 = 0$ at D the stationary value is nevertheless a maximum.

Example 3. One end A of a uniform rod AB of length $2a$ and weight W can turn freely about a fixed smooth hinge; the other end B is attached by a light elastic string of unstretched length a to a fixed support at the point O vertically above and distant $4a$ from A. If the equilibrium of the vertical position of the rod with B above A is stable, find the minimum modulus of elasticity of the string.

Consider the case in which $O\hat{A}B = \theta$, Fig 7.24. The potential energy of the rod (taking A as the level of zero energy) is $Wa\cos\theta$ and the potential energy of the elastic string is

$$\lambda(OB - a)^2/(2a) = \tfrac{1}{2}\lambda a\{2\sqrt{(5 - 4\cos\theta)} - 1\}^2,$$

where λ is the modulus of the string. The string is always stretched since the least value of OB is $2a$. Therefore the potential energy of the system is

$$V = Wa\cos\theta + \tfrac{1}{2}\lambda a[21 - 16\cos\theta - 4\sqrt{(5 - 4\cos\theta)}]$$

$$\Rightarrow \frac{1}{a}\frac{dV}{d\theta} = \left[8\lambda - W - \frac{4\lambda}{\sqrt{(5 - 4\cos\theta)}}\right]\sin\theta. \tag{1}$$

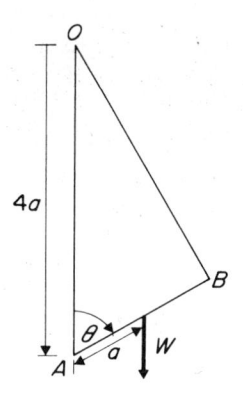

FIG. 7.24.

In order that the vertical position with B above A, $\theta = 0$, should be stable V must be a minimum at $\theta = 0$, i.e. $dV/d\theta$ must change from negative to positive as θ passes through 0. Since $\sin \theta$ changes sign from negative to positive as θ passes through 0; this is equivalent to the condition

$$8\lambda - W - \frac{4\lambda}{\sqrt{(5 - 4\cos\theta)}} > 0 \qquad (2)$$

for θ small but non-zero. Since the function on the left-hand side of inequality (2) increases as $|\theta|$ increases from zero, the required condition is obtained from the condition that the l.h. side of (2) is greater than or equal to zero when $\theta = 0$

$$\Rightarrow 8\lambda - W - 4\lambda \geq 0$$
$$\Leftrightarrow \lambda \geq \tfrac{1}{4} W.$$

The minimum modulus of the string is therefore $\tfrac{1}{4} W$.

We now discuss, by means of an illustrative example the stability of equilibrium of a rigid body which rolls on a fixed surface so that each point of it moves parallel to a vertical plane. The theory covers the two-dimensional motion of a rocking stone. In the example below, both the rolling body and the fixed surface on which it rolls have circular cross-sections.

Example. Find the distance from the centre of a uniform solid sphere, of radius a, to the centre of mass of the smaller of the segments into which it is divided by a plane distant $\tfrac{1}{2}a$ from its centre.

This segment rests with its plane face horizontal and uppermost and its curved surface in contact with the highest point of a fixed sphere of radius a. Assuming that the surfaces are sufficiently rough to prevent sliding, show that the segment may be rolled through an angle $2\cos^{-1}(20/27)$ into a position of unstable equilibrium.

(L.)

Integration shows that the centre of mass of the segment is at a distance $27a/40$ from the centre of the sphere from which it is cut. In the oblique position shown in Fig. 7.25 the segment has been turned through an angle 2θ. If W is the weight of the segment, the potential energy, V, is given by

$$\frac{V}{W} = 2a\cos\theta - \frac{27a}{40}\cos 2\theta$$

$$\Rightarrow \frac{1}{W}\frac{dV}{d\theta} = -2a\sin\theta + \frac{27a}{20}\sin 2\theta = \frac{27a}{10}\sin\theta\left(\cos\theta - \frac{20}{27}\right)$$

$$\Rightarrow \frac{1}{W}\frac{d^2V}{d\theta^2} = -2a\cos\theta + \frac{27a}{10}\cos 2\theta.$$

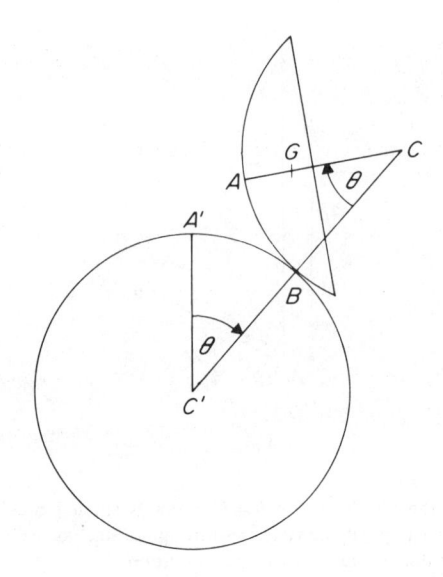

FIG. 7.25.

Hence $dV/d\theta$ vanishes for $\theta = 0$ and $\cos\theta = 20/27$. When $\theta = 0$,

$$d^2V/d\theta^2 = W\left(-2a + \frac{27a}{10}\right) > 0.$$

Also, when $\cos\theta = 20/27$,

$$d^2V/d\theta^2 = Wa\left(-\frac{40}{27} + \frac{27}{5}\cdot\frac{400}{(27)^2} - \frac{27}{10}\right) < 0.$$

Therefore the position $\theta = 0$ is a position of stable equilibrium (V has a minimum there) and the oblique position, $\cos\theta = 20/27$, is a position of unstable equilibrium (V has a maximum there).

Exercise 7.6

1. A uniform rod AB of weight W and length a, is free to rotate about A, which is fixed. To B is attached one end of an elastic string, of modulus W and natural length a. The rod is hung vertically downwards and the free end of the string is then attached to a fixed point, at a height b vertically above A. Show that the rod is in stable equilibrium if $b < a$. (O.C.)

2. A uniform rod AB, of weight W and length $2a$, is free to rotate in a vertical plane about the point A. A light elastic string, of modulus kW and natural length a, has one end attached to B and the other to a fixed point O which is vertically above A. If $OA = 2a$ show that when AB makes an angle θ with the downward vertical, the potential energy of the system, when the string is stretched, may be expressed in the form

$$Wa\{(4k-1)\cos\theta - 4k\cos(\theta/2)\} + \text{constant}.$$

Deduce that, if $k > \frac{1}{3}$, the equilibrium position in which the rod is vertical, with B below A, is unstable, and that there is an oblique position of equilibrium which is stable. (L.)

3. Two uniform rods AB, BC each of length a and of the same weight are smoothly hinged together at B and rest in equilibrium over two small smooth pegs in a horizontal line at a distance c ($< a$) apart. Prove that the angle of inclination of each rod to the vertical is

$$\sin^{-1}\left(\frac{c}{a}\right)^{1/3}.$$

Show also that the equilibrium is stable for displacements in which A moves in a vertical line passing through the middle point of the line joining the pegs. (N.)

4. Three equal uniform bars AB, BC, CD, each of length $2a$, are smoothly jointed at B, C and rest with BC horizontal and AB, CD each on small smooth pegs at the same level at a distance $2(a+b)$ apart.
   Show that, if $2a > 3b$, there are two positions of equilibrium, and determine which of them is stable.
   If $2a = 3b$, show that there is only one position of equilibrium and that it is unstable. (N.)

5. Four equal uniform rods, each of weight W and length $2a$, are smoothly jointed to form a rhombus $ABCD$ which is smoothly pivoted to a fixed point at A. A uniform disc of radius $a/2$ and weight $8W$ is in the same plane as the rods and rests in smooth contact with the rods BC, CD, the whole hanging in a vertical plane. Show that there is an equilibrium position in which the rods are inclined at an angle $\tan^{-1}(\frac{1}{2})$ to the vertical. Discuss the stability of this position. (N.)

6. Four uniform rods, each of weight W and length $2a$, are smoothly jointed together to form a rhombus $ABCD$, and a light elastic string of modulus $2W$ and natural length a connects A and C. The vertex A is smoothly pinned to a fixed support and the system hangs at rest. Show that there is a position of stable equilibrium in which the angle BCD of the rhombus is $120°$. (L.)

7. A rod AB of mass m and length $2a$ can move freely about an end A which is fixed. An elastic string of natural length a and modulus $\frac{1}{2}mg$ is attached to B and to a small ring which can move freely on a smooth horizontal wire at a height $3a$ above A and in the vertical plane through A. Find the positions of equilibrium and examine their stability. (N.)

8. A heavy rod AB can turn freely in a vertical plane about one end A which is fixed. To the other end B is tied a light elastic string of natural length $\frac{3}{4}AB$ and of modulus equal to half the weight of the rod. The other end of the string is attached to a light ring which can slide on a smooth horizontal bar, which is fixed at a height equal to twice AB above A and in the vertical plane through AB. Find the equilibrium positions of the rod and discuss the stability in each case. (L.)

9. A smooth ring P of mass m is free to slide on a smooth fixed vertical circular wire of radius a and centre O. A light elastic string, of natural length $2a$ and modulus kmg, passes through the ring, its ends being fixed to the ends A and B of the horizontal diameter of the wire. Find the potential energy of the system when OP makes an angle 2θ with AB, the ring being below AB. Show that if $k > 2 + \sqrt{2}$ the ring is in unstable equilibrium when at the lowest point of the wire. Investigate the stability in this position when $k = 2 + \sqrt{2}$. (L.)

10. A uniform cube of edge $2a$ rests in equilibrium on the top of a fixed rough cylinder of radius b whose axis is horizontal. Two of the faces of the cube are horizontal. By considering the potential energy when it is rolled over through an angle θ, show that the equilibrium is stable if a is less than b.
    Show also that, in this case, the cube can be rolled into another position of equilibrium, which is unstable.

11. A uniform solid consists of a right circular cone of height h and a hemisphere with their bases, each of radius a, in contact and with their circular boundaries coinciding. The solid stands with its axis vertical and vertex upwards on the top of a rough sphere of radius $3a$. Prove that the equilibrium is stable if h is less than $\frac{1}{2}(\sqrt{5}-1)a$, but is unstable if h is greater than *or equal to* this value. (L.)

7:7 A UNIFORM FLEXIBLE INELASTIC STRING HANGING UNDER GRAVITY

We define a *flexible string* as a continuous line of particles in which the only action which one element can exert on a neighbouring element is a force (the tension) along the line joining the elements, i.e. along the tangent to the string. There is no couple and no transverse force; the string offers no resistance to bending. In this section we consider the equilibrium of such a string, fixed at its ends and hanging freely under gravity as shown in Fig. 7.26.

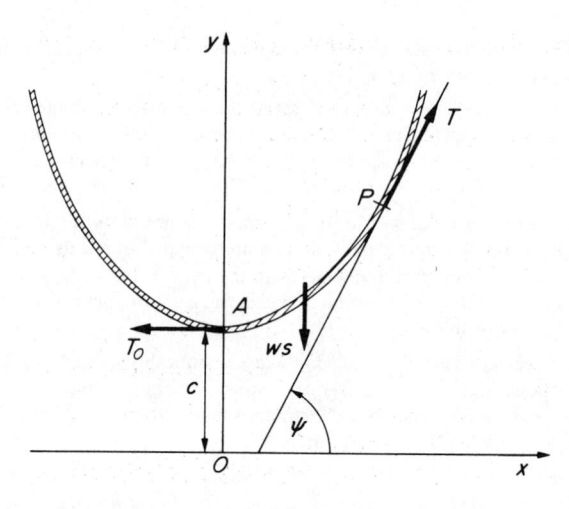

FIG. 7.26.

In this figure A is the lowest point of the string, P is another point of the string and the tangent at P to the string makes an angle ψ with the horizontal. We suppose that the weight per unit length of the string is uniform and equal to w. Then the forces acting on the portion AP of the string are,

(i) its weight ws,
(ii) the tension T exerted by the part of the string to the right of P along the tangent to the string at P,
(iii) the tension T_0 exerted by the part of the string to the left of A, *horizontally* along the tangent to the string at A.

Then, for the equilibrium of this part of the string,

$$T_0 = T \cos \psi, \tag{7.7}$$

$$ws = T \sin \psi. \tag{7.8}$$

Equation (7.7) shows that the horizontal component of the tension is constant throughout the string. We write $T_0 = wc$, where c is constant. Then, by division,

$$s = c \tan \psi. \tag{7.9}$$

This equation (7.9) is the intrinsic equation of the curve of the string. The curve is called the *catenary* (Latin—*catena*, a chain).

We now select as axes of cartesian coordinates the vertical line through A as y-axis and the horizontal line at a distance c below A as x-axis. The x-axis is sometimes called the *directrix* of the catenary (not associated with the directrix of a conic), and c is called the *parameter* of the catenary.

In order to obtain y in terms of ψ for the catenary we use the relationship

$$\frac{dy}{d\psi} = \frac{dy}{ds} \cdot \frac{ds}{d\psi}.$$

But $dy/ds = \sin\psi$, and, from equation (7.9), $ds/d\psi = c\sec^2\psi$,

$$\Rightarrow \frac{dy}{d\psi} = c\sec^2\psi \sin\psi = c\sec\psi \tan\psi$$

$$\Leftrightarrow y = \int c\sec\psi \tan\psi \, d\psi$$

$$\Leftrightarrow y = c\sec\psi + C_1,$$

where C_1 is constant. With the axes we have chosen, $y = c$ when $\psi = 0$. Therefore $C_1 = 0$ and so

$$y = c\sec\psi.$$

Therefore,
$$y^2 - s^2 = c^2(\sec^2\psi - \tan^2\psi)$$
$$\Leftrightarrow y^2 = c^2 + s^2. \qquad (7.11)$$

Also
$$\frac{dx}{d\psi} = \frac{dx}{ds} \cdot \frac{ds}{d\psi}.$$

$$\Rightarrow \frac{dx}{d\psi} = \cos\psi . c\sec^2\psi$$

$$\Leftrightarrow x = \int c\sec\psi \, d\psi$$

$$\Leftrightarrow x = c\ln(\sec\psi + \tan\psi) + C_2,$$

where C_2 is constant. We have chosen axes so that $x = 0$ when $\psi = 0 \Rightarrow C_2 = 0$

$$\Rightarrow x = c\ln(\sec\psi + \tan\psi). \qquad (7.12)$$

Also,
$$\frac{dy}{dx} = \tan\psi = \sqrt{(\sec^2\psi - 1)} = \sqrt{\left(\frac{y^2}{c^2} - 1\right)}$$

$$\Leftrightarrow x = \int \frac{c\,dy}{\sqrt{(y^2 - c^2)}} = c\cosh^{-1}\left(\frac{y}{c}\right) + K,$$

where K is constant, and from the condition $y = c$ when $x = 0$ it follows that $K = 0$

$$\Rightarrow y = c\cosh(x/c). \qquad (7.13)$$

This is the cartesian equation of the catenary. The student will recognise, in the shape of the catenary, the shape of the curve $y = \cosh x$ discussed in Volume 2, Part I.

From (7.11) and (7.13),
$$s = \sqrt{(y^2 - c^2)} = \sqrt{[c^2\cosh^2(x/c) - c^2]}$$
$$\Rightarrow s = c\sinh(x/c). \qquad (7.14)$$

[*Alternatively*, equations (7.13) and (7.14) can be obtained by writing equation (7.12) in the form

$$\sec\psi + \tan\psi = e^{x/c}$$

from which
$$\sec\psi - \tan\psi = e^{-x/c}.$$

Then addition and subtraction lead to the equations

$$y = c\cosh(x/c), \qquad s = c\sinh(x/c).]$$

From
$$T_0 = T\cos\psi,$$
we have
$$T = T_0 \sec\psi = wc\sec\psi$$
$$\Rightarrow T = wy. \tag{7.15}$$

This equation shows that the tension at any point of the string is proportional to the ordinate at that point.

Example 1. A uniform heavy chain of length 32 m hangs symmetrically over two smooth pegs at the same level so that the lowest point of the portion of the chain between the pegs is 2 m below the level of the pegs. Find the length of either vertical portion of the chain and show that the distance between the pegs is 16 ln 2 m.

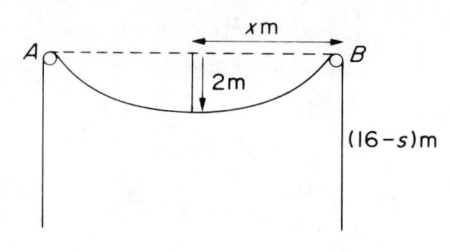

FIG. 7.27.

It is a characteristic of a chain hung freely over smooth pegs in this way that the ends of the vertical portions of the chain are on the x-axis (the directrix of the catenary). The reason for this is apparent in the work which follows.

If $2s$ m is the length of chain between the pegs, Fig. 7.27, the length of each vertical portion of the chain is $(16-s)$ m. The tension in the curved portion of the chain at B is equal to the tension in the vertical portion of the chain at B. Therefore
$$wy = w(16 - s),$$
where y is the ordinate of B referred to the usual axes for the catenary and the weight of 1 metre of the chain is equal to the weight of w kilograms.

Thus, $\qquad y = 16 - s.$
But also, $\qquad y = 2 + c$
and $\qquad y^2 = s^2 + c^2,$

where c m is the parameter of the catenary
$$\Rightarrow (16-s)^2 = s^2 + (14-s)^2$$
$$\Leftrightarrow s^2 + 4s - 60 = 0.$$

Therefore, since s is positive, $s = 6$, $c = 8$, and the length of the vertical portion of the chain is 10 m.

Also at B, $\qquad y = c\cosh(x/c)$
$$\Rightarrow 10 = 8\cosh(x/8)$$
$$\Rightarrow x = 8\cosh^{-1}\left(\frac{5}{4}\right) = 8\ln\left\{\frac{5}{4} + \sqrt{\left(\frac{25}{16} - 1\right)}\right\} = 8\ln 2.$$

The distance between the pegs is $2x$ m $= 16\ln 2$ m.

Example 2. The ends of a uniform flexible chain of length 56 m are attached to two points A and B, where B is 14 m higher than A. The tension at B is $39/25$ times the tension at A. If the lowest point C of the chain lies between A and B, find the lengths of chain CA and CB and show that the parameter c of the catenary in which the chain hangs is 15 m.

AN INELASTIC STRING HANGING UNDER GRAVITY

Let the ordinates of A and B referred to the usual axes be y_A and y_B, and let the tensions at A and B be T_A and T_B, Fig. 7.28.

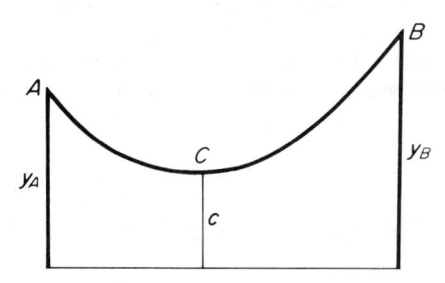

FIG. 7.28.

At A, $$T_A = wy_A,$$

At B, $$T_B = wy_B = w(14 + y_A)$$

$$\Rightarrow \frac{y_A}{14 + y_A} = \frac{25}{39} \Leftrightarrow y_A = 25.$$

At A, $\quad s_A^2 = y_A^2 - c^2 \quad \Leftrightarrow s_A^2 = 25^2 - c^2.$
At B, $\quad (56 - s_A)^2 = y_B^2 - c^2 \quad \Leftrightarrow (56 - s_A)^2 = 39^2 - c^2.$

Hence, equating values of c^2, we find $s_A = 20$, $c = 15$.
AC is 20 m long, CB is 36 m long and the parameter of the catenary is 15 m.

Example 3. A uniform heavy flexible string AB, of length a and weight wa, will break if subjected to a tension greater than $3wa$. The string is attached to a fixed point at A and hangs in equilibrium under the action of a horizontal force nwa applied at B. If the string is about to break, show that (1) $n = 2\sqrt{2}$, (2) the tangent to the string at A makes an angle $\sin^{-1}(\frac{1}{3})$ with the horizontal, (3) the height of A above the level of B is $(3 - 2\sqrt{2})a$.

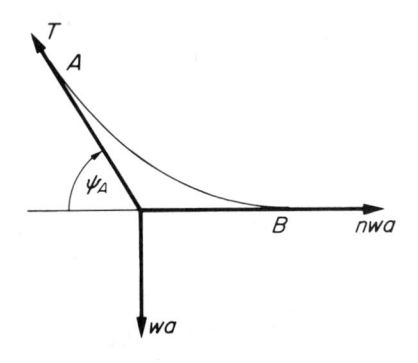

FIG. 7.29.

From equation (7.15) the tension in the string is greatest at A and therefore, since the string is just about to break, the force acting on the string at A, equal and opposite to the tension in the string at A, is $3wa$. This force acts along the tanget to the string Fig. 7.29.

The whole string is in equilibrium under the action of this force, the horizontal force nwa at B, and its weight wa. The lines of action of these three forces will therefore be concurrent as shown in Fig. 7.29.

From the triangle of forces,
$$(3wa)^2 = (wa)^2 + (nwa)^2$$
and hence
$$n = 2\sqrt{2}. \tag{1}$$

The tension at B is nwa; but, since the string is horizontal at B, the tension at B is $wy_B = wc$, where y_B is the ordinate of B and c is the parameter of the catenary
$$\Rightarrow cw = nwa$$
$$\Leftrightarrow c = 2a\sqrt{2}.$$
At A, since $T_A = 3wa$,
$$y_A = 3a$$
and since $y = c \sec \psi$, where ψ is the angle made by the tangent to the string with the horizontal,
$$3a = c \sec \psi_A \Leftrightarrow \sec \psi_A = \frac{3}{2\sqrt{2}} \Leftrightarrow \sin \psi_A = \tfrac{1}{3}. \tag{2}$$

The vertical distance of A above B is
$$y_A - c = 3a - 2a\sqrt{2} = (3 - 2\sqrt{2})a. \tag{3}$$

The tightly stretched wire

We consider a uniform wire of length $2l$ tightly stretched between two points A and B on the same horizontal level and distance $2a$ apart; k is the *sag* at the mid-point of the wire and c is the parameter of the catenary, Fig. 7.30. Since chord $BC <$ arc BC, it follows from the right-angled triangle BCD that
$$k^2 = BC^2 - BD^2 < l^2 - a^2.$$
But $l - a$ is small and therefore
$$k \approx \sqrt{[2l(l-a)]}$$
is small compared with l.

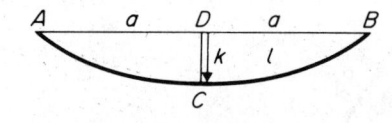

FIG. 7.30.

Then, for the point B, from equation (7.13),
$$k + c = c \cosh(a/c),$$
and from equation (7.14),
$$l = c \sinh(a/c).$$
Hence
$$\left(\frac{k+c}{c}\right)^2 - \left(\frac{l}{c}\right)^2 = 1$$
$$\Leftrightarrow c = \frac{l^2}{2k} - \frac{k}{2}.$$

It follows that, since k is small compared with l or a, c is large and approximately equal to $l^2/(2k)$, and a/c is approximately equal to $2k/l$ and is therefore small. We now obtain a good approximation to l from the formula $l = c \sinh(a/c)$ by expanding $\sinh(a/c)$ and neglecting terms involving higher powers of a/c than the second. Thus,

$$l \approx c\left(\frac{a}{c} + \frac{a^3}{6c^3}\right),$$

$$l \approx a + \frac{a^3}{6c^2}.$$

Also, if y_B is the ordinate at B,

$$y_B = c \cosh \frac{a}{c}$$

$$\Rightarrow y_B \approx c\left(1 + \frac{a^2}{2c^2}\right)$$

$$\Leftrightarrow y_B \approx c + \frac{a^2}{2c}.$$

(This approximation shows the shape of the curve, when the sag is small compared with the span, to be approximately parabolic.)

But
$$k = y_B - c$$

$$\Rightarrow k \approx \frac{a^2}{2c}.$$

The difference between the length and the span

$$2(l-a) \approx \frac{a^3}{3c^2} = \frac{8}{3}\left(\frac{a^2}{2c}\right)^2 \bigg/ 2a \approx \frac{8}{3} \cdot \frac{k^2}{2a}.$$

Therefore, the difference between the length and the span

$$\approx (8/3)\{(\text{sag})^2/\text{span}\}.$$

Exercise 7.7

1. A uniform flexible chain, of length $54a$, hangs over two small smooth pegs, the lengths of the vertical portions being $20a$ and $13a$ respectively. Show that the parameter of the catenary between the pegs is $12a$, and that the horizontal distance between the pegs is $12a \ln(9/2)$. (L.)

2. A uniform chain, of length 14 m and mass 28 kg, is attached at its ends to two fixed points A and B, B being 2 m above the level of A. The tension at A is equal to the weight of 26 kg. Show that the length of chain between A and the lowest point C (which lies between A and B) is 5 m, and find the tension at C. Show also that the horizontal distance between A and B is 12 ln 3 m.

3. A uniform chain, of length $2l$ and weight $2wl$, hangs in equilibrium with its ends attached to two points B and C at the same level. The depth of the middle point A of the chain below BC (the central sag) is k and the tangents to the chain at B and C are each inclined at an angle α to the horizontal. Show that $k = l \tan(\alpha/2)$. If the tension at B is double the tension at A, show that $k = l/\sqrt{3}$. If now the distance BC is altered so that the central sag is halved, show that the tension at A is increased in the ratio 11/4. (N.)

4. A uniform heavy flexible string AB, of length $2a$, has its ends A, B attached to two light rings which can slide on a fixed straight horizontal wire. The coefficient of friction between each ring and the wire is $1/\sqrt{3}$. Show that, when equilibrium is limiting at A and B, the tangent to the string at A makes an angle $\pi/3$ with the horizontal. Find the depth of the mid-point of the string below AB and show that the distance AB is

$$(2a/\sqrt{3}) \ln(2 + \sqrt{3}).$$ (N.)

5. A uniform heavy flexible string AB, of length $2l$, has its ends A, B attached to light rough rings which can slide on a fixed horizontal rod. The coefficient of friction between the ring and the rod is $1/\sqrt{3}$. Show that when equilibrium is limiting at both ends the depth of the mid-point M of AB below the rod is $l/\sqrt{3}$.

If a particle of the same weight as the string is now attached to M and the string takes up a new symmetrical position of limiting equilibrium, show that M falls a distance $l(3-\sqrt{7})/\sqrt{3}$. (L.)

6. The ends of a uniform string of length $2l$ are attached to small light rings which are threaded on a rough straight horizontal wire. Prove that the greatest possible distance between the rings in a position of equilibrium is $2l \tan \lambda \ln \cot \lambda/2$, where λ is the angle of friction. (L.)

7. A uniform heavy chain AB is of length $8l$ and weight w per unit length. A smooth ring of weight wl fixed to the end A is free to slide on a fixed vertical wire, and the chain is slung over a smooth peg C so that a length $5l$ of the chain hangs vertically. Find the parameter of the catenary AC, and show that the distance of the peg C from the vertical wire is $3l \ln (\sqrt{10}-1)$. (L.)

8. A uniform chain of length 23 m is attached to a fixed point A and passes over a small smooth peg at B, the portion beyond B hanging vertically. If in the equilibrium position the catenary in which the chain hangs has its vertex between A and B and the tangents to the catenary at A and B make angles $\tan^{-1}(3/4)$ and $\tan^{-1}(12/5)$ respectively with the horizontal, find the parameter c of the catenary and the vertical and horizontal distances of B from A. (L.)

9. The sag of a telegraph wire stretched between two poles at a distance $2a$ apart is k. Show that the tension at each end of the wire is approximately

$$w\left(\frac{a^2}{2k}+\frac{7k}{6}\right),$$

where w is the weight per unit length of the wire. (L.)

10. A uniform chain of length $2l$ hangs over two smooth pegs at the same level and at a distance $2kl$ apart. Prove that for a position of equilibrium in which the chain hangs in a catenary $y = c \cosh(x/c)$ between the pegs, l/c is a root of the equation $k\xi = \ln \xi$. Hence show that equilibrium is only possible if $k \leqslant e^{-1}$. (L.)

7:8 A ROPE IN CONTACT WITH A ROUGH SURFACE

It is possible to make practical use of the friction caused by contact with a rough surface by taking a number of turns of a rope round a cylinder such as a bollard. We now consider the mechanics of this situation.

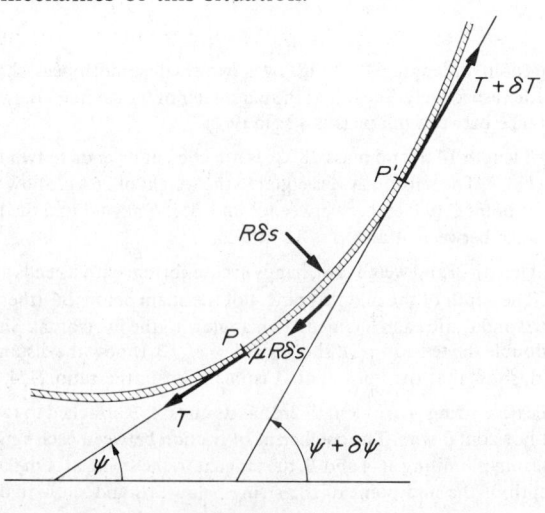

Fig. 7.31.

Figure 7.31 shows the forces on a small portion PP' of a light rope in contact with a rough surface. Since the rope is acted upon by external forces, the tension will not be constant throughout the rope; suppose that the tensions at P, P' are $T, T+\delta T$ respectively. Friction and normal reaction will be proportional to the length δs of PP'; if the normal reaction is $R\delta s$, and the friction is limiting, the frictional force will be $\mu R\delta s$ in the opposite sense to that in which T is increasing, μ being the coefficient of friction. Resolving in the direction of T, which makes the angle ψ with the initial direction of the rope,

$$T + \mu R \delta s = (T + \delta T)\cos\delta\psi = T + \delta T + O[(\delta\psi)^2]$$
$$\Rightarrow \mu R \delta s \approx \delta T. \qquad (7.16)$$

Resolving in the normal direction,

$$R\delta s = T\sin\delta\psi = T\delta\psi + O[(\delta\psi)^3] \qquad (7.17)$$
$$\Rightarrow R\delta s \approx T\delta\psi.$$

From equations (7.16) and (7.17),

$$\mu\delta\psi \approx \frac{\delta T}{T} \Rightarrow \frac{dT}{d\psi} = \mu T$$

$$\Rightarrow \int_0^\psi \mu\,d\psi = \int_{T_0}^T \frac{1}{T}\,dT,$$

where T_0 is the tension at one end of the rope,

$$\Rightarrow \mu\psi = \ln(T/T_0) \Leftrightarrow T = T_0 e^{\mu\psi}. \qquad (7.18)$$

This result shows that the tension increases rapidly as ψ increases; if, for instance, $\mu = 0.5$, each complete turn of the rope round a bollard multiplies the tension by a factor e^π. A ship of mass 2000 tonnes with a speed of $5\,\mathrm{m\,s^{-1}}$ has a momentum of $2000 \times 10^3 \times 5 = 10^7\,\mathrm{kg\,m\,s^{-1}}$; since $e^{3\pi} > 10^4$, a man needs only to exert a pull of 100 N on a rope with the three turns round a bollard to bring the ship to rest in 10 seconds—provided of course that the rope can withstand the resulting tension!

Exercise 7.8

1. A light rough rope passes over a circular cylinder in a plane normal to its axis, and is about to slip. Show that, if θ is the angle subtended at the axis by two points A, B on the rope, $T_A = e^{\pm\mu\theta}T_B$, where the sign depends upon the direction of slip, μ is the coefficient of friction between the rope and the cylinder, and T_A, T_B are the tensions of the rope at A and B.

A horizontal rough circular beam is fixed in position; a parallel circular beam of weight W is suspended by a rope which is fixed to the bottom of the upper beam and passes half-way round both beams twice; the distance between the beams is sufficiently large to allow all connecting sections of the rope to be regarded as vertical. Calculate the tension on the free end of the rope,
   (a) when the lower beam is just rising,
   (b) when it is just falling.

2. A light rope AB passes over a rough circular cylinder, which is fixed with its axis horizontal. Bodies of masses M and m are attached at A and B respectively so that the ends of the rope hang in a vertical plane perpendicular to the axis of the cylinder. If A is about to descend, find the greatest mass that can be added to B before B starts to descend.

3. If, in the initial situation of question 2, the coefficient of friction μ between the rope and the cylinder is small enough for A to descend, find the acceleration of A.

Miscellaneous Exercise 7

1. Forces $(-9\mathbf{i}+\mathbf{j}-2\mathbf{k})$, $(3\mathbf{i}+2\mathbf{j}-3\mathbf{k})$ and $(6\mathbf{i}-3\mathbf{j}+5\mathbf{k})$ act through points with position vectors $(-11\mathbf{i}+2\mathbf{j}-5\mathbf{k})$, $(\mathbf{i}-4\mathbf{j}+5\mathbf{k})$ and $(-8\mathbf{i}+4\mathbf{j}-8\mathbf{k})$ respectively. Prove that these forces are equivalent to a couple, and find the moment of this couple. (L.)

2. The position vectors of the vertices A, B, C, D of a tetrahedron $ABCD$ are $\mathbf{a, b, c, d}$ respectively relative to an origin O.

Forces $\lambda\overrightarrow{AB}$, $\lambda\overrightarrow{AC}$, $\lambda\overrightarrow{AD}$, $\mu\overrightarrow{OB}$, $\mu\overrightarrow{OC}$, $\mu\overrightarrow{OD}$, where λ and μ are constants, act along the edges AB, AC, AD, OB, OC, OD respectively. Find the resultant of this system of forces.

If $\lambda = 2$, $\mu = 3$, $\mathbf{a} = (15\mathbf{i}+5\mathbf{j})$, $\mathbf{b} = (\mathbf{i}-5\mathbf{k})$, $\mathbf{c} = (3\mathbf{i}+2\mathbf{j}-3\mathbf{k})$, $\mathbf{d} = (-4\mathbf{i}-8\mathbf{j}+2\mathbf{k})$, find the moment of the couple required to maintain equilibrium if the tetrahedron is smoothly hinged along the edge BC. (L.)

3. $\mathbf{i, j, k}$ are unit vectors in the direction of three mutually perpendicular axes forming a right-handed set. Write down the values of $\mathbf{i} \times \mathbf{j}$, $\mathbf{j} \times \mathbf{k}$ and $\mathbf{i} \times \mathbf{k}$.

A cube has edges of length a. There are six edges which do not meet a given diagonal, joined together in a chain. Forces, each of magnitude F, act along these six edges. Find the force or couple needed to keep the cube in equilibrium if the senses in which these forces act round the chain are, consecutively, (i) $+ + + + + +$, (ii) $+ - + - + -$, (iii) $+ + - + + -$. (O.C.S.M.P.)

4. The force $\mathbf{F}_1 = (7\mathbf{i}+5\mathbf{j}+2\mathbf{k})$ acts at the point whose position vector is $(-2\mathbf{i}-3\mathbf{j}+3\mathbf{k})$ and the force $\mathbf{F}_2 = (-\mathbf{i}+4\mathbf{j}+\mathbf{k})$ acts at the point whose position vector is $(3\mathbf{i}+10\mathbf{j}+7\mathbf{k})$. Show that the lines of action of these forces intersect. Find the magnitude of their resultant and the vector and cartesian equations of the line of action of this resultant.

Show that the cosine of the angle between the lines of action of the forces $\mathbf{F}_1$ and $\mathbf{F}_2$ is $(5\sqrt{39})/78$. (L.)

5. A non-uniform rigid beam AB, of length $3a$ and weight nW, rests on supports P and Q at the same level, where $AP = PQ = QB = a$. When a load of weight W is hung from A, the beam is on the point of tilting about P. Find the distance of the centre of gravity of the beam from A. When an additional load of weight W_1 is hung from B, the forces exerted on the supports at P and Q are equal. Find W_1 in terms of n and W.

If a couple, of moment L and acting in the vertical plane through AB, is now applied to the loaded beam, the reaction at P is increased in the ratio 3:2. Show that

$$L = \tfrac{1}{3}(n+1)Wa.$$ (N.)

6. A uniform circular cylinder of weight W and radius a rests between two fixed rough planes each inclined to the horizontal at an angle α. Prove that the least couple required to rotate the cylinder about its axis has moment

$$\frac{\mu Wa}{(1+\mu^2)\cos\alpha},$$

where μ is the coefficient of friction between the cylinder and each plane. (L.)

7. A chain of four equal uniform rods, AB, BC, CD, DE, each of weight W and freely jointed together at B, C, D, hangs symmetrically from two fixed points A and E in the same horizontal line. The angles of inclination of AB and BC to the vertical are θ and ϕ respectively. Prove that $\tan\phi = 3\tan\theta$.

Show also that the resultant reaction at B is

$$W\sqrt{(1+\tfrac{1}{4}\tan^2\phi)}.$$ (L.)

8. Three equal uniform rods AB, BC, CD, each of length $2c$ and weight W, are smoothly jointed at B and C and rest with AB, CD in contact with two smooth pegs at the same level. In the position of equilibrium AB and CD are inclined at an angle α to the vertical, BC being horizontal. Prove that the distance between the pegs is $2c(1+\tfrac{2}{3}\sin^3\alpha)$.

If β is the angle which the reaction at B makes with the vertical, prove that $\tan\alpha\tan\beta = 3$. (O.C.)

9. Three equal smooth circular cylinders each of weight W are placed with their axes parallel, each cylinder in contact with the other two. A light smooth continuous elastic band, whose modulus of elasticity is λ, is placed round them in a plane perpendicular to the axes. In this position the band is just unstretched. The cylinders are then placed so that the curved surfaces of two of them rest on a smooth horizontal table. If T is the tension in the elastic band in the position of equilibrium, θ the inclination to the vertical of the plane passing through the axis of the upper cylinder and one of the lower cylinders, show that

$$T = \tfrac{1}{2} W \tan \theta$$

and
$$\frac{W}{\lambda} = \frac{2 \cot \theta (2 \sin \theta - 1)}{3 + \pi}.$$ (N.)

*10. A cylinder of weight W and radius $4a$ lies on a smooth floor and against a smooth wall. A smooth cylinder of radius a lies on the floor in contact with the first cylinder along a generator. A horizontal force P is applied symmetrically to the smaller cylinder so as to push the system against the wall, its line of action being at a height a above the floor. Prove that as P is increased, the large cylinder will cease to exert pressure on the floor when $P = \tfrac{4}{3} W$.

The two cylinders are removed and bound together by an endless cord at tension T passing round them in a plane perpendicular to their axes. The system is placed on the floor away from the wall. Show that each cylinder exerts on the floor a pressure equal to its own weight. (O.C.)

*11. Two equal uniform ladders AB, BC, each of length $2l$ and weight W, are smoothly hinged together at B, while the ends A and C rest on a rough horizontal plane. The coefficient of friction is μ at A and C, and the angle ABC is 2α. Prove that a man of weight w can ascend a distance x given by

$$x = \frac{2lW(2\mu - \tan \alpha)}{w(\tan \alpha - \mu)},$$

provided that $x < 2l$ and $\mu < \tan \alpha < 2\mu$. (L.)

*12. A rhombus $ABCD$ is formed by four uniform freely jointed rods each of weight W and length $2a$, and kept in shape by a light rod BD of length $2a$. The rhombus is suspended from A, and a particle of weight w is suspended from two inextensible strings each of length $\sqrt{3}a$ attached to B and D. Prove that the thrust in BD is $\{8\sqrt{3}W + (2\sqrt{3} + 3\sqrt{2})w\}/12$.

*13. Four uniform rods each of weight W and length $2a$ are freely jointed at their extremities to form a rhombus $ABCD$. This framework is freely suspended from A and is maintained in the form of a square by a light inextensible string which joins C to the mid-point of AB. Use the principle of virtual work to show that the tension in the string is $W\sqrt{10}$.

The framework, with string attached, is now placed flat on a smooth horizontal table and the rod AD is rigidly clamped to the table. A couple of moment M, in the plane of the framework, is applied to the rod AB in such a sense as to tighten the string. Find the tension in the string.

*14. AB and BC are two rods fixed in a vertical plane, AB being inclined at an angle $\alpha (> \pi/4)$ to the horizontal with BC perpendicular to AB and B below the levels of A and C. A uniform rod PQ of weight W and length $4a$ has smooth light rings fixed to its ends so that P can slide on AB and Q on BC. Prove that the height y of the centre of gravity of the rod PQ above B is given by

$$y = 2a \sin(\alpha + \theta),$$

where θ is the angle BPQ.

Use the principle of virtual work to show that if the rod PQ is in equilibrium $\theta = \tfrac{1}{2}\pi - \alpha$.

If a weight W is attached to the rod PQ at a distance a from P, prove that in the new position of equilibrium

$$5 \tan \theta = 3 \cot \alpha.$$

*15. Three light rods OA, OB, OC, each of length a, are freely jointed at O and their other ends are joined by light inextensible strings BC, CA, AB, each of length b. The tripod is placed with the ends A, B, C on smooth horizontal ground with the strings taut and a weight W is suspended from O. Prove that
  (i) the height h of O above the ground is given by

$$h^2 = a^2 - b^2/3,$$

  (ii) the thrust in each rod is $Wa/3h$,
  (iii) the tension in each string is $Wb/9h$.

16. A heavy uniform rectangular lamina $ABCD$, in which the length of the diagonals is $16a$ and the angle

ADB is α ($\cos^{-1}\frac{4}{5} < \alpha < 45°$), rests in a vertical plane with the sides AB, AD on two small smooth horizontal pegs E and F. EF has length $5a$ and makes an angle 2α with the upward vertical.

Prove that the lamina is in equilibrium when AC is vertical and when AC makes an angle $\cos^{-1}\frac{4}{5}$ with the vertical on either side. Prove that the first of these positions is stable and the others unstable. (L.)

17. Two equal uniform rods AB and AC, each of length $2b$, are freely jointed at A. The rods rest on a smooth circular cylinder of radius a with their plane perpendicular to the axis of the cylinder, which is horizontal. Prove that if 2θ is the angle BAC when the rods are in equilibrium with A vertically above the axis of the cylinder, then $b \sin^3 \theta = a \cos \theta$, and show that this equation has only one root in the range $0 < \theta < \pi/2$.

Show, also, that if the rods are constrained to move in a vertical plane and A is constrained to move vertically the position of equilibrium is stable. (L.)

18. A uniform bar AB of length l and weight W_1 is smoothly hinged to a fixed point A. The bar is supported by a light inextensible string attached to B and passing over a small light smooth pulley at a point C, which is vertically above A at a height $a(> l)$. The string carries a freely hanging weight W_2 at its other end. Show that there are three positions of equilibrium, provided that

$$\frac{2a}{a+l} < \frac{W_1}{W_2} < \frac{2a}{a-l}.$$

If $W_1 = 2W_2$ show that two of the positions of equilibrium are stable and one is unstable. (L.)

19. A hemispherical cup, of radius r and mass m, has a particle of mass m attached to a point on its rim. The cup is placed in equilibrium with a point of its curved surface in contact with the highest point of a fixed rough sphere of radius R. Prove that the equilibrium is stable, if

$$R > r(4\sqrt{5/5} - 1).$$ (N.)

20. Each end of a uniform chain, of length $4a$ and weight $8W$, is attached to a ring of weight W. The rings are threaded on a rough fixed horizontal bar and the coefficient of friction between each ring and the bar is $3/5$. If the rings are in limiting equilibrium, prove that the sag in the chain is a and that the distance between the rings is $3a \ln 3$. (N.)

21. A uniform flexible chain, of length $2l$ and weight W, hangs from two fixed points H, K on the same level, so that the lowest point of the chain is at a depth h below the line HK. Show that the parameter of the catenary in which the chain hangs is given by $2hc = l^2 - h^2$, and deduce that the tension at either point H or K is $\frac{1}{4}W(l/h + h/l)$. (L.)

22. A uniform heavy chain BC, of length $2l$ and weight w per unit length, hangs symmetrically over two fixed smooth pegs H, K at the same level, and the lowest point A of the portion of the chain between the pegs is at a depth $l/18$ below the level of the pegs. Find the lengths of the portions of the chain which hang vertically, and the distance between the pegs. Show further that the thrust on either peg is $\frac{1}{6}\sqrt{13}$ of the weight of the chain. (L.)

23. A force system consists of three forces $\mathbf{F}_1 = \mathbf{i} - \mathbf{j} + 2\mathbf{k}$, $\mathbf{F}_2 = \mathbf{i} + 3\mathbf{j} - \mathbf{k}$, $\mathbf{F}_3 = s\mathbf{i} + t\mathbf{j} + 2\mathbf{k}$, where s and t are constants. They act respectively through the points whose position vectors are $3\mathbf{i} - \mathbf{j} + \mathbf{k}, \mathbf{j} + 2\mathbf{k}, \mathbf{k}$. Obtain, in terms of s and t, the equivalent force system consisting of a single force $\mathbf{F}$ acting through the origin and a couple of moment $\mathbf{G}$.

Determine s and t so that $\mathbf{G}$ is parallel to $\mathbf{F}$. (L.)

24. In a regular tetrahedron $OABC$ each edge is of length l. The centroid of the triangle ABC is G and M is the mid-point of OG. With O as origin the position vectors of A, B, C are $\mathbf{a}, \mathbf{b}, \mathbf{c}$ respectively. Show that the forces $\lambda \overrightarrow{MA}$, $\lambda \overrightarrow{MB}$, $\lambda \overrightarrow{MC}$ are mutually perpendicular and show also that the resultant of these forces is $\frac{1}{2}\lambda(\mathbf{a} + \mathbf{b} + \mathbf{c})$. Find the position vector of the point where the line of action of this resultant meets the plane ABC.

25. A uniform sphere of radius b and weight W is placed on a fixed rough sphere of radius a and a weight w is attached to the upper sphere at the point of contact. Show that the potential energy when the upper sphere has rolled into a position where the common radius of the spheres is inclined at an angle θ to the vertical is

$$V = (W + w)(a + b) \cos \theta - wb \cos \frac{a+b}{b}\theta + \text{constant}.$$

Hence show that the sphere can rest in stable equilibrium with $\theta = 0$ if $wa > Wb$.

26. One end of a uniform chain of length l is attached to the back of a car at a height h above the ground ($h < l$). The car is travelling at constant speed along a straight level road, the coefficient of friction between the road and the chain being μ. Prove that the length of chain in contact with the ground is

$$\mu h + l - \sqrt{[(\mu^2 + 1)h^2 + 2\mu h l]}.$$

Prove that when $l = 5h$ and $\mu = \frac{1}{2}$ the tension in the chain nowhere exceeds half its own weight. (L.)

27. Forces $\mathbf{F}_1 = \mathbf{i} + 2\mathbf{j} + 3\mathbf{k}$ and $\mathbf{F}_2 = 2\mathbf{i} + \mathbf{k}$ act at points with position vectors $2\mathbf{i} + 5\mathbf{j} + c\mathbf{k}$ and $5\mathbf{i} + c\mathbf{j} + 2\mathbf{k}$ respectively. Given that these forces meet at a point, find the value of c and determine a vector equation of the line of action of the resultant of forces $\mathbf{F}_1$ and $\mathbf{F}_2$.
Show that the sum of the moments of these forces about the origin is $12\mathbf{i} - 4\mathbf{j} - 7\mathbf{k}$. (L.)

28. Forces $\mathbf{F}_1 = 2\mathbf{i} - \mathbf{j} + 3\mathbf{k}$, $\mathbf{F}_2 = 4\mathbf{i} + \mathbf{j} + 5\mathbf{k}$, $\mathbf{F}_3 = -6\mathbf{i} - 8\mathbf{k}$ act respectively at points with position vectors

$$\mathbf{i} - 2\mathbf{j} - \mathbf{k}, \qquad 7\mathbf{i} - 2\mathbf{j} + 7\mathbf{k}, \qquad 3\mathbf{i} - 3\mathbf{j} + \mathbf{k}.$$

Show that the lines of action of $\mathbf{F}_1$ and $\mathbf{F}_2$ meet at the point with position vector $3\mathbf{i} - 3\mathbf{j} + 2\mathbf{k}$.
Show that this system of forces is equivalent to a couple, and find the magnitude of the couple.
If the force $\mathbf{F}_3$ is now replaced by a force $\mathbf{F}_4$ such that $\mathbf{F}_1$, $\mathbf{F}_2$ and $\mathbf{F}_4$ are in equilibrium, find a vector equation of the line of action of $\mathbf{F}_4$. (L.)

29. A uniform rectangular lamina $ABCD$ rests in a vertical plane with A in contact with a smooth vertical wall and B attached by a light inextensible string to a point P of the wall vertically above A. The plane of the lamina is perpendicular to the wall. $AB = BP = 2a$ and $AD = 2b$. If the inclination of the string to the wall is θ, show that the depth of the centre of mass of the lamina below P is $3a \cos \theta + b \sin \theta$.
Hence, or otherwise, show that in equilibrium $\tan \theta = b/(3a)$ and determine whether the equilibrium is stable or unstable. (L.)

30. The resultant of three forces $\mathbf{F}, \mathbf{G}, \mathbf{H}$ is $\mathbf{R}$, the resultant of $\mathbf{F}, \mathbf{G}, -\mathbf{H}$ is $\mathbf{S}$ and the resultant of $\mathbf{F}, -\mathbf{G}, \mathbf{H}$ is $\mathbf{T}$. Prove that if $\mathbf{S}$ is perpendicular to both $\mathbf{R}$ and $\mathbf{T}$ then $\mathbf{S}$ is also perpendicular to $\mathbf{G}$. (O.)

31. A uniform rod AB, of mass m and length $2a$, can turn about a fixed smooth pivot at A. A light inextensible string is attached to the other end B of the rod, and passes through a smooth ring fixed at a point C vertically above A so that $AC = 2a$. A particle P of mass $m/2$ hangs from the other end of the string. The angle CAB is θ, where $0 \leqslant \theta \leqslant \pi$. Show that the potential energy, V, of the system is given by

$$V = 2mga \sin(\theta/2) + mga \cos \theta + \text{constant}.$$

Show also that the system is in equilibrium when $\theta = \pi/3$ and when $\theta = \pi$ and investigate the stability of these positions of equilibrium. (L.)

32. An endless uniform flexible string, of length greater than $2\pi a$, hangs over a smooth pulley of radius a and is in contact with two-thirds of the circumference of the pulley. Show that the total length of the string is

$$4\pi a/3 + 3a/\ln(2 + \sqrt{3}). \qquad \text{(L.)}$$

33. A fixed smooth wire is in the form of the parabola $y^2 = 4ax$, where Oy is vertical. A small ring of mass m slides on the wire, and is attached by a light spring, of natural length $2a$ and modulus kmg, to a pin fixed at the focus of the parabola. Show that, when the ring is at a point with coordinates $(at^2, 2at)$, the potential energy of the system is given by

$$V = \tfrac{1}{4}kmga(t^2 - 1)^2 + 2mgat + \text{constant}.$$

Show that there is at least one position of equilibrium for all positive values of k, and obtain the set of values of k for which there is only one equilibrium position. [Assume that the spring remains straight when in compression.] (L.)

8 Further Dynamics of a Particle with one Degree of Freedom

8:1 DAMPED HARMONIC OSCILLATIONS

When a particle moving in a straight line, in what otherwise would be SHM, is also subject to a resistance which is proportional to its speed, the equation of motion becomes

$$\frac{d^2x}{dt^2} = -\omega^2 x - 2\lambda \frac{dx}{dt},$$

where x is the directed distance of the particle from the fixed point and ω, λ are constants.

This equation of motion is still a linear equation of the second order with constant coefficients. [The method of solution of such equations is discussed in Volume 2, Part I, § 18:7–8.] The nature of the motion of the particle as determined by its equation of motion is dependent on the nature of the roots of the auxiliary equation used in solving the equation of motion. The example below illustrates the different cases which can arise.

Example. A particle of mass m moves in a horizontal straight line under the action of a force directed towards a fixed point O of the line. This force varies as the distance of the particle from O and is equal to $4mx$ at a distance x from O. There is a resistance to the motion of the particle which is proportional to its speed and which is equal to $2\lambda mv$ when the speed of the particle is v. The particle starts from rest at the point $x = a$. Discuss the motion of the particle in each of the three cases

(a) $\lambda = 2\frac{1}{2}$,   (b) $\lambda = 2$,   (c) $\lambda = 1$.

The force $4mx$ is equivalent to a force $-4mx$ in the direction of x increasing on whichever side of O the particle is. The resistance to the motion of the particle is equivalent to a force $-2\lambda m\dot{x}$ in the direction of x increasing in whichever direction the particle is moving. Hence the equation of motion of the particle is

$$\frac{d^2x}{dt^2} + 2\lambda \frac{dx}{dt} + 4x = 0.$$

(a) When $\lambda = 2\frac{1}{2}$, this equation becomes

$$\frac{d^2x}{dt^2} + 5\frac{dx}{dt} + 4x = 0.$$

The auxiliary equation is $m^2 + 5m + 4 = 0$ and the roots of this equation are $m_1 = -4$, $m_2 = -1$. The solution of the equation of motion is therefore

$$x = Ae^{-4t} + Be^{-t}.$$

Since $x = a$ when $t = 0$, $A + B = a$. Also

$$\frac{dx}{dt} = -4Ae^{-4t} - Be^{-t}.$$

and since $dx/dt = 0$ when $t = 0$, $-4A - B = 0$

$$\Rightarrow A = -\frac{a}{3}, \quad B = \frac{4a}{3}$$

$$\Rightarrow x = -\tfrac{1}{3}a(e^{-4t} - 4e^{-t})$$

$$\Leftrightarrow x = \tfrac{1}{3}ae^{-4t}(4e^{3t} - 1).$$

This expression for x is positive for all values of t and $\lim_{t \to \infty} x = 0$.

It follows from this analysis that the particle never passes through the origin $x = 0$ but approaches it asymptotically. The x-t graph of the motion is shown in Fig. 8.1(i).

(b) When $\lambda = 2$, the equation of motion becomes

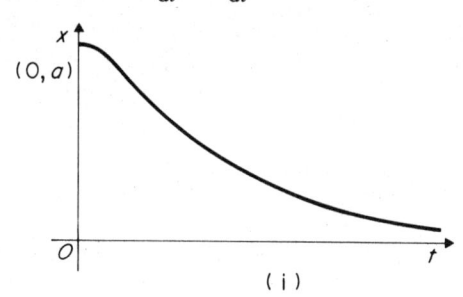

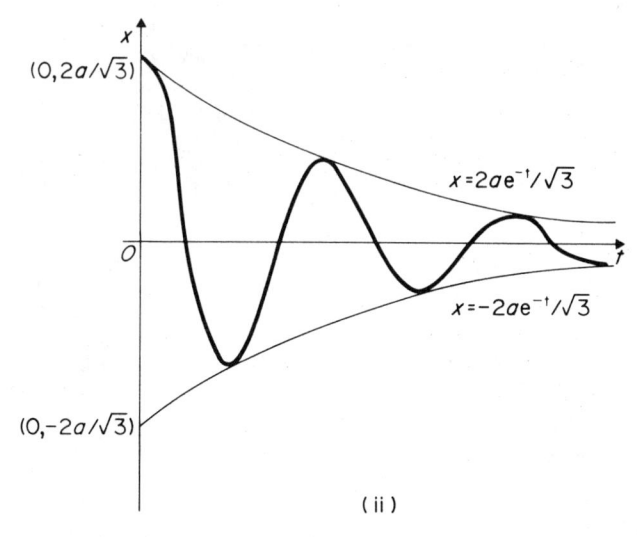

$$\frac{d^2x}{dt^2} + 4\frac{dx}{dt} + 4 = 0.$$

FIG. 8.1.

The auxiliary equation is $m^2 + 4m + 4 = 0$, which has a repeated root $m = -2$. The complete solution of the equation of motion is therefore

$$x = e^{-2t}(A + Bt),$$

where A and B are constants.
$$x = a \text{ when } t = 0 \Rightarrow A = a.$$

Also
$$\frac{dx}{dt} = -2e^{-2t}(A + Bt) + Be^{-2t}$$

and $dx/dt = 0$ when $t = 0 \Rightarrow -2A + B = 0$, $B = 2a$
$$\Rightarrow x = ae^{-2t}(1 + 2t).$$

Again x is positive for all values of t and
$$\lim_{t \to \infty} x = a \lim_{t \to \infty} \left[e^{-2t} + 2t/(1 + 2t + 2t^2 + \ldots) \right]$$
$$= a \lim_{t \to \infty} \left[e^{-2t} + 1 \Big/ \left(\frac{1}{2t} + 1 + 2t + \ldots \right) \right]$$
$$= 0.$$

The motion is similar to that of case (a).

(c) When $\lambda = 1$ the equation of motion becomes
$$\frac{d^2x}{dt^2} + 2\frac{dx}{dt} + 4 = 0.$$

The auxiliary equation is $m^2 + 2m + 4 = 0$, which has the complex roots
$$m_1 = -1 + i\sqrt{3}, \quad m_2 = -1 - i\sqrt{3}.$$

The complete solution of the equation of motion is therefore
$$x = e^{-t}[A \cos(t\sqrt{3}) + B \sin(t\sqrt{3})].$$

$x = a$ when $t = 0 \Rightarrow A = a$.
Also $dx/dt = -e^{-t}[A \cos(t\sqrt{3}) + B\sin(t\sqrt{3})] + e^{-t}[-\sqrt{3}A\sin(t\sqrt{3}) + \sqrt{3}B\cos(t\sqrt{3})]$ and $dx/dt = 0$ when $t = 0$

$$\Rightarrow -A + \sqrt{3}B = 0, \quad B = \frac{a\sqrt{3}}{3}$$

$$\Rightarrow x = ae^{-t}\left[\cos(t\sqrt{3}) + \frac{\sqrt{3}}{3}\sin(t\sqrt{3}) \right]$$

$$\Leftrightarrow x = ae^{-t}\frac{2}{\sqrt{3}}\sin\left[(t\sqrt{3}) + \frac{\pi}{3} \right].$$

Therefore x has zeros given by
$$t\sqrt{3} + \frac{\pi}{3} = n\pi, \quad (n = 1, 2, 3, \ldots), \quad \text{i.e. by } t = \frac{\pi}{\sqrt{3}}\left(\frac{3n - 1}{3} \right).$$

Also
$$\frac{dx}{dt} = \frac{2ae^{-t}}{\sqrt{3}}\left[\sqrt{3}\cos\left(t\sqrt{3} + \frac{\pi}{3} \right) - \sin\left(t\sqrt{3} + \frac{\pi}{3} \right) \right]$$

and thus $\frac{dx}{dt}$ is zero when $\tan\left(t\sqrt{3} + \frac{\pi}{3} \right) = \sqrt{3}$

$$\Rightarrow t\sqrt{3} + \frac{\pi}{3} = n\pi + \frac{\pi}{3}, \quad (n = 0, 1, 2, \ldots),$$

$$\Leftrightarrow t = \frac{n\pi}{\sqrt{3}}.$$

Therefore x has stationary values (which are alternately maxima and minima) for values of t given by $t = n\pi/3$. Because of the factor e^{-t} in the expression for x, the numerical values of these maxima and minima decrease progressively as t increases and the decrease is in geometrical progression. For all values of t, $|x| \leqslant 2ae^{-t}/\sqrt{3}$. For values of t given by $t\sqrt{3} + \pi/3 = \frac{1}{2}(4n+1)\pi$, $(n = 0, 1, 2, \ldots)$, $x = 2ae^{-t}/\sqrt{3}$ and for values of t given by $t\sqrt{3} + \pi/3 = \frac{1}{2}(4n+3)\pi$, $(n = 0, 1, 2, \ldots)$, $x = -2ae^{-t}/\sqrt{3}$. The x-t graph of the function is therefore as shown in Fig. 8.1(ii).

At the common points of the graph of the function and each of the graphs $x = \pm 2ae^{-t}/\sqrt{3}$ the gradients of the two curves are equal and the curves therefore touch. These common points do not coincide with the maxima and minima of x.

We conclude that, in this case, the particle oscillates with constantly decreasing amplitude of oscillation, but always passing through the "centre" $x = 0$.

[Note that in this example x, a, t etc. are numbers (without dimensions).]

In general, it can be shown by the method applied to the particular cases of the above example, that if the equation of the motion is

$$\frac{d^2x}{dt^2} + 2\lambda\frac{dx}{dt} + \omega^2 x = 0,$$

then (a) when $\lambda \geqslant \omega$, the particle approaches the position $x = 0$ asymptotically (the particle, in some cases, passes through the origin before returning to approach the origin asymptotically),

(b) when $\lambda < \omega$, the particle oscillates about $x = 0$ with oscillations, the amplitudes of which decrease in geometrical progression.

Motion of a simple pendulum. When a simple pendulum oscillates in a medium of which the resistance varies as the speed, its equation of motion can be reduced to the form

$$\frac{d^2\theta}{dt^2} + \lambda\frac{d\theta}{dt} + \frac{g}{l}\sin\theta = 0$$

which, when θ is small so that terms involving the third and higher powers of θ can be neglected, becomes

$$\frac{d^2\theta}{dt^2} + \lambda\frac{d\theta}{dt} + \frac{g}{l}\theta = 0.$$

Forced oscillations. If the particle is subjected to a force which is a function of the time, its equation of motion takes the form

$$\frac{d^2x}{dt^2} + \lambda\frac{dx}{dt} + \omega^2 x = f(t).$$

In these cases the solution of the equation of motion involves the finding of a particular integral in addition to the complementary function.

Example 1. A particle of mass m is attached to one end, B, of a light spring, AB, of natural length l and modulus mln^2. At the time $t = 0$ the spring and particle are lying at rest on a smooth horizontal table, with the spring straight but unstretched. The end A is then moved in a straight line in the direction BA with constant acceleration f, so that, after t seconds, its distance in this direction from its initial position is $\frac{1}{2}ft^2$. Show that the distance, x, of the particle at time t in the direction BA from its initial position satisfies the equation

$$\frac{d^2x}{dt^2} + n^2 x = \frac{1}{2}n^2 ft^2.$$

Find the value of x at time t and show that the tension in the spring never exceeds $2mf$.

278   **FURTHER DYNAMICS OF PARTICLE**   Ch. 8 §8:1

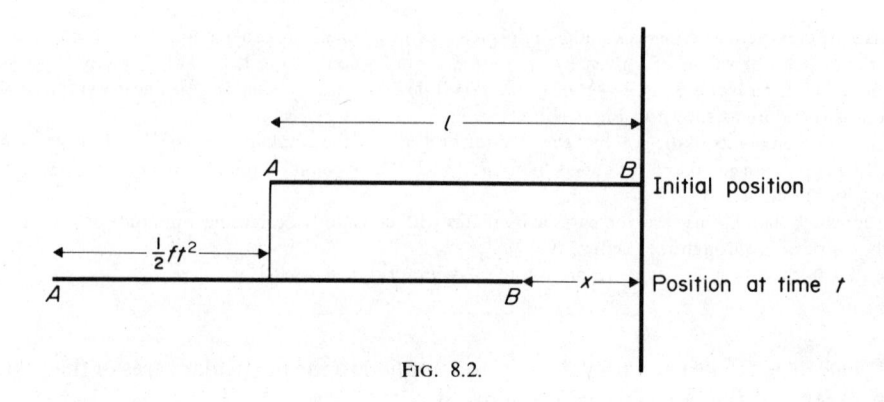

FIG. 8.2.

At time t the length of spring is $l - x + \frac{1}{2}ft^2$ (Fig. 8.2). The only force acting in the direction of motion on the particle is the elastic tension in the spring. The equation of motion for the particle is therefore:

$$m\frac{d^2x}{dt^2} = mln^2\left(\frac{1}{2}ft^2 - x\right)\Big/l$$

$$\Leftrightarrow \frac{d^2x}{dt^2} + n^2x = \frac{1}{2}n^2ft^2.$$

The general solution of this equation is

$$x = \frac{1}{2}ft^2 - A\cos nt - B\sin nt - \frac{f}{n^2}.$$

$x = 0$ when $t = 0$ $\Rightarrow A = -(f/n^2)$.
Also, $\dot{x} = 0$ when $t = 0$ $\Rightarrow B = 0$, and the value of x at time t is given by

$$x = \frac{1}{2}ft^2 + \frac{f}{n^2}\cos nt - \frac{f}{n^2}.$$

Therefore, if T is the tension in the spring, and considering the motion of the particle,

$$T = m\frac{d^2x}{dt^2} \quad \Rightarrow T = m(f - f\cos nt).$$

The maximum value of T occurs when $\cos nt = -1$, and then $T = 2mf$.

Example 2. A particle P of mass m moves along a straight line so that $OP = x$, where O is a fixed point on the line. The forces acting on P are

(i) a force mn^2x directed towards O,
(ii) a resistance $2mnv$, where v is the speed of P,
(iii) a force $F = ma\cos nt$ acting along the direction x increasing.

Write down the differential equation satisfied by x and solve this equation given that at time $t = 0$ the particle is at rest at O.
Find the rate at which the force F is doing work at time t. (L.)

The equation of motion for the particle is

$$m\ddot{x} = -mn^2x - 2mn\dot{x} + ma\cos nt$$

$$\Rightarrow \ddot{x} + 2n\dot{x} + n^2x = a\cos nt.$$

The complementary function is $e^{-nt}(A + Bt)$, where A and B are constants.

Particular integral. If x = the real part of $C\,e^{int}$ is a solution,

$$-Cn^2 + 2in^2 C + n^2 C = a$$

$$\Leftrightarrow C = \frac{a}{2in^2}.$$

The particular integral is, therefore, the real part of $(a/2in^2)e^{int}$

$$\Rightarrow \frac{a}{2n^2}\sin nt$$

$$\Rightarrow x = e^{-nt}(A + Bt) + \frac{a}{2n^2}\sin nt.$$

But $x = 0$ when $t = 0 \Rightarrow A = 0$.
Also $\dot{x} = 0$ when $t = 0 \Rightarrow B = -a/(2n)$

$$\Rightarrow x = -\frac{a}{2n}t\,e^{-nt} + \frac{a}{2n^2}\sin nt.$$

At time t, the force is doing work at the rate $F\dot{x}$, i.e. at the rate

$$\frac{ma^2}{2n}\cos nt(\cos nt + nt\,e^{-nt} - e^{-nt}).$$

[That part of the motion of the particle which is represented by $-(a/2n)te^{-nt}$ in the equation for x, and which dies away as t increases, is called the *transient* part of the motion. That part which is represented by $(a/2n^2)\sin nt$ is the *forced oscillation*.]

Exercise 8.1

In questions 1–4 the differential equation is the equation of motion of a particle moving in a straight line under the action of a restoring force directed towards a fixed point O of the line and proportional to the distance of the particle from O and also of a resistance to its motion which is proportional to its speed. In each case state the nature of the motion.

1. $\ddot{x} + 2\dot{x} + 2x = 0$; $x = 1$, $\dot{x} = 0$ when $t = 0$.
Calculate x when $t = \frac{1}{4}\pi$.

2. $\ddot{x} + 2\dot{x} + x = 0$; $x = 2$, $\dot{x} = 0$ when $t = 0$.
Calculate each of x and $\dot{x}$ when $t = 1$, giving your answers correct to 2 significant figures.

3. $\ddot{x} + 2\dot{x} + 5x = 0$; $x = 2$, $\dot{x} = 0$ when $t = 0$.
Calculate each of x and $\dot{x}$ when $t = \frac{1}{2}\pi$.

4. $\ddot{x} + 3\dot{x} + 2x = 0$; $x = 1$, $\dot{x} = 0$ when $t = 0$.
Show that x is always positive and that for $x < \frac{1}{10}$, $t > \ln(10 + 3\sqrt{10})$.

5. A particle of mass m moves in a straight line and x is its distance at time t from a fixed point of the line. Explain the nature of the forces acting on the particle, given that the equation of motion is

$$\frac{d^2x}{dt^2} + 2k\frac{dx}{dt} + 5k^2 x = 0,$$

where k is a positive constant.
Solve the equation of motion given that $x = 0$, $dx/dt = u$ when $t = 0$.
Show that, when x is next zero,

$$\frac{dx}{dt} = -u\,e^{-\pi/2}$$

and find the corresponding value of d^2x/dt^2. (N.)

6. A particle moves in a straight line so that its distance x from a fixed point in the line satisfies the

differential equation

$$\frac{d^2x}{dt^2} + 4\frac{dx}{dt} + 4x = 0.$$

The particle starts from rest at time $t = 0$ when $x = a$. Prove that its greatest speed in the ensuing motion is $2a e^{-1}$.

7. A particle P of mass m moves in a straight line under the action of a force $mn^2 OP$, which is always directed towards a fixed point O in the line. If the resistance to motion is $2\lambda mnv$, where v is the speed and $0 < \lambda < 1$, write down the equation of motion, and find x in terms of the time t, given that when $t = 0$, $x = 0$ and

$$\frac{dx}{dt} = u, \quad \text{where} \quad x = OP. \tag{N.}$$

8. A simple pendulum whose period of small oscillations *in vacuo* is $\pi/2$ s is made to perform small oscillations under gravity in a fluid which offers resistance to the motion of the bob. The force of resistance is $2mkv$ where m is the mass and v the speed of the bob: k is a constant whose value depends on the fluid being used. Prove that the angular displacement, θ, of the pendulum from the vertical during the small oscillations satisfies the following differential equation,

$$\frac{d^2\theta}{dt^2} + 2k\frac{d\theta}{dt} + 16\theta = 0.$$

If the resistance is such that $k = 3$, show that

$$\theta = a e^{-3t} \sin \sqrt{7} t$$

gives a possible motion, where a is arbitrary but small. If the fluid is such that the resistance is greater, $k = 5$, show that $\theta = a e^{-5t} \sinh 3t$ is a possible motion. By means of rough graphs of θ against the time t point out the chief characteristics of these two motions. (O.C.)

9. A body is attached by a spring to a point which is oscillating with sinusoidal motion. The displacement s of the body from a fixed reference point at any time t is given by the differential equation

$$\frac{d^2s}{dt^2} + 4s = \lambda \sin t,$$

where λ is a constant. Given that $s = 0$ when $t = 0$, and $ds/dt = 2\lambda/3$ when $t = 0$, solve the equation for s in terms of t. (L.)

8:2 MOTION IN A STRAIGHT LINE UNDER VARIABLE FORCES

In the following examples a relationship between two of the quantities s, v, t (in the usual notation) is known and necessary boundary conditions are specified, thereby enabling us to calculate the remaining quantities.

Example 1. A particle moves in a straight line so that at time t from the beginning of the motion its distance from a fixed point O in the line is s and its speed is v where $v = ks$, k being constant, and $s = a$ when $t = 0$. Find
  (a) the initial acceleration of the particle,
  (b) the value of s when $t = 2$ seconds,
  (c) the equation of motion of the particle relating v and t.

(a) From $v = ks$, $dv/ds = k$ and the acceleration of the particle is

$$v\frac{dv}{ds} = k^2 s.$$

Therefore, since $s = a$ when $t = 0$, the initial acceleration of the particle is $k^2 a$.

(b)
$$v = \frac{ds}{dt} = ks$$

$$\Leftrightarrow \int \frac{ds}{s} = \int k\,dt \qquad \Leftrightarrow \ln s = kt + A,$$

where A is constant.
But $s = a$ when $t = 0 \Rightarrow A = \ln a \qquad \Rightarrow \ln s - \ln a = kt$

$$\Leftrightarrow \ln\left(\frac{s}{a}\right) = kt \qquad \Leftrightarrow s = a e^{kt}.$$

Therefore, when $t = 2$ seconds, $s = a e^{2k}$.

(c) *First Method.* $v = ks$ and $s = a e^{kt}$.
Eliminating s we have $v = ka e^{kt}$.

Second Method. $s = a e^{kt} \Rightarrow v = \dfrac{ds}{dt} = ka e^{kt}$.

Example 2. In the usual notation, the equation of motion of a particle moving in a straight line is $f = k/s^2$, for $s \geqslant 1$, and $s = 1$, $v = 0$ when $t = 0$. Show that v cannot exceed $\sqrt{(2k)}$ and find the time taken by the particle to move from $s = 2$ to $s = 4$.

$$v\frac{dv}{ds} = \frac{k}{s^2} \qquad \Leftrightarrow \int v\,dv = \int \frac{k}{s^2}\,ds$$

$$\Leftrightarrow \frac{v^2}{2} = A - \frac{k}{s},$$

where A is constant.
But $s = 1$ when $v = 0 \Rightarrow A = k \qquad \Rightarrow v^2 = 2k\left(1 - \frac{1}{s}\right)$

$$\Leftrightarrow v = \sqrt{\left\{2k\left(1 - \frac{1}{s}\right)\right\}} \Leftrightarrow v < \sqrt{(2k)} \text{ for all values of } s.$$

Also
$$v = \frac{ds}{dt} = \sqrt{\left\{2k\left(1 - \frac{1}{s}\right)\right\}}.$$

Hence the time to move from $s = 2$ to $s = 4$ is T, where

$$T = \frac{1}{\sqrt{(2k)}} \int_2^4 \sqrt{\left(\frac{s}{s-1}\right)}\,ds.$$

Consider
$$I = \int \sqrt{\left(\frac{s}{s-1}\right)}\,ds.$$

Put $s = \cosh^2 u$, then $ds/du = 2 \cosh u \sinh u$

$$\Rightarrow I = \int \frac{\cosh u}{\sinh u} 2\cosh u \sinh u\,du = \int 2\cosh^2 u\,du$$

$$= \int (1 + \cosh 2u)\,du = u + \tfrac{1}{2}\sinh 2u$$

$$\Rightarrow T = \frac{1}{\sqrt{(2k)}}\left[u + \frac{1}{2}\sinh 2u\right]_{s=2}^{s=4}.$$

When $s = 4$, $\quad \cosh^2 u = 4 \quad \Rightarrow \sinh u = \sqrt{3} \quad \Leftrightarrow \tfrac{1}{2}\sinh 2u = 2\sqrt{3}$.
When $s = 2$, $\quad \cosh^2 u = 2 \quad \Rightarrow \sinh u = 1 \quad \Leftrightarrow \tfrac{1}{2}\sinh 2u = \sqrt{2}$.

When $s = 4$, $\quad u = \cosh^{-1} \sqrt{s} = \ln(2 + \sqrt{3})$.
When $s = 2$, $\quad u = \ln(1 + \sqrt{2})$.

$$T = \frac{1}{\sqrt{(2k)}} \left\{ 2\sqrt{3} - 2 + \ln\left(\frac{2+\sqrt{3}}{1+\sqrt{2}}\right) \right\}.$$

Example 3. A particle is projected with speed V from a fixed point O, and moves in a straight line OX. The retardation is proportional to v^3, where v is the speed of the particle when its distance from O is x. Show that

(a) $\dfrac{1}{v} = \dfrac{1}{V} + ax$, where a is a positive constant,

(b) $t = \dfrac{x}{V} + \dfrac{1}{2}ax^2$, where t is the time taken to travel the distance x. (N.)

The equation of motion of the particle is

$$v \frac{dv}{dx} = -bv^3,$$

where b is a positive constant,

$$\Leftrightarrow \int \frac{v \, dv}{v^3} = \int -b \, dx \quad \Leftrightarrow \quad -\frac{1}{v} = -bx + k,$$

where k is constant.
But $v = V$ when $x = 0 \Rightarrow k = -1/V \quad \Rightarrow 1/v = (1/V) + bx$

$$\Leftrightarrow \frac{dt}{dx} = \frac{1}{V} + bx \quad \Leftrightarrow \quad t = \int \left(\frac{1}{V} + bx\right) dx$$

$$\Leftrightarrow t = \frac{x}{V} + \frac{bx^2}{2} + C,$$

where C is constant.
But $x = 0$ when $t = 0 \Rightarrow C = 0$

$$\Leftrightarrow t = \frac{x}{V} + \frac{bx^2}{2}$$

$\Rightarrow t = (x/V) + (ax^2/2)$, where a is a positive constant.

Exercise 8.2(a)

In each of questions 1–7 the equation of motion of a particle moving in a straight line and the boundary conditions for the motion are given in the usual notation.

1. $v^2 = 36 - 4s^2$; $s = 3$ when $t = 0$.
Calculate f when $s = 1$ and calculate f when $t = \frac{1}{3}\pi$.

2. $f = k/s$ where k is a positive constant; $v = 0$ when $s = 1$. Find s in terms of v and k.

3. $v = 2 - 3s$, $s = 0$ when $t = 0$. Calculate f when $s = \frac{1}{2}$ and calculate s when $t = 1$. Show that s cannot exceed the value $\frac{2}{3}$.

4. $f = -(a + bv^2)$ when a and b are positive constants; $s = 0$ and $v = v_0$ when $t = 0$. Prove that $v = 0$ when

$$t = \frac{1}{\sqrt{(ab)}} \tan^{-1}\left(v_0 \sqrt{\frac{b}{a}}\right), \quad s = \frac{1}{2b} \ln\left(1 + \frac{bv_0^2}{a}\right).$$

5. $f = -\lambda v^3$, where λ is a positive constant; $v = 20$ and $s = 0$ when $t = 0$; $v = 10$ when $t = 15$. Show that $\lambda = \frac{1}{4000}$ and find v when $s = 600$.

6. $f = \frac{1}{100}v$; $v = 1$ when $t = 0$. Find v when $t = 10$ and find t when $v = 100$.

7. $f = 2 + \frac{1}{5}v$; $v = 0$, $s = 0$ when $t = 0$. Find v and s when $t = 1$.

8. A particle moves in a straight line so that its distance s from a fixed point of the line at time t is given by the equation $s^2 = a^2 + V^2 t^2$, where a and V are constants. Find in terms of s its velocity and acceleration at time t. (O.C.)

9. A car is travelling along a straight road. When it is passing a certain position O the engine is switched off. At time t after the car has passed O the speed v is given by the formula

$$\frac{1}{v} = A + Bt,$$

where A and B are positive constants. Show that the retardation is proportional to the square of the speed.
If when $t = 0$ the retardation is 1 ms^{-2} and $v = 80$ ms^{-1}, find A and B.
If x is the distance moved from O in time t, express
(i) x in terms of t, (ii) v in terms of x.

10. A particle is moving along a straight line away from a fixed point O in the line so that when its distance from O is x its speed v is given by $v = k/x$, where k is a constant. Show that the particle has a retardation which is inversely proportional to the cube of its distance from O.
If P, Q, R, S are points in that order on the straight line such that the distances PQ, QR, RS are all equal, show that the times taken to traverse these successive distances increase in arithmetical progression.

Miscellaneous problems

The examples which follow illustrate problems in which the equation of motion of a body is derived from an analysis of the force system acting upon the body.

Example 1. The motion of a body of mass m (moving in a horizontal straight line) is resisted by a constant force ma and a variable force mv^2/k, where v is the speed. Show that the body is brought to rest by the resistances from speed u in a distance $(k/2)\ln\{1 + u^2/(ak)\}$ and in a time $\sqrt{(k/a)}\tan^{-1}\{u/\sqrt{(ak)}\}$.

From the equation $P = mf$ for the body, we derive the equation of motion either as

$$\text{(a)} \quad v\frac{dv}{ds} = -a - \frac{v^2}{k}$$

or as

$$\text{(b)} \quad \frac{dv}{dt} = -a - \frac{v^2}{k}.$$

Each of these equations is a differential equation with variables separable.
From (a), the distance in which the body is brought to rest is

$$k\int_u^0 \frac{-v\,dv}{ka+v^2} = k\int_0^u \frac{v\,dv}{ka+v^2} = \frac{1}{2}k\left[\ln(ka+v^2)\right]_0^u$$

$$= \frac{1}{2}k\{\ln(ka+u^2) - \ln(ka)\} = \frac{1}{2}k\ln\left(\frac{ka+u^2}{ka}\right)$$

$$= \frac{1}{2}k\ln\left(1 + \frac{u^2}{ka}\right).$$

From (b), the time in which the body is brought to rest is

$$k\int_0^u \frac{dv}{ka+v^2} = \frac{k}{\sqrt{(ka)}}\left[\tan^{-1}\left\{\frac{v}{\sqrt{(ka)}}\right\}\right]_0^u = \sqrt{\left(\frac{k}{a}\right)}\tan^{-1}\left\{\frac{u}{\sqrt{(ka)}}\right\}.$$

Example 2. A body is projected vertically upwards from the earth's surface with speed u. Neglecting atmospheric resistance and assuming the gravitational attraction between the body and the earth to be inversely proportional to the square of the distance of the body from the centre of the earth, calculate the smallest value of u for which the body does not return to the earth.

The equation of motion for the body is

$$mv\frac{dv}{ds} = -\frac{mgR^2}{s^2}.$$

where R is the radius of the earth and s is the distance of the body from the centre of the earth. Therefore

$$v\frac{dv}{ds} = -\frac{gR^2}{s^2}$$

$$\Leftrightarrow \frac{v^2}{2} = \frac{gR^2}{s} + A,$$

where A is constant.
Since $v = u$ when $s = R$, $A = (u^2/2) - gR$

$$\Leftrightarrow \frac{v^2}{2} = \frac{gR^2}{s} + \frac{u^2}{2} - gR,$$

so that v will never become zero if $u^2 > 2gR$, i.e. the required least value of u is $\sqrt{(2gR)}$.

Taking R as 6400 km and g as 9·8 ms^{-2}, this result gives about 40 000 km h^{-1} as the minimum "speed of escape".

Motion under gravity in a resisting medium

Physical experience tends to indicate that for bodies moving under gravity in the atmosphere, or in fluids generally, the magnitude of the resistance to motion is dependent on the speed of the body. We consider here two cases in each of which the resistance is assumed to be a simple function of the speed.

1. Resistance proportional to speed

Let the resistance $= mkv$, where k is a positive constant, m is the mass of the body and v its speed.

(a) For a particle of mass m falling vertically from rest, one form of the equation of motion is

$$m\frac{dv}{dt} = mg - mkv.$$

Here we measure v and s *downward* from the point of release.

[We note that, if and when v reached the value g/k the acceleration would become zero and the particle would continue to descend with uniform speed.] Integrating this equation of motion, we have

$$\int \frac{dv}{g - kv} = t \quad \Leftrightarrow \quad t = -\frac{1}{k}\ln(g - kv) + A,$$

where A is constant.

But $v = 0$ when $t = 0 \Leftrightarrow A = \dfrac{1}{k}\ln g$

$$\Leftrightarrow t = \dfrac{1}{k}\ln\left(\dfrac{g}{g-kv}\right)$$

$$\Leftrightarrow v = \dfrac{g}{k}(1-e^{-kt}). \qquad (8.1)$$

Since $\lim\limits_{t\to\infty}(1-e^{-kt}) = 1$, v increases continuously with t and, for sufficiently large values of t, v can take any positive value less than g/k however near to g/k. The quantity g/k is called the *terminal velocity* of the particle.
(Note that the *dimensions* of k in this discussion are time^{-1} $[T^{-1}]$.)

A second form of the equation of motion of the particle is

$$mv\dfrac{dv}{ds} = mg - mkv$$

$$\Leftrightarrow \int\dfrac{v\,dv}{g-kv} = s \quad \Leftrightarrow s = -\dfrac{1}{k}\int\left(1 - \dfrac{g}{g-kv}\right)dv$$

$$\Leftrightarrow s = -\dfrac{v}{k} - \dfrac{g}{k^2}\ln(g-kv) + B,$$

where B is a constant.
But $v = 0$ when $s = 0 \Rightarrow B = (g/k^2)\ln g$

$$\Leftrightarrow s = \dfrac{g}{k^2}\ln\left(\dfrac{g}{g-kv}\right) - \dfrac{v}{k}. \qquad (8.2)$$

Also from (8.1),

$$\dfrac{ds}{dt} = \dfrac{g}{k}(1-e^{-kt})$$

$$\Leftrightarrow s = \dfrac{gt}{k} + \dfrac{g}{k^2}e^{-kt} + C,$$

where C is constant.
But $s = 0$ when $t = 0 \Rightarrow C = -(g/k^2)$

$$\Leftrightarrow s = \dfrac{g}{k^2}(kt + e^{-kt} - 1). \qquad (8.3)$$

(b) The equation of motion for a particle projected vertically upwards with speed u in such a medium is

$$\dfrac{dv}{dt} = -g - kv$$

[Here v and s are measured *upward* from the point of projection.]

$$\Leftrightarrow t = -\int\dfrac{dv}{g+kv} \quad \Leftrightarrow t = -\dfrac{1}{k}\ln(g+kv) + C,$$

where C is a constant.

$v = u$ when $t = 0 \Rightarrow C = (1/k) \ln (g + ku)$

$$\Leftrightarrow t = \frac{1}{k} \ln \left(\frac{g + ku}{g + kv} \right). \tag{8.4}$$

The time to reach the highest point of the path, where $v = 0$, is thus

$$\frac{1}{k} \ln \left(1 + \frac{ku}{g} \right). \tag{8.5}$$

Another form of the equation of motion is

$$v \frac{dv}{ds} = -g - kv,$$

from which the greatest height reached is given by

$$-\int_u^0 \frac{v \, dv}{g + kv} = \int_0^u \frac{v \, dv}{g + kv}$$

$$= \frac{1}{k} \int_0^u \left(1 - \frac{g}{g + kv} \right) dv = \frac{1}{k} \left[v - \frac{g}{k} \ln (g + kv) \right]_0^u$$

$$= \frac{1}{k} \left\{ u - \frac{g}{k} \ln \left(1 + \frac{ku}{g} \right) \right\}. \tag{8.6}$$

2. *Resistance proportional to the square of the speed*

Let the resistance $= mkv^2$, where m is the mass of the particle and v its speed.

(a) For a particle of mass m falling vertically from rest, one form of the equation of motion is

$$m \frac{dv}{dt} = mg - mkv^2.$$

Here v and s are measured *downward* from the point of release.

[We note that, if and when v reached the value $\sqrt{(g/k)}$, the acceleration would become zero and the particle would continue to descend with uniform speed $\sqrt{(g/k)}$.]

Integrating the equation of motion, we have

$$t = \int \frac{dv}{g - kv^2} \quad [(v < \sqrt{(g/k)}]$$

$$\Leftrightarrow t = \frac{1}{2\sqrt{(kg)}} \ln \left(\frac{\sqrt{\left(\frac{g}{k}\right)} + v}{\sqrt{\left(\frac{g}{k}\right)} - v} \right) + A, \tag{8.7}$$

where A is constant.

But $v = 0$ when $t = 0 \Rightarrow A = 0$

$$\Leftrightarrow e^{2t\sqrt{(kg)}}\left\{\sqrt{\left(\frac{g}{k}\right)} - v\right\} = \sqrt{\left(\frac{g}{k}\right)} + v$$

$$\Leftrightarrow v = \sqrt{\left(\frac{g}{k}\right)}\left\{\frac{e^{2t\sqrt{(kg)}} - 1}{e^{2t\sqrt{(kg)}} + 1}\right\} = \sqrt{\left(\frac{g}{k}\right)}\tanh\{t\sqrt{(kg)}\}. \qquad (8.7a)$$

Since $\displaystyle\lim_{t\to\infty}\left\{\frac{e^{2t\sqrt{(kg)}} - 1}{e^{2t\sqrt{(kg)}} + 1}\right\} = \lim_{t\to\infty}\left\{\frac{1 - e^{-2t\sqrt{(kg)}}}{1 + e^{-2t\sqrt{(kg)}}}\right\} = 1,$

v increases continuously with t, and, for sufficiently large values of t, v can take any positive value less than $\sqrt{(g/k)}$, however near to $\sqrt{(g/k)}$. The quantity $\sqrt{(g/k)}$ is the *terminal speed* of the particle. [Note: the phrase *terminal velocity* is frequently used.]
(Note that the *dimensions* of k in this discussion are L^{-1}.)

A second form of the equation of motion is

$$mv\frac{dv}{ds} = mg - mkv^2$$

$$\Leftrightarrow s = \int\frac{v\,dv}{g - kv^2}$$

$$\Leftrightarrow s = -\frac{1}{2k}\ln(g - kv^2) + A,$$

where A is constant.

$s = 0$ when $v = 0 \Rightarrow A = \dfrac{1}{2k}\ln g \Leftrightarrow s = \dfrac{1}{2k}\ln\left(\dfrac{g}{g - kv^2}\right) \qquad (8.8)$

$$\Leftrightarrow v^2 = \frac{g}{k}(1 - e^{-2ks}). \qquad (8.9)$$

(b) The equation of motion for a particle projected vertically upwards in such a medium with speed u is

$$\frac{dv}{dt} = -g - kv^2,$$

where v and s are measured *upward* from the point of projection,

$$\Leftrightarrow t = -\int\frac{dv}{g + kv^2}$$

$$\Leftrightarrow t = -\frac{1}{\sqrt{(kg)}}\tan^{-1}\left\{v\sqrt{\left(\frac{k}{g}\right)}\right\} + B,$$

where B is constant.

$v = u$ when $t = 0 \Rightarrow B = \dfrac{1}{\sqrt{(kg)}}\tan^{-1}\left\{u\sqrt{\left(\dfrac{k}{g}\right)}\right\}$

$$\Leftrightarrow \sqrt{(kg)}\,t = \tan^{-1}\left\{u\sqrt{\left(\frac{k}{g}\right)}\right\} - \tan^{-1}\left\{v\sqrt{\left(\frac{k}{g}\right)}\right\}.$$

The time to reach the highest point of the path (where $v = 0$) is thus

$$\frac{1}{\sqrt{(kg)}} \tan^{-1}\left\{u\sqrt{\left(\frac{k}{g}\right)}\right\}. \tag{8.10}$$

Also
$$v\frac{dv}{ds} = -g - kv^2$$

$$\Leftrightarrow -\int \frac{v\,dv}{g + kv^2} = s \Leftrightarrow s = -\frac{1}{2k} \ln(g + kv^2) + C,$$

where C is constant.

$s = 0$ when $v = u \Rightarrow C = \{1/(2k)\} \ln(g + ku^2)$

$$\Leftrightarrow s = \frac{1}{2k} \ln\left(\frac{g + ku^2}{g + kv^2}\right).$$

The greatest height reached by the particle is thus

$$\frac{1}{2k} \ln\left(1 + \frac{ku^2}{g}\right). \tag{8.11}$$

[When the resistance to the motion is proportional to the speed, the particle is projected vertically upwards and the upward direction is taken as the positive direction, the whole of the motion *both upwards and downwards* is described by the equation

$$\frac{dv}{dt} = -g - kv.$$

When the resistance to the motion is proportional to the square of the speed, *separate equations for the upward and downward motions are required.*]

Example 1. A particle is projected vertically upwards with speed nu. Show that, if the resistance of the medium in which it is projected is assumed to vary as the square of the speed of the particle and if the terminal velocity when the particle is allowed to fall from rest in this medium is u, then (a) the particle returns to the ground with its original kinetic energy reduced in the ratio $1/(1 + n^2)$, (b) the time taken by the particle to return to its starting point is T, where

$$T = \frac{u}{g}\left(\tan^{-1} n + \ln\left[\sqrt{(n^2 + 1)}n\right]\right).$$

(To avoid repetition, results obtained above are used in the work which follows. The reader is not, however, advised to quote these results in his own work, but to obtain them from first principles whenever necessary.)

An equation of motion for the particle projected upwards is

$$\frac{dv}{dt} = -g - kv^2,$$

where $u = \sqrt{(g/k)}$ and $v = nu$ when $t = 0$, from which we have obtained the result (8.10),

$$t_1 = \frac{1}{\sqrt{(kg)}} \tan^{-1}\left\{nu\sqrt{\left(\frac{k}{g}\right)}\right\} = \frac{u}{g} \tan^{-1} n,$$

where t_1 is the time taken to reach the highest point of the path.

Another equation of motion for the particle projected vertically upwards is
$$v\frac{dv}{ds} = -g - kv^2,$$
from which we obtain the result (8.11)
$$s_1 = \frac{1}{2k}\ln\left(1 + \frac{kn^2u^2}{g}\right) = \frac{u^2}{2g}\ln(1+n^2)$$
for the greatest height s_1 reached by the particle.

An equation of motion for the particle on its return fall to the earth is
$$v\frac{dv}{ds} = g - kv^2,$$
from which we obtain the result (8.8),
$$s = \frac{1}{2k}\ln\left(\frac{g}{g-kv^2}\right) = \frac{u^2}{2g}\ln\left(\frac{u^2}{u^2-v^2}\right)$$
relating the distance s fallen from the highest point to the speed v at that point. Therefore, when the particle strikes the ground,
$$\frac{u^2}{2g}\ln(1+n^2) = \frac{u^2}{2g}\ln\left(\frac{u^2}{u^2-v^2}\right)$$
$$\Leftrightarrow 1+n^2 = \frac{u^2}{u^2-v^2} \Leftrightarrow v^2 = \frac{n^2u^2}{1+n^2}.$$

The initial kinetic energy of the particle was $\tfrac{1}{2}mn^2u^2$ and the final kinetic energy of the particle is $\tfrac{1}{2}mn^2u^2/(1+n^2)$. Therefore the ratio in which the kinetic energy is reduced is $1/(1+n^2)$.

Another equation of motion for the particle on its return fall to earth is
$$\frac{dv}{dt} = g - kv^2.$$
from which we obtain the result (8.7),
$$t_2 = \frac{1}{2\sqrt{(kg)}}\ln\left[\frac{\sqrt{\left(\frac{g}{k}\right)}+v}{\sqrt{\left(\frac{g}{k}\right)}-v}\right] = \frac{u}{2g}\ln\left(\frac{u+v}{u-v}\right)$$
for t_2 the time taken on the downward journey, where $v^2 = n^2u^2/(1+n^2)$,
$$\Leftrightarrow t_2 = \frac{u}{2g}\ln\left[\frac{\sqrt{(n^2+1)}+n}{\sqrt{(n^2+1)}-n}\right] = \frac{u}{g}\ln\left[\sqrt{(n^2+1)}+n\right]$$
$$\Leftrightarrow T = t_1 + t_2 = \frac{u}{g}\left(\tan^{-1}n + \ln\left[\sqrt{(n^2+1)}+n\right]\right).$$

Example 2. An engine pulls a train along a level track against a resistance which at any time is k times the momentum. The engine works at constant power $9mkv_0^2$ where m is the total mass of the engine and the train. Show that the speed increases from v_0 to $2v_0$ in a time $\{\ln(8/5)\}/(2k)$.

The power is cut off when the train has speed U and a constant braking force F then acts in addition to the resistance. Show that the train will stop after a further time $(1/k)\ln\{(F+mkU)/F\}$. (N.)

The total force exerted by the engine at time t, when the speed is v and the acceleration is dv/dt, is
$$mkv + m\frac{dv}{dt}.$$

Therefore the power equation of the motion is

$$9mkv_0^2 = \left(mkv + m\frac{dv}{dt}\right)v \iff v\frac{dv}{dt} = k(9v_0^2 - v^2)$$

$$\Rightarrow \frac{1}{k}\int_{v_0}^{2v_0} \frac{v\,dv}{9v_0^2 - v^2} = T_1,$$

where T_1 is the time during which the speed increases from v_0 to $2v_0$,

$$\iff T_1 = \frac{1}{2k}\left[\ln(9v_0^2 - v^2)\right]_{2v_0}^{v_0} = \frac{1}{2k}[\ln(8v_0^2) - \ln(5v_0^2)] = \frac{1}{2k}\ln\left(\frac{8}{5}\right).$$

After the power is cut off the equation of motion of the train becomes

$$-(F + mkv) = m\frac{dv}{dt} \Rightarrow T_2 = -\int_U^0 \frac{m\,dv}{F + mkv},$$

where T_2 is the further time required for the train to stop,

$$\iff T_2 = \frac{1}{k}\left[\ln(F + mkv)\right]_0^U = \frac{1}{k}\ln\{(F + mkU)/F\}.$$

Exercise 8.2(b)

1. A particle of mass of 10 kg starts from rest and is acted upon by a force in a constant direction and increasing uniformly in 5 seconds from zero to 6 N. Prove that t seconds after the start of the motion the acceleration of the particle is $0.12t$ ms^{-2}. Find the distance the particle moves in the first 5 seconds and show that when it has moved a distance x m its speed is v ms^{-1}, where $50v^3 = 27x^2$.

2. A load W is raised through a vertical height h, starting from rest and coming to rest under gravity, by a chain incapable of supporting a load greater than P where $P > W$. Show that the shortest time in which this can be done is

$$\sqrt{\left\{\frac{2hP}{g(P - W)}\right\}}.$$

If a large cargo is to be discharged in this way, show that the time occupied by upward journeys from rest to rest will be least if the cargo is sent up in loads of $\tfrac{2}{3}P$. (L.)

3. The net accelerating force acting on a train of mass W varies with the speed, being $F(v)$ when the speed is v. Show that the time t required to reach a speed of V from rest is given by

$$t = \int_0^V \frac{5W\,dv}{18F(v)}.$$

4. A train of mass W kg moves on the level under the action of a pull P newtons against a resistance of R newtons. The speed of the train at any instant is v ms^{-1}. Show that the distance travelled by the train whilst its speed varies from v_0 ms^{-1} to v_1 ms^{-1} is

$$W\int_{v_0}^{v_1} v\,dv/(P - R) \text{ metres}.$$

If $W = 3 \times 10^5$ and $R = 9000 + 70v^2$ show that the distance travelled by the train in slowing down, with the engine shut off, from 72 km h^{-1} to 54 km h^{-1} is about 860 m.

5. A car of mass M moves from rest on a horizontal road against a constant resistance R. The pull exerted by the engine decreases uniformly with the distance from the starting point, being initially P and falling to R

after the car has travelled a distance a. Derive an expression for the accelerating force when the car has moved a distance x from rest ($x \leqslant a$), and show that, if t is the time taken to describe the distance x,

$$\left(\frac{dx}{dt}\right)^2 = k^2(2ax - x^2),$$

where $k^2 = (P - R)/(Ma)$.
    Verify by differentiation that $x = a(1 - \cos kt)$ is a solution of the above equation, and show that it satisfies the initial conditions.

6. A particle of mass m can move in a resisting medium in which the resistance varies as the square of the speed, the magnitude of the terminal speed being V. If it is projected vertically upwards with a speed $V \tan \alpha$, show that it will return to the point of projection with a speed $V \sin \alpha$.
    Show also that the amount of energy, kinetic and potential together, which is lost in its ascent is

$$\tfrac{1}{2}mV^2(\tan^2 \alpha - 2 \ln \sec \alpha).\qquad\text{(L.)}$$

7. A particle of mass m falls from rest under gravity and the resistance to its motion is kmv^2 when its speed is v, the factor k being a constant. Prove that, if the distance fallen is then x,

$$v^2 = \frac{g}{k}(1 - e^{-2kx}).$$

If as x increases from d to $2d$ the speed increases from V to $5V/4$, find, in terms of V only, the greatest possible speed of the particle. (N.)

8. A particle is released from rest at a height a above a horizontal plane. The resistance to the motion of the particle is kv^2 per unit mass, where v is its speed and k is a constant. Show that the particle strikes the plane with speed V where

$$kV^2 = g(1 - e^{-2ka}).$$

If the particle rebounds from the plane without loss of energy and reaches a maximum height b in the subsequent motion, show that

$$e^{-2ka} + e^{2kb} = 2.\qquad\text{(L.)}$$

9. A particle moves under gravity in a medium in which the resistance to motion per unit mass is k times the speed where k is a constant. The particle is projected vertically upwards with speed g/k. Show that the speed v and height x reached after time t are given by

$$kv = g(2e^{-kt} - 1) \quad\text{and}\quad k^2 x = g(2 - 2e^{-kt} - kt).$$

Show that the greatest height H the particle can attain is given by

$$k^2 H = g(1 - \ln 2).\qquad\text{(N.)}$$

10. A particle of mass m is projected with speed V vertically upwards from a point on horizontal ground. Its subsequent motion is subject to gravity and to a resistance kmv^2, where v is the speed and k is a constant. Show that the greatest height attained is

$$\frac{1}{2k}\ln\left(1 + k\frac{V^2}{g}\right).$$

Find an expression for the work done against the resistance during the whole of the upward motion. (N.)

11. The engine of a train of mass M pulls the train against a variable resistance $M kv$, where k is constant and v is the speed of the train. The time taken to reach the speed v from rest is

$$\frac{1}{2k}\ln\left(\frac{H}{H - Mkv^2}\right),\quad \text{where}\quad v < \sqrt{\left(\frac{H}{Mk}\right)}$$

and H is constant. Prove that the engine is working at a constant rate.

12. A train of mass M is being pulled by its engine against a constant resistance R. The engine works at a constant rate H. Prove that the time taken to reach a speed $v(< H/R)$ from rest is

$$\frac{MH}{R^2}\ln\left(\frac{H}{H-Rv}\right) - \frac{Mv}{R}.$$

*8:3 THE RECTILINEAR MOTION OF BODIES WITH VARIABLE MASS

In all problems considered so far in these volumes the bodies involved have been assumed to be of constant mass. This assumption in most cases allows of a close approximation to the correct equation of motion, but there are many cases in which the equation of motion would be appreciably altered if the assumption were not made, e.g. in a motor car, consumption of fuel and escape of exhaust gases with a velocity differing from that of the car, influence the equation of motion.

The advent of the jet engine and the rapid development of the large rockets required to put earth satellites and space probes into orbit have given rise to a variety of problems involving bodies of changing mass or bodies which pick up and eject the material of the medium through which they move. Here we are concerned primarily with the rectilinear motion of such a body but the methods are applicable to more complicated problems.

In these cases we do not write down the equations of motion of the separate parts of a system but consider the change in momentum of the system as a whole. The system in question consists of the body in motion, i.e. an aircraft or a rocket, together with any matter which is absorbed and any matter which is ejected in a short interval of time δt. At the beginning of this interval all these different parts of the system have their separate states of motion; at the end of the interval each part has, usually, a different state of motion. We use the fact that the change of linear momentum of the *whole* system in the interval δt is given by the impulse of the *external* forces acting on the *whole* system. Having written down the equation expressing this momentum balance for an interval of time δt and noting that any product of small quantities such as $\delta m \cdot \delta v \to 0$ as $\delta t \to 0$, we obtain a differential equation by dividing by δt and then taking the limit as $\delta t \to 0$.

In solving these problems it is advisable in each case to proceed from first principles and derive the differential equation of the motion rather than to quote standard results.

We first consider two general forms of motion in a straight line during which the mass of the body varies.

1. A body falls under constant gravity picking up matter from rest as it falls, so that at time t its mass is m and speed is v. Consider the system which consists of the body at a given instant and the material, of mass δm, which it picks up in the subsequent interval δt; this material is initially at rest. The initial momentum of the system is

$$p = mv + \delta m.0.$$

Finally the body and the additional material are moving with speed $v + \delta v$, and the final momentum is

$$p + \delta p = (m + \delta m)(v + \delta v).$$

The external force acting on the *whole* system throughout this interval is the total weight $mg + (\delta m)g$

$$\Rightarrow (m + \delta m)g\,\delta t = \delta p = (m + \delta m)(v + \delta v) - mv$$
$$\Leftrightarrow mg\,\delta t + g\,\delta m.\delta t = m\,\delta v + v\,\delta m + \delta m.\delta v$$
$$\Leftrightarrow mg + g\,\delta m = m\frac{\delta v}{\delta t} + v\frac{\delta m}{\delta t}$$
$$\Rightarrow \lim_{\delta t \to 0}(mg + g\,\delta m) = \lim_{\delta t \to 0} m\frac{\delta v}{\delta t} + \lim_{\delta t \to 0} v\frac{\delta m}{\delta t}$$
$$\Rightarrow mg = m\frac{dv}{dt} + v\frac{dm}{dt} = \frac{d}{dt}(mv). \tag{8.12}$$

If the matter picked up had been moving downwards with speed u, then in place of equation (8.12), the momentum balance would have been

$$(m + \delta m)g\,\delta t = \delta p = (m + \delta m)(v + \delta v) - (mv + \delta m.u)$$

and this leads to the differential equation

$$mg = m\frac{dv}{dt} + v\frac{dm}{dt} - u\frac{dm}{dt} = \frac{d}{dt}(mv) - u\frac{dm}{dt}. \tag{8.13}$$

Note that equation (8.12) expresses the second law of motion in the form

rate of change of downward momentum = downward force

but equation (8.13) shows that this result must be modified when the additional mass is not picked up from rest.

2. A body moves vertically upwards under gravity so that at time t its mass is m and its speed is v. The body ejects material at the rate of k units of mass per second vertically downwards with a speed u *relative to the body*. In this case the system considered is the body and the material it ejects in the short interval δt. At the start of the interval the body has mass m and speed v; at the end of the interval it has mass $m + \delta m$ (a mass $-\delta m = k\,\delta t$ having been ejected) with a speed $v + \delta v$, and the ejected mass has a speed between $v + \delta v - u$ and $v - u$. The upward momentum of the ejected mass therefore lies between $(-\delta m)(v - u)$ and $(-\delta m)(v + \delta v - u)$ and can therefore be taken as $-\delta m(v - u)$. The external force acting on the whole system is the weight mg. The momentum equation is therefore

$$(m + \delta m)(v + \delta v) + (-\delta m)(v - u) - mv = -mg\,\delta t$$
$$\Leftrightarrow m\,\delta v + u\,\delta m = -mg\,\delta t$$
$$\Rightarrow \lim_{\delta t \to 0} m\frac{\delta v}{\delta t} + \lim_{\delta t \to 0} u\frac{\delta m}{\delta t} = -mg$$
$$\Leftrightarrow m\frac{dv}{dt} + u\frac{dm}{dt} = -mg, \tag{8.14}$$

where $dm/dt = -k$.

In the following examples a wide variety of simple problems is considered. We repeat that in every case the student should derive the equations of motion from first principles as illustrated above.

*Example 1. A spherical raindrop of initial radius a falls from rest under gravity. Its radius increases with time at a constant rate μ owing to condensation from a surrounding cloud which is at rest. Find the distance fallen by the raindrop after time t.

At time t the radius of the raindrop is r and its downward speed is v. But equation (8.12) above holds and therefore

$$\frac{d}{dt}\left(\frac{4}{3}\pi r^3 \varrho g\right) = \frac{4}{3}\pi r^3 \varrho g, \tag{1}$$

where ϱ is the density. Also because of condensation

$$\frac{dr}{dt} = \mu$$
$$\Rightarrow r = a + \mu t.$$

Substitution in (1) gives

$$\frac{d}{dt}\{(a+\mu t)^3 v\} = (a+\mu t)^3 g.$$

Integration and use of the initial condition, $v = 0$ at $t = 0$

$$\Rightarrow (a+\mu t)^3 v = \frac{g}{4\mu}\{(a+\mu t)^4 - a^4\}. \tag{2}$$

If the distance fallen in time t is x, $v = dx/dt$ and (2)

$$\Rightarrow \frac{dx}{dt} = \frac{g}{4\mu}\left[a + \mu t - \frac{a^4}{(a+\mu t)^3}\right].$$

Integration and use of the initial condition, $x = 0$ at $t = 0$

$$\Rightarrow x = \frac{g}{4\mu}\left[at + \frac{1}{2}\mu t^2 + \frac{a^4}{2\mu(a+\mu t)^2} - \frac{a^2}{2\mu}\right]. \tag{3}$$

[Note that the usual result of free fall without condensation can be obtained by letting $\mu \to 0$ in (2) and (3).]

---

*Example 2. A rocket continuously ejects matter backwards with speed c relative to itself. Show that if gravity is neglected the speed v and total mass m of the rocket are related by the equation

$$m\frac{dv}{dt} + c\frac{dm}{dt} = 0.$$

Deduce that whatever the rate of burning of the rocket, v and m are related by the formula

$$v = c \ln(M/m),$$

where M is a constant.

Assuming that m decreases at a constant rate k, show that the distance the rocket travels from rest before the mass has fallen from the initial value m_0 to m_1 is

$$c(m_1/k)[m_0/m_1 - 1 - \ln(m_0/m_1)].$$

In this case the analysis of case (2) above is valid and the first equation follows at once from equation (8.14) on putting $g = 0$, $u = c$. The variables of this equation are separable

$$\Rightarrow c\int_{m_0}^{m}\frac{1}{m}dm = -\int_{V}^{v} 1\,dv, \tag{1}$$

where m_0 is the initial mass and V is the initial speed of the rocket,

$$\Leftrightarrow v = V + c \ln(m_0/m).$$

But, since V is constant, we can write $V = c \ln(M/m_0)$, where M is constant, and then

$$v = c \ln(M/m_0) + c \ln(m_0/m)$$
$$\Leftrightarrow v = c \ln(M/m). \qquad (2)$$

If $V = 0$ and the rocket travels a distance x whilst the mass falls from m_0 to m, we can write (2) in the form

$$\frac{dx}{dt} = \frac{dx}{dm}\frac{dm}{dt} = c \ln\left(\frac{m_0}{m}\right). \qquad (3)$$

But m decreases at a constant rate k and therefore $dm/dt = -k$

$$\Rightarrow \frac{dx}{dm} = -\frac{c}{k}(\ln m_0 - \ln m).$$

Hence the distance travelled by the rocket before the mass has fallen from m_0 to m_1 is

$$X = -\frac{c}{k}\int_{m_0}^{m_1}(\ln m_0 - \ln m)dm$$

$$= -\frac{c}{k}\Big[m \ln m_0 - m \ln m + m\Big]_{m_0}^{m_1}$$

$$= -\frac{c}{k}(m_1 \ln m_0 - m_1 \ln m_1 + m_1 - m_0)$$

$$= \frac{m_1 c}{k}\left[\frac{m_0}{m_1} - 1 - \ln\left(\frac{m_0}{m_1}\right)\right].$$

Note that here we have used m as the independent variable. We could have found x in terms of t by integrating (3) with $m = m_0 - kt$ and finally eliminating t but in many problems of rocketry the mass ratio m_0/m_1 is of prime importance and it is very convenient to use m as the independent variable.

Example 3. A rocket in rectilinear motion is propelled by ejecting all the products of combustion of the fuel from the tail at a constant rate and at a constant speed relative to the rocket. Show that, for a given initial total mass M, the final kinetic energy of the rocket is greatest when the initial mass of fuel is $(1 - e^{-2})M$.

In this case, and again neglecting gravity, result (2) of Example 2 above shows that the speed, when the initial mass M has fallen to λM ($\lambda < 1$), is $c \ln(1/\lambda)$. Therefore the kinetic energy is

$$T = \tfrac{1}{2}\lambda M(-c \ln \lambda)^2 = \tfrac{1}{2}Mc^2\lambda(\ln \lambda)^2.$$

Consider

$$f(\lambda) = \lambda(\ln \lambda)^2.$$

Then

$$f'(\lambda) = (\ln \lambda)(\ln \lambda + 2)$$

and therefore $f(\lambda)$ has a maximum when $\ln \lambda = -2$, i.e. when $\lambda = e^{-2}$. Therefore T is a maximum when the final mass of the rocket is $e^{-2}M$, i.e. when the initial mass of fuel is $(1 - e^{-2})M$.

Example 4. A rocket of initial total mass M propels itself by ejecting mass at a constant rate μ per unit time with speed u relative to the rocket. If the rocket is at rest directed vertically upwards, show that it will not initially leave the ground unless $\mu u > Mg$, and assuming this condition to hold show that its speed after time t is given by

$$-u \ln(1 - \mu t/M) - gt.$$

Show also that when the mass of the rocket has been reduced to half the initial value, its height above ground level will be

$$\frac{uM}{2\mu}(1 - \ln 2 - Mg/4\mu u) \qquad \text{(L.)}$$

In this case the equation of motion is given by equation (8.14) which takes the form

$$(M - \mu t)\frac{dv}{dt} - \mu u = -(M - \mu t)g$$

$$\Leftrightarrow \frac{dv}{dt} = \frac{\mu u - (M - \mu t)g}{M - \mu t}. \qquad (1)$$

For the rocket to leave the ground initially we must have $dv/dt > 0$ at $t = 0$, i.e. $\mu u > Mg$. [If this condition is not satisfied the rocket will not lift off the ground until the right hand side of (1) increases to zero, i.e. until $t = (Mg - \mu u)/(\mu g)$. It may even happen that, if the exhaust velocity is too low and/or the final mass of the rocket is too large, the rocket cannot leave the ground at all.]

Integration of (1) subject to the initial conditions $v = 0$ at $t = 0$ gives

$$v = \int_0^t \left\{ \frac{\mu u}{M - \mu t} - g \right\} dt = \left[-u \ln(M - \mu t) - gt \right]_0^t$$

$$= -u \ln(1 - \mu t/M) - gt.$$

Writing $v = dx/dt$ and integrating again we find the distance risen whilst the mass of the rocket is reduced to half its initial value, i.e. during a time $M/(2\mu)$, is

$$X = \int_0^X 1 \, dx = \int_0^{M/(2\mu)} \{-u \ln(1 - \mu t/M) - gt\} \, dt$$

$$= \left[-\mu t \ln\left(1 - \frac{\mu t}{M}\right) + ut + \frac{uM}{\mu} \ln\left(1 - \frac{\mu t}{M}\right) - \frac{1}{2}gt^2 \right]_0^{M/(2\mu)}$$

$$= \frac{uM}{2\mu}\left(1 - \ln 2 - \frac{Mg}{4\mu u}\right).$$

---

*Example 5.** Two buckets of water each of total mass M_0 are suspended at the ends of a cord passing over a smooth pulley and are initially at rest. Water begins to leak from a small hole in the side of one of the buckets at a steady slow rate of m units of mass per second. Establish the equations of motion, and prove that the speed V of the bucket when a mass M_1 of water has escaped is given by

$$V = \frac{2M_0 g}{m} \ln\left(\frac{2M_0}{2M_0 - M_1}\right) - \frac{gM_1}{m}.$$

Suppose that at time t after release the mass of the leaking bucket is M and its upward speed is v. Let T be the tension in the string at this instant. Then the equation of motion (downwards) of the other bucket is

$$M_0 g - T = M_0 \frac{dv}{dt}. \qquad (1)$$

The change in momentum of the leaking bucket and its contents from time t to time $t + \delta t$ is

$$(M + \delta M)(v + \delta v) + (-\delta M)v - Mv$$
$$= M\delta v + \text{terms of order } (\delta t)^2, \text{ etc.}$$

Equating this to the impulse in time δt of the upward forces, i.e. to $(T - Mg)\delta t + \text{terms of order } (\delta t)^2$,

$$\Rightarrow M\frac{dv}{dt} = T - Mg. \qquad (2)$$

Thus the usual equations of motion are unaffected by the leak. This is because the water which leaks out has no speed *relative to the bucket* at the moment when it leaves the bucket.

Equations (1) and (2) give

$$\frac{dv}{dt} = \frac{(M_0 - M)g}{M_0 + M}$$

$$\Rightarrow \frac{dv}{dt} = \frac{mgt}{2M_0 - mt}$$

$$\Rightarrow v = \int_0^t \frac{mgt}{2M_0 - mt} dt$$

$$= g\int_0^t \left(-1 + \frac{2M_0}{2M_0 - mt}\right) dt$$

$$= g\left[-t - \frac{2M_0}{m} \ln(2M_0 - mt)\right]_0^t$$

$$\Rightarrow v = g\left[-t - \frac{2M_0}{m} \ln\left(1 - \frac{mt}{2M_0}\right)\right].$$

The required result is obtained by writing $t = M_1/m$.

---

*Exercise 8.3

*1. A spherical hailstone, falling under gravity in still air, increases its radius r by condensation according to the law $dr/dt = \lambda r$, where λ is constant. If air resistance is neglected, prove that the hailstone approaches a limiting velocity $g/(3\lambda)$. (L.)

*2. A raindrop falls from rest through an atmosphere containing water vapour at rest. The mass of the raindrop, which is initially m_0, increases by condensation uniformly with time in such a way that after a given time T it is equal to $2m_0$. The motion is opposed by a frictional force $\lambda m_0/T$ times the speed of the drop, where λ is a positive constant. Show that after time T the speed is

$$\frac{gT}{2+\lambda}(2 - 2^{-(1+\lambda)}).$$

*3. A raindrop falls through a stationary cloud, its mass m increasing by accretion uniformly with the distance fallen, $m = m_0(1 + kx)$. The motion is opposed by a resisting force $m_0 k \lambda v^2$ proportional to the square of the speed v. If $v = 0$ when $x = 0$ prove that

$$v^2 = \frac{2g}{(3+2\lambda)k}\left\{1 + kx - \frac{1}{(1+kx)^{2+2\lambda}}\right\}.$$

*4. A small body is projected vertically upwards in a cloud, the speed of projection being $\sqrt{(2gh)}$. During the ascent the body picks up moisture from the cloud, its mass at height x above the point of projection being $m_0(1 + \alpha x)$, where α is a positive constant, and the added mass is picked up from rest. Prove that the greatest height reached is h', where

$$h' = [\sqrt[3]{(1+3\alpha h)} - 1]/\alpha. \qquad (L.)$$

*5. A rocket driven car, of total initial mass M, loses mass at a constant rate m per unit time at constant ejection speed V relative to the car. If the total resistance to motion is kv when the speed is v, show that the acceleration of the car along a straight horizontal road is

$$\frac{mV - kv}{M - mt}$$

at time t from the start, and hence that the speed from rest is then

$$\frac{mV}{k}[1-(1-mt/M)^{k/m}].\qquad\text{(L.)}$$

*6. A train of mass M is moving with speed V when it begins to pick up water from rest at a uniform rate. The power is constant and equal to H. If after time t a mass m of water has been picked up, find the speed and show that the loss of energy is

$$\frac{m(Ht+MV^2)}{2(m+M)}.$$

*7. A mass m of water issues per unit time from a pipe with uniform speed u, and strikes a pail which retains it, there being no elasticity. Initially the pail is at rest, and at a subsequent instant is moving in the direction of the stream with speed V. Prove that

$$\frac{dV}{dt}=\frac{m(u-V)^3}{Mu^2},$$

and that the loss of energy up to this instant is

$$\tfrac{1}{2}MuV,$$

where M is the mass of the pail, and gravity is omitted from consideration.

Miscellaneous Exercise 8

1. A light spring AB of natural length a and modulus of elasticity $3amn^2$, initially lies straight and at its natural length on a horizontal table, and a particle of mass m is attached to it at B. Starting at time $t=0$, the end A is moved in the direction AB with constant speed V. The motion of the particle is resisted by a force equal to $4mn$ times the speed of the particle. If x is the extension of the spring at time t, prove that

$$\frac{d^2x}{dt^2}+4n\frac{dx}{dt}+3n^2x=4nV.$$

Find x in terms of t, and deduce the value of the force applied at A at time t. (N.)

2. A particle of mass m is suspended from a fixed point by a spring of natural length a and modulus λ, and a periodic force equal to $mf\sin pt$ acts downwards on it. The motion of the particle is resisted by a force equal to k times its speed. If x is the downward distance from the equilibrium position at time t, show that

$$\frac{d^2x}{dt^2}+\frac{k}{m}\frac{dx}{dt}+\frac{\lambda}{ma}x=f\sin pt.$$

If $k=4m$ and $\lambda=5ma$, show that the amplitude of the periodic forced oscillations is

$$\frac{|f|}{\sqrt{\{(5-p^2)^2+16p^2\}}}.$$

Show that this amplitude decreases steadily as p^2 increases. (N.)

3. A body is projected vertically upwards with speed U. Assuming that the air resistance is proportional to the square of the speed, show that the body will rise to a height

$$\frac{V^2}{2g}\ln\left(\frac{U^2+V^2}{V^2}\right),$$

where V is the limiting speed in downward motion, and find the corresponding time. (L.)

4. A particle P of mass m moves in a horizontal straight line. The only force acting on P is a resistance of magnitude mkv^3, where k is a positive constant and v is the speed of P. At time t the distance of P from a fixed point O of the line is x. If $x=0$ and $v=c$ when $t=0$, show that

$$v=\frac{c}{\sqrt{(1+2kc^2t)}}=\frac{c}{(1+kcx)}.\qquad\text{(L.)}$$

5. A particle falls from rest under gravity in a medium in which the resistance is proportional to the square of the speed and in which the terminal speed is U. Show that the distance fallen after time t is

$$\frac{U^2}{g}\ln\cosh\left(\frac{gt}{U}\right),$$

where g is the acceleration due to gravity.

If the particle is projected upwards with speed V in the same medium, prove that the time taken to reach the greatest height is

$$\frac{U}{g}\arctan\left(\frac{V}{U}\right). \tag{L.}$$

6. A particle moves in a resisting medium under gravity, where the resistance is proportional to the speed of the particle and the terminal speed of the particle is V. The particle is projected vertically upwards with speed u and reaches a greatest height h above the point of projection in time t_1. If u/V is small show that

$$gt_1 = u - \frac{1}{2}\frac{u^2}{V}, \qquad gh = \frac{1}{2}u^2 - \frac{1}{3}\frac{u^3}{V},$$

approximately, when terms involving higher powers of u/V are neglected. (L.)

7. A particle of mass m moves in a straight line under the action of a force of magnitude $|25\,mx|$ directed towards a fixed point O of the line, where x is the displacement of the particle from O at time t. The particle experiences a resistance to motion whose magnitude is km times the speed of the particle. Show that the equation of motion of the particle is

$$\frac{d^2x}{dt^2} + k\frac{dx}{dt} + 25x = 0.$$

Given that, when $t = 0$, $x = a\,(>0)$ and $\dfrac{dx}{dt} = 0$, find x in terms of a and t in the two cases (i) $k = 10$, (ii) $k = 8$.

Prove that in the first case the particle moves *towards* O for all $t > 0$. In the second case prove that the particle first passes through O when $t = \frac{1}{3}(\pi - \alpha)$, where α is the acute angle $\sin^{-1}\frac{3}{5}$. (N.)

8. A train of mass m moves along a straight horizontal track. The engine works at the constant rate P and the resistance to motion is constant and of magnitude R. Find the maximum speed of the train.

The train starts from rest. Show that the time which elapses before the train has attained half its maximum speed is

$$mP(\ln 2 - \tfrac{1}{2})/R^2. \tag{L.}$$

9. A particle of mass m moves in a straight line under no forces except a resistance mkv^3, where v is the speed and k is a positive constant. If the initial speed is $u > 0$ and x is the distance covered in time t, prove that

$$kx = \frac{1}{v} - \frac{1}{u}, \qquad t = \frac{x}{u} + \frac{1}{2}kx^2. \tag{L.}$$

10. The retardation of a train moving with power cut off is $k(v^2 + c^2)$, where v is the speed and k and c are constants. Prove that, if initially $v = 2c$, the speed will be halved in a distance

$$\frac{1}{2k}\ln\frac{5}{2} \text{ in time } \frac{1}{ck}\tan^{-1}\frac{1}{3}.$$

Prove also that the train will come to rest in a further distance $\dfrac{1}{2k}\ln 2$ in additional time $\pi/(4kc)$.

11. A particle of unit mass is projected vertically upwards under gravity with speed u and reaches its maximum height H in time T. Given that the air resistance is kv, where k is constant and v is the speed of the particle, find H and T and show that

$$kH = u - gT.$$

*12. If a rocket, originally of mass M, throws off every unit of time a mass eM with relative speed V, and if M' be the mass of the case, etc., show that it cannot rise at once unless $eV > g$, nor at all unless

$$\frac{eMV}{M'} > g.$$

If it just rises vertically at once, show that its greatest speed is

$$V \ln \frac{M}{M'} - \frac{g}{e}\left(1 - \frac{M'}{M}\right),$$

and that the greatest height it reaches is

$$\frac{V^2}{2g}\left(\ln \frac{M}{M'}\right)^2 + \frac{V}{e}\left(1 - \frac{M'}{M} - \ln \frac{M}{M'}\right). \tag{L.}$$

*13. A body consists of equal masses M of inflammable and non-inflammable material. It descends freely under gravity from rest whilst the combustible part burns at the uniform rate of λM, where λ is a constant. If the burning material is ejected vertically upwards with a constant upward speed u relative to the body, and the air resistance is neglected, show, from considerations of momentum, or otherwise, that

$$\frac{d}{dt}[(2 - \lambda t)v] = \lambda(u - v) + g(2 - \lambda t)$$

where v is the speed of the body at time t.

Hence show that the body descends a distance

$$g/(2\lambda^2) + (1 - \ln 2)u/\lambda$$

before all the inflammable material is burnt. (L.)

14. A particle P, of unit mass, is projected vertically downwards from a point O with speed V in a medium which exerts on the particle a resistance to motion of magnitude kv, where k is a positive constant and v is the speed of the particle. Show that the speed tends to a value which is independent of V and that the distance fallen by P in time t is

$$[gt - (V - g/k)(e^{-kt} - 1)]/k.$$

After a time T an identical particle Q is projected vertically downwards from O with speed $2V$. Show that, provided that $T > V/g$, ultimately P is a distance $(gT - V)/k$ below Q. (L.)

15. A particle P of unit mass moves along a straight line so that $OP = x$, where O is a fixed point on the line. The forces acting on P are (a) a force $5k^2 x$ directed towards O, (b) a resistance $2kv$, where v is the speed of P. Write down the differential equation satisfied by x, and solve this equation given that,

$$\text{when } t = 0, \; x = 0 \text{ and } \frac{dx}{dt} = 2.$$

Sketch the graph of x against t for $t \geq 0$ when $k = 1$. (L.)

16. The only force acting on a particle of mass m moving in a straight line is a resistance $mk(V^2 + 3v^2)$, where k is a constant, V is the initial speed and v is the speed of the particle at time t. When $t = T$, $v = \frac{1}{3}V$ and the particle has travelled a distance X. Show that

$$6kX = \ln 3 \text{ and } 6VkT\sqrt{3} = \pi. \tag{L.}$$

17. A particle P of mass m is attached to one end of a light elastic string of natural length l and modulus of elasticity $6mn^2l$. The other end of the string is fastened to a fixed point O on a horizontal table and P is initially at rest on the table with $OP = l$. A variable force is now applied to P in the direction OP, the magnitude of the force being $mn^2l\,e^{-nt}$ where t is the time measured from the initial application of the force. The motion of P is opposed by a resistance of magnitude $5mn$ times the speed. Show that when the extension of the string is x then

$$\frac{d^2x}{dt^2} + 5n\frac{dx}{dt} + 6n^2 x = n^2 l\, e^{-nt}.$$

Show that the greatest extension of the string is $2l/27$. (N.)

18. A rod is fixed at an angle of 30° to the horizontal. A bead of mass m can move on the rod and is connected to the lowest point O of the rod by a light spring of natural length a and modulus of elasticity $5mn^2 a$, where n is a positive constant. There is a force of magnitude $2mn|v|$, where v is the velocity of the bead, which acts along the rod and opposes the motion of the bead. The bead is released from rest at a distance a from O and at time t after release it has moved a distance x down the rod from its initial position. Show that

$$\frac{d^2 x}{dt^2} + 2n\frac{dx}{dt} + 5n^2 x = \tfrac{1}{2}g.$$

Find x at time t and show that, when t is large, $x \approx g/(10n^2)$. (N.)

*19. A particle of mass $\tfrac{1}{2}$ kg is released from rest at a height of 10 m above the surface of the earth. At time t after release it has fallen a distance x and has velocity v.

Assuming an air resistance cv^2 where $c = \tfrac{1}{10}$ kg m^{-1} and taking g to be 10 m s^{-2}, write down the equation of motion in terms of the variables v and x. Show by integration that

$$v^2 = 50(1 - e^{-0.4x}).$$

By substituting this value for v^2 in the equation of motion, show that

$$\ddot{x} = 10 e^{-0.4x}.$$

Given that $x = 0.05$ m when $t = 0.1$ second, use a step-by-step process with step length 0.1 second to estimate the distance fallen in half a second. [Tabulate your working to two decimal places only.]

Find the difference between this distance and the distance fallen in the same time under free fall (without air-resistance). (N.)

20. A particle is projected vertically upwards with speed $2c$ in a medium in which the resistance per unit mass, when the speed is v, is gv^2/c^2. Show that the particle returns to the point of projection with speed $2c/\sqrt{5}$.

Find, in terms of c and g, the time of ascent of the particle. (L.)

21. A particle of unit mass moves on a straight line subject to a force $k^2/2x^2$ towards a point O on the line, where x is the distance of the particle from O and k is a positive constant.

The particle is projected from the point $x = \tfrac{1}{4}a(>0)$ away from O with a speed $(3k^2/a)^{1/2}$. Show that during the motion

$$\left(\frac{dx}{dt}\right)^2 = k^2\left(\frac{1}{x} - \frac{1}{a}\right)$$

and that the particle comes to rest at the point $x = a$.

Express the time T for the particle to reach the point $x = a$ as a definite integral. By evaluating this integral, using the substitution $x = a\sin^2\theta$ or otherwise, show that

$$T = \left(\frac{\pi}{3} + \frac{\sqrt{3}}{4}\right)\frac{a^{3/2}}{k}.$$

(N.)

22. A light spring AB of natural length a and modulus of elasticity $3amn^2$ lies straight and at its natural length at rest on a horizontal table. A particle of mass m is attached to the end A. From time $t = 0$ onwards the end B is caused to move in the direction AB with constant acceleration f. The resulting motion of the particle is resisted by a force of magnitude $4mn$ times the speed of the particle.

Prove that if x is the extension of the spring at time t the velocity of the particle, in the sense of the motion of B, is

$$ft - \frac{dx}{dt}.$$

Deduce that the extension of the spring is given by

$$\frac{d^2 x}{dt^2} + 4n\frac{dx}{dt} + 3n^2 x = 4nft + f,$$

and find x in terms of t. (N.)

23. A bead, P, of mass m can slide on a long straight rod AB and is attached to one end of a spring of natural length a and modulus of elasticity $2mn^2a$, where n is a positive constant. The other end of the spring is fastened to A. The system is at rest with AB horizontal and $AP = a$. The rod is now made to move horizontally along its length so that at time t its displacement in the direction AB is $a \sin nt$. The bead slides along the rod and its motion is opposed by a resistance of magnitude $3 mn$ times its speed relative to the rod. Denoting the extension of the spring at time t by x, prove that

$$\frac{d^2 x}{dt^2} + 3n \frac{dx}{dt} + 2n^2 x = an^2 \sin nt.$$

Find x in terms of a, n and t. (N.)

24. A particle of mass 1 kg moves in a straight line under the action of a force directed towards a fixed point O of the line. This force is proportional to the distance x m of the particle from O and is 2N when $x = 1$. The resistance to the motion of the particle is proportional to its speed V m/s and is 3N when $V = 1$. The particle starts from rest when $x = a$. Show that, after t s,

$$x = Ae^{-t} + Be^{-2t},$$

where A and B are constants. Find the values of A and B.

Give a sketch of x against t indicating clearly both the behaviour of x when t is small and the behaviour of x when t is large. (L.)

25. A particle moves along the x-axis so that at time t its displacement from the origin O satisfies the differential equation

$$\frac{d^2 x}{dt^2} + 4 \frac{dx}{dt} + 13x = 40 \cos 3t.$$

Find a particular integral of this differential equation in the form $p \cos 3t + q \sin 3t$, where p and q are constants. Obtain further the complementary function of this differential equation.

Hence express x in terms of t, given that

$$x = 1, \frac{dx}{dt} = 12 \text{ when } t = 0.$$

Show that, when t is large, the motion approximates to a simple harmonic oscillation about O with period $2\pi/3$ and with amplitude $\sqrt{10}$. (L.)

*26. A raindrop falls from rest through an atmosphere containing water vapour at rest. The mass of the raindrop is initially M and it increases uniformly with time by condensation, doubling after a time T. The motion is opposed by a frictional resistance equal to (M/T) times the speed. Show that the speed, v, of the raindrop satisfies the differential equation

$$(T+t)\frac{dv}{dt} + 2v = (T+t)g.$$

Hence show that after time T the raindrop has fallen a distance $gT^2/3$. (L.)

*27. Initially the total mass of a rocket is M, of which kM is the mass of the fuel. Starting from rest, the rocket gives itself a constant vertical acceleration of magnitude g by ejecting fuel with constant speed u relative to itself. If m denotes its remaining mass at time t, show that the rate of decrease of m with respect to t is $2mg/u$, and deduce that

$$m = Me^{-2gt/u}.$$

Find in terms of M, u and k an expression for the kinetic energy of the rocket when the fuel is exhausted. Find the value of k for which this energy is a maximum. (Assume that the height reached is sufficiently small for g to be considered constant.) (L.)

9 Dynamics of a Particle with two Degrees of Freedom

9:1 FURTHER DIFFERENTIATION OF VECTORS—APPLICATIONS TO KINEMATICS

In § 1:1 and § 1:2 we defined differentiation and integration of a vector with respect to a scalar parameter, usually the time t. The product rule for differentiation leads at once to the result

$$\frac{d}{dt}(\mathbf{a} \times \mathbf{b}) = \frac{d\mathbf{a}}{dt} \times \mathbf{b} + \mathbf{a} \times \frac{d\mathbf{b}}{dt}. \tag{9.1}$$

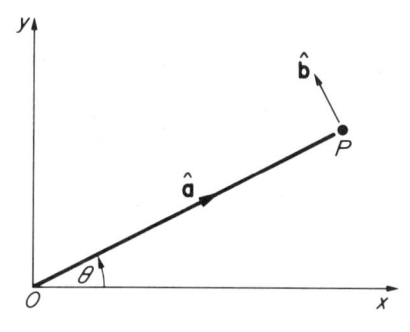

FIG. 9.1.

If $\hat{\mathbf{a}}$ is a unit vector in *two* dimensions, Fig. 9.1, we can write

$$\hat{\mathbf{a}} = \mathbf{i}\cos\theta + \mathbf{j}\sin\theta \tag{9.2}$$

$$\Rightarrow \frac{d\hat{\mathbf{a}}}{dt} = -\mathbf{i}\dot{\theta}\sin\theta + \mathbf{j}\dot{\theta}\cos\theta = \dot{\theta}\hat{\mathbf{b}}, \tag{9.3}$$

where $\hat{\mathbf{b}} = -\mathbf{i}\sin\theta + \mathbf{j}\cos\theta = \mathbf{i}\cos(\theta + \tfrac{1}{2}\pi) + \mathbf{j}\sin(\theta + \tfrac{1}{2}\pi)$. Therefore $\hat{\mathbf{b}}$ is another unit vector obtained by rotating $\hat{\mathbf{a}}$ through one right angle in the positive sense. If $\mathbf{k}$ is a unit vector perpendicular to the plane of $\mathbf{i}, \mathbf{j}$ and forms a right handed set with them, we can also write (9.3) in the form

$$\frac{d\hat{\mathbf{a}}}{dt} = \dot{\theta}(\mathbf{k} \times \hat{\mathbf{a}}). \tag{9.4}$$

303

Note that, since

$$\hat{\mathbf{b}} = -\mathbf{i}\sin\theta + \mathbf{j}\cos\theta, \tag{9.2a}$$

$$\frac{d\hat{\mathbf{b}}}{dt} = -\mathbf{i}\dot\theta\cos\theta - \mathbf{j}\dot\theta\sin\theta = -\dot\theta\hat{\mathbf{a}}. \tag{9.5}$$

Special representations

Here we give some formulae for the components of the velocity and acceleration of a point in terms of different representations of its position.

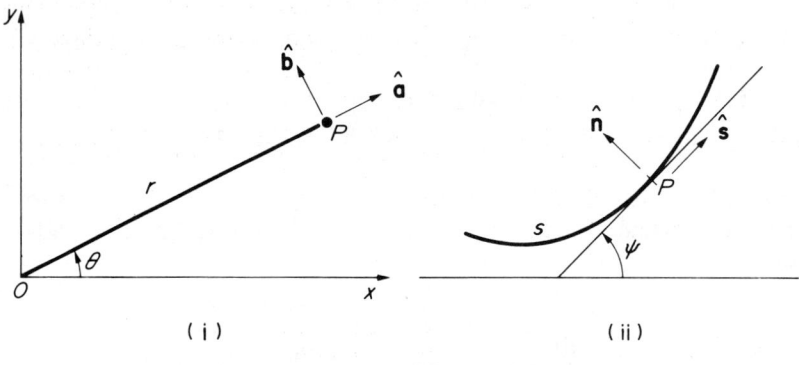

FIG. 9.2.

1. Polar coordinates

In Fig. 9.2 (i) the coordinates of P referred to initial line Ol (i.e. Ox) and pole O are (r, θ). The vector $\overrightarrow{OP}$ can be represented as $\mathbf{r} = r\hat{\mathbf{a}}$, where $\hat{\mathbf{a}}$ is the unit vector in the direction $\overrightarrow{OP}$ and therefore $\hat{\mathbf{a}} = \mathbf{i}\cos\theta + \mathbf{j}\sin\theta$. Then, by differentiation,

$$\mathbf{v} = \frac{dr}{dt}\hat{\mathbf{a}} + r\frac{d\hat{\mathbf{a}}}{dt}$$

and therefore from (9.3)

$$\mathbf{v} = \dot r\hat{\mathbf{a}} + r\dot\theta\hat{\mathbf{b}}, \tag{9.6}$$

where $\hat{\mathbf{b}}$ is the unit vector in the direction at right angles to $\overrightarrow{OP}$, i.e. in the *transverse* direction.

By differentiating equation (9.6) again we obtain the acceleration

$$\mathbf{f} = \frac{d\mathbf{v}}{dt} = \ddot r\hat{\mathbf{a}} + \dot r\frac{d\hat{\mathbf{a}}}{dt} + (\dot r\dot\theta + r\ddot\theta)\hat{\mathbf{b}} + r\dot\theta\frac{d\hat{\mathbf{b}}}{dt}.$$

But, from (9.5) $d\hat{\mathbf{b}}/dt = -\dot\theta\hat{\mathbf{a}}$, and hence

$$\mathbf{f} = (\ddot r - r\dot\theta^2)\hat{\mathbf{a}} + (2\dot r\dot\theta + r\ddot\theta)\hat{\mathbf{b}}. \tag{9.7}$$

The radial component of this acceleration is

$$\ddot r - r\dot\theta^2. \tag{9.8}$$

and the transverse component is

$$2\dot{r}\dot{\theta} + r\ddot{\theta} = \frac{1}{r}\frac{d}{dt}(r^2\dot{\theta}). \tag{9.9}$$

Note that when the particle is moving in a circle the radial component of the acceleration is $-r\dot{\theta}^2$ and the transverse (in this case tangential) component of acceleration is $r\ddot{\theta}$.

2. Intrinsic coordinates

This representation is used when a particle moves in a plane along a curve, Fig. 9.2 (ii). One of the intrinsic coordinates is the arc length s along the curve from some fixed point to P, and the other is the angle ψ between the tangent and a fixed direction. The direction of the tangent at P is denoted by the unit vector $\hat{s}$ and the direction of the normal towards the centre of curvature (see Volume 2, Part I, § 20:10) by $\hat{n}$. [In Fig. 9.2 (ii) the curvature κ is such that ψ increases as s increases, i.e. $\dfrac{d\psi}{ds} = \kappa = \dfrac{1}{\varrho} > 0.$]

The velocity vector is given by

$$\mathbf{v} = v\hat{s} = \frac{ds}{dt}\hat{s}. \tag{9.10}$$

Differentiation with respect to t gives the acceleration

$$\mathbf{f} = \frac{d\mathbf{v}}{dt} = \frac{dv}{dt}\hat{s} + v\frac{d\hat{s}}{dt}.$$

But from (9.3)

$$\frac{d\hat{s}}{dt} = \frac{d\psi}{dt}\hat{n} = \frac{ds}{dt}\cdot\frac{d\psi}{ds}\hat{n} = \frac{v}{\varrho}\hat{n}$$

$$\Rightarrow \mathbf{f} = \frac{dv}{dt}\hat{s} + \frac{v^2}{\varrho}\hat{n}. \tag{9.11}$$

The *normal* component of this acceleration is v^2/ϱ towards the centre of curvature and the *tangential* component is

$$\frac{dv}{dt}\left(=\frac{d^2s}{dt^2} = v\frac{dv}{ds}\right).$$

If the curvature is in the opposite sense so that $d\psi/ds < 0$, provided that $\hat{n}$ is directed towards the centre of curvature, the same formula holds.

Example 1. A particle describes the curve $r = a\sin 2\theta$ in such a manner that the radius vector from the origin rotates with uniform angular speed ω. Find, in terms of ω, expressions for the radial and transverse components of acceleration of the particle and find the component of acceleration perpendicular to the initial line when $\theta = \tfrac{1}{4}\pi$.

Because the radius vector rotates with uniform angular speed ω, $\dot{\theta} = \omega = $ constant and therefore, since $r = a\sin 2\theta$,

$$\dot{r} = 2a\dot{\theta}\cos 2\theta = 2a\omega\cos 2\theta \quad \text{and} \quad \ddot{r} = -4a\omega^2\sin 2\theta.$$

Therefore, from (9.8) the radial component of acceleration is $-4a\omega^2 \sin 2\theta - r\omega^2$, and from (9.9) and since $\ddot{\theta} = 0$ the transverse component of acceleration is $4a\omega^2 \cos 2\theta$. The component of acceleration perpendicular to the initial line is therefore

$$(-4a\omega^2 \sin 2\theta - r\omega^2) \sin\theta + 4a\omega^2 \cos 2\theta \cos\theta.$$

When $\theta = \pi/4, r = a$. Hence when $\theta = \pi/4$ this component of acceleration is $-5a\omega^2/\sqrt{2}$. The negative sign implies that this acceleration component is in fact towards the initial line.

Example 2. A particle describes the curve (catenary) whose intrinsic equation is $s = c \tan\psi$ with uniform speed u. Find expressions for the normal and tangential components of the acceleration of the particle at the point $(c, \pi/4)$ of the curve.

The tangential component of acceleration is zero since the particle is describing the curve with uniform speed [equation (9.11)]. For this curve $\varrho = |ds/d\psi| = c \sec^2\psi$, and therefore ϱ at the point $(c, \pi/4)$ is $2c$. Hence the normal component of acceleration at $(c, \pi/4)$ is $u^2/\varrho = u^2/(2c)$.

Example 3. A particle moves so that its position vector $\mathbf{r}$ at time t is given by

$$\mathbf{r} = \mathbf{a} \cos\omega t + \mathbf{b} \sin\omega t,$$

where $\mathbf{a}$ and $\mathbf{b}$ are constant vectors and ω is constant. Obtain the velocity $\dot{\mathbf{r}}$ and acceleration $\ddot{\mathbf{r}}$ of the particle and describe the motion briefly.

From the definition of differentiation it follows that

$$\dot{\mathbf{r}} = -\mathbf{a}\omega \sin\omega t + \mathbf{b}\omega \cos\omega t,$$
$$\ddot{\mathbf{r}} = -\mathbf{a}\omega^2 \cos\omega t - \mathbf{b}\omega^2 \sin\omega t$$
$$\Rightarrow \ddot{\mathbf{r}} = -\omega^2 \mathbf{r}.$$

The acceleration, $\ddot{\mathbf{r}}$, of the particle P is always directed towards the origin O and is of magnitude $\omega^2 OP$. Further, since

$$\mathbf{r} \cdot (\mathbf{a} \times \mathbf{b}) = 0,$$

the position vector $\overrightarrow{OP} = \mathbf{r}$ is always perpendicular to the fixed vector $\mathbf{a} \times \mathbf{b}$ and so $\overrightarrow{OP}$ lies in a plane, the plane of $\mathbf{a}$ and $\mathbf{b}$. This is otherwise obvious since $\mathbf{a}, \mathbf{b}$ form a basis for the vector $\mathbf{r}$.

The path of P is an ellipse.

Exercise 9.1

1. A vector $\mathbf{v}$ varies with time so that, at time t, $d\mathbf{v}/dt = \mathbf{i} \cos t + \mathbf{j} \sin t + \mathbf{k}$. When $t = 0$,

$$\mathbf{v} \times \frac{d\mathbf{v}}{dt} = \mathbf{0} \text{ and } \mathbf{v} \cdot \mathbf{v} = 18.$$

Find the two possible values of $\mathbf{v}$ at time t.

2. A vector $\mathbf{u}$ is given by

$$\mathbf{u} = x\mathbf{a} + y\mathbf{b}$$

where $\mathbf{a}$ and $\mathbf{b}$ are constant vectors with $\mathbf{a} \times \mathbf{b} \neq \mathbf{0}$ and x and y are functions of the time t such that at all times

$$\mathbf{u} \times \frac{d\mathbf{u}}{dt} = \mathbf{0}.$$

Show that y/x is constant. (N.)

3. A unit vector $\hat{\mathbf{a}}$ lies in the xy plane and its direction at time t makes an angle θ with the x-axis. Prove that

$$\frac{d\hat{\mathbf{a}}}{dt} = \hat{\mathbf{b}} \frac{d\theta}{dt},$$

where $\hat{\mathbf{b}}$ is the unit vector in the xy plane whose direction at time t makes an angle $\pi/2 + \theta$ with the x-axis. Deduce that

$$\frac{d\hat{\mathbf{b}}}{dt} = -\hat{\mathbf{a}}\frac{d\theta}{dt}.$$

Express $\dfrac{d^2\hat{\mathbf{a}}}{dt^2}$ in terms of $\hat{\mathbf{a}}$, $\hat{\mathbf{b}}$, $\dfrac{d\theta}{dt}$ and $\dfrac{d^2\theta}{dt^2}$. (N.)

4. A particle describes the curve $r = ae^{\theta\sqrt{3}}\cosh 2\theta$ in such a manner that the angular speed about the origin is constant. Show that the resultant acceleration of the particle at any instant makes an angle of $30°$ with the radius vector from the origin. (L.)

5. A particle describes the cardioid $r = a(1 - \cos\theta)$ in such a manner that the radius vector from the origin rotates with constant angular speed ω. Show that its acceleration can be resolved into a component $2a\omega^2$ parallel to the initial line and a component $(4r - 3a)\omega^2$ towards the origin.

9:2 MOTION REFERRED TO CARTESIAN AXES

In two dimensions the position vector $\mathbf{r}$ of a point P can be represented in the form $x\mathbf{i} + y\mathbf{j}$ and therefore the velocity of P can be represented in the form

$$\mathbf{v} = \dot{\mathbf{r}} = \dot{x}\mathbf{i} + \dot{y}\mathbf{j}$$

since $\mathbf{i}$ and $\mathbf{j}$ are constant. Also, the acceleration of P can be represented in the form

$$\mathbf{f} = \dot{\mathbf{v}} = \ddot{x}\mathbf{i} + \ddot{y}\mathbf{j}.$$

We give below an illustrative example; most problems of this kind reduce to the solution of simultaneous differential equations.

Example. A particle moves in a plane so that its accelerations parallel to cartesian axes of coordinates in its plane are λa parallel to Ox and $-\mu y$ parallel to Oy where λ, μ, a are positive constants. The particle starts from rest at the point $(0, a)$. Find the equation of the path of the particle.

$$\ddot{x} = \lambda a$$
$$\Leftrightarrow \dot{x} = \lambda a t + C, \quad \text{where } C \text{ is constant.}$$
$$\dot{x} = 0 \quad \text{when} \quad t = 0 \Rightarrow C = 0$$
$$\Rightarrow \dot{x} = \lambda a t$$
$$\Leftrightarrow x = \tfrac{1}{2}\lambda a t^2 + K, \quad \text{where } K \text{ is constant.}$$
$$x = 0 \quad \text{when} \quad t = 0 \Rightarrow K = 0$$
$$\Rightarrow x = \tfrac{1}{2}\lambda a t^2. \tag{1}$$

Also
$$\ddot{y} = -\mu y$$

and the general solution of this differential equation is

$$y = A\cos(t\sqrt{\mu}) + B\sin(t\sqrt{\mu}),$$

where A and B are constants.

$$y = a \quad \text{when} \quad t = 0 \Rightarrow A = a.$$

Also,
and
$$\dot{y} = -A\sqrt{\mu}\sin(t\sqrt{\mu}) + B\sqrt{\mu}\cos(t\sqrt{\mu})$$
$$\dot{y} = 0 \quad \text{when} \quad t = 0 \Rightarrow B = 0.$$

The particular solution of the differential equation is therefore

$$y = a\cos(t\sqrt{\mu}). \tag{2}$$

Eliminating t between (1) and (2) we find
$$y = a \cos \sqrt{\{2\mu x/(a\lambda)\}}$$
is the equation of the path of the particle.

Exercise 9.2

In each of questions **1–4** the motion of a particle is defined in a conventional notation. In each case, find the equation of the path.

1. $\ddot{x} = 0$, $\ddot{y} = -g$; $x = 0$, $y = 0$, $\dot{x} = v$, $\dot{y} = u$ when $t = 0$.

2. $\ddot{x} = 2$, $\ddot{y} = 1$; $x = 0$, $y = 0$, $\mathbf{v} = \hat{\mathbf{s}}u\sqrt{2}$ when $t = 0$, where the direction of $\hat{\mathbf{s}}$ bisects the angle between the axes.

3. $\ddot{x} = \omega^2 a \sin \omega t$, $\ddot{y} = \omega^2 a \cos \omega t$; $x = 0$, $y = -a$, $\dot{y} = 0$, $\dot{x} = -\omega a$ when $t = 0$.

4. $\ddot{x} = -\omega^2 x$, $\ddot{y} = -\omega^2 y$; $x = a$, $y = 0$, $\dot{x} = 0$, $\dot{y} = \omega b$ when $t = 0$.

5. Given that $\overrightarrow{OP} = \mathbf{r}$ and $\mathbf{r} = t^2\mathbf{i} + 2kt\mathbf{j}$, show that the locus of P is a parabola. [Resolve $\mathbf{r}$ into its components parallel to the cartesian coordinate axes.]

6. Given that $\overrightarrow{OP} = \mathbf{r}$ and $\mathbf{r} = t\mathbf{i} + (1/t)\mathbf{j}$, find the equation of the locus of P.

7. A particle starts from the origin with velocity 2 in the direction of the x-axis and moves with unit accelerations in the directions of the x- and y-axes respectively. Find an expression for the distance travelled by the particle in the first second of its motion. [Use $(ds/dt)^2 = (dx/dt)^2 + (dy/dt)^2$.]

8. The motion of a particle P, whose coordinates are (x, y) referred to a pair of fixed axes through a point O, satisfies the equations
$$\frac{d^2x}{dt^2} = -\omega^2 x, \quad \frac{d^2y}{dt^2} = -\omega^2 y;$$
the initial conditions are
$$x = a, \quad y = 0, \quad \frac{dx}{dt} = 0 \quad \text{and} \quad \frac{dy}{dt} = b\omega \quad \text{when} \quad t = 0,$$
where ω, a and b are positive constants. Prove that the path of the particle is the ellipse $(x/a)^2 + (y/b)^2 = 1$.

Find, in terms of OP, the magnitude of the component perpendicular to OP of the velocity of P. Hence, or otherwise, show that the angular speed of OP is
$$\frac{\omega ab}{OP^2}. \tag{N.}$$

9. A particle starts from a point whose position vector is $3\mathbf{j}$ with initial velocity vector $4\omega\mathbf{i}$, where ω is a constant. Throughout its motion it has an acceleration $-\omega^2(x\mathbf{i} + y\mathbf{j})$, where $x\mathbf{i} + y\mathbf{j}$ is the position vector of the particle at time t. Find the equation of the path of the particle in vector and cartesian form.

A second particle is projected from the origin at time $t = 0$ with velocity vector $\mathbf{i} + \mathbf{j}$. If this particle has a constant acceleration $(\mathbf{i} + \mathbf{j})/5$, find the least value of ω for which a collision is possible. Find also the value of t when the collision occurs. (L.)

10. A particle moves in a plane so that its acceleration vector is $-x\mathbf{i} - 4y\mathbf{j}$ when its position vector is $x\mathbf{i} + y\mathbf{j}$. At time $t = 0$ the position vector of the particle is $\mathbf{i} + \mathbf{j}$ and its velocity vector is $\mathbf{i} - 2\mathbf{j}$. Find the position vector of the particle at time t.

When $t = \frac{3}{4}\pi$ the forces acting on the particle are removed. Prove that the position vector of the particle when $t = \frac{7}{4}\pi$ is $-\pi\sqrt{2}\mathbf{i} + (2\pi + 1)\mathbf{j}$. (L.)

11. A particle of mass m moves in a horizontal plane under the action of a variable force $\mathbf{F}$ so that the position vector of the particle at time t is $\mathbf{r} = 4\cos kt\, \mathbf{i} + 3\sin kt\, \mathbf{j}$, where k is a constant. Find
  (a) the period of the motion,
  (b) the greatest magnitude of $\mathbf{F}$.

If the force $\mathbf{F}$ ceases to act when $t = \pi/(3k)$, find the position vector of the particle when $t = 4\pi/(3k)$. (L.)

12. A particle of mass m moves in such a way that at time t its position vector is
$$t^2\mathbf{i} + (2t^2 - 5t)\mathbf{j} + (t^2 + 5t)\mathbf{k}.$$
Find the force required to maintain this motion.
  This force is the resultant of a force $m(14\mathbf{i} - 2\mathbf{k})$ and another force $\mathbf{R}$. Find $\mathbf{R}$ and show that it is always perpendicular to the direction of motion of the particle. (N.)

13. A particle of mass m moves in the plane of rectangular axes Ox, Oy under the influence of a force whose components when the particle is at the point (x, y) are $-4mn^2 y$, $-mn^2(5y - x)$ respectively, where n is a positive constant. At time $t = 0$ the particle leaves the point $(0, 0)$ with a velocity whose components are $na, 0$. By finding a differential equation for $x - y$, prove that, at any time t, $x = y + a \sin nt$. Deduce that $y = \tfrac{1}{3}a \sin nt - \tfrac{1}{6}a \sin 2nt$ and find the corresponding expression for x.
  Verify that when $t = 2\pi/n$ the position and velocity of the particle are the same as at $t = 0$. (N.)

9:3 MOTION OF A PARTICLE ON A SMOOTH CURVE

When a particle moves on a smooth curve (fixed in a vertical plane) whose intrinsic equation is $s = f(\psi)$, the velocity and acceleration are given by equations (9.10), (9.11) respectively.

Figure 9.3 shows the forces acting on a particle P of mass m which is free to move under gravity on the inside of a smooth curve which is fixed in a vertical plane. These forces are the weight of the particle and the normal reaction R of the curve on the particle. The particle is at the point $P(x, y)$ at time t from the beginning of the motion.

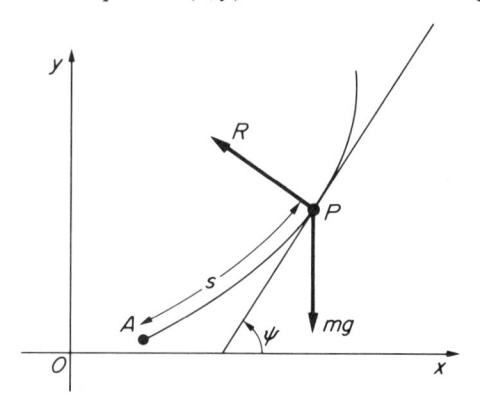

FIG. 9.3.

The equations which determine the motion of the particle are
$$R - mg \cos \psi = \frac{mv^2}{\varrho} \qquad (9.12)$$
and the energy equation
$$\tfrac{1}{2}mv^2 + mgy = k, \qquad (9.13)$$
where v is the speed of the particle at P and k is a constant depending upon the boundary conditions of the motion.

The energy equation could be replaced by the equation of motion for the particle in the *tangential* direction

$$-mg \sin \psi = m\ddot{s}, \tag{9.14}$$

where the length of the arc AP is s. This equation can be obtained directly from the energy equation by differentiation with respect to s, thus:

$$\tfrac{1}{2}mv^2 + mgy = k$$

$$\Leftrightarrow mv\frac{dv}{ds} + mg\frac{dy}{ds} = 0.$$

But

$$v\frac{dv}{ds} = \ddot{s} \quad \text{and} \quad \frac{dy}{ds} = \sin \psi$$

$$\Rightarrow \ddot{s} + g \sin \psi = 0. \tag{9.15}$$

Example 1. A particle moves under gravity on a smooth cycloid whose intrinsic equation is $s = 4a \sin \psi$; $s = 0$ when $\psi = 0$, Fig. 9.4. The particle starts from rest at the point A, $\psi = \tfrac{1}{4}\pi$. Show that the reaction between the particle and the curve is positive throughout the motion from A to O, the lowest point of the curve, and calculate
  (a) the normal reaction between the particle and the curve at O,
  (b) the time taken by the particle to move from A to O.

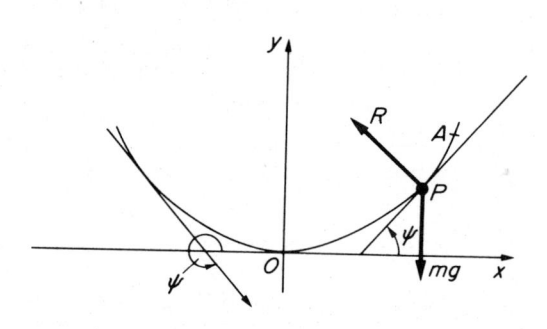

FIG. 9.4.

If the particle is at $P(s, \psi)$ at time t from the beginning of the motion, the equations of motion for the particle are:

(1) $$R - mg \cos \psi = \frac{mv^2}{\varrho},$$

(2) $$v^2 = 2g(y_A - y_P),$$

where v is the speed of the particle at P and y_A, y_P are respectively the heights of A and P above the horizontal through O.

Also

$$dy/ds = \sin \psi$$

$$\Rightarrow y = \int \sin \psi \, ds = \int \sin \psi \frac{ds}{d\psi} d\psi = \int \sin \psi \cdot 4a \cos \psi \, d\psi$$

$$= -a \cos 2\psi + K,$$

where K is constant. At A,

$$\psi = \tfrac{1}{4}\pi$$

$$\Rightarrow y_A - y_P = a \cos 2\psi$$

$$\Rightarrow v^2 = 2ga \cos 2\psi.$$

But
$$\varrho = \frac{ds}{d\psi} = 4a\cos\psi$$

$$\Rightarrow R = mg\cos\psi + \frac{2mga\cos 2\psi}{4a\cos\psi} = \frac{mg(4\cos^2\psi - 1)}{2\cos\psi}.$$

But, for $0 \leqslant \psi \leqslant \frac{1}{4}\pi$, $4\cos^2\psi > 1$ and therefore R is positive for all values of ψ in this range.

[The symmetry of the motion about O shows that when the particle is on the negative side of O, R remains positive. This result is apparent from the value of R obtained above since the definitions involved here give $-\frac{1}{4}\pi \leqslant \psi \leqslant 0$ for values of ψ on the x negative side of the origin.]

At O, $\psi = 0$ and so the normal reaction there is $3mg/2$.

Also, for the tangential component of the motion of the particle at P,
$$m\ddot{s} = -mg\sin\psi$$

$$\Rightarrow \ddot{s} = -\frac{gs}{4a}.$$

The general solution of this equation is
$$s = A\cos\left\{t\sqrt{\left(\frac{g}{4a}\right)}\right\} + B\sin\left\{t\sqrt{\left(\frac{g}{4a}\right)}\right\}$$

and the boundary conditions $s = 2a\sqrt{2}$, $\dot{s} = 0$ when $t = 0$ give $A = 2a\sqrt{2}$, $B = 0$

$$\Rightarrow s = 2a\sqrt{2}\cos\left\{t\sqrt{\left(\frac{g}{4a}\right)}\right\}.$$

Therefore, when $s = 0$,
$$0 = \cos\left\{t\sqrt{\left(\frac{g}{4a}\right)}\right\} \Leftrightarrow t = \pi\sqrt{\left(\frac{a}{g}\right)}.$$

The particle moves from A to O in time $\pi\sqrt{(a/g)}$.

[*Note.* It is an important characteristic of the cycloid that the motion discussed here is a simple harmonic motion whatever its amplitude.]

Example 2. Show that the radius of curvature of the parabola $x^2 = 4ay$ at the point $(2ap, ap^2)$ is $2a(1 + p^2)^{3/2}$.

A small heavy bead can slide freely along a fixed smooth wire in the form of this parabola, the axis of y being directed vertically upwards. The bead is projected from the origin with speed U. Show that when the direction of motion has turned through an angle ψ the speed of the bead is

$$\sqrt{(U^2 - 2ga\tan^2\psi)}$$

and that the force exerted by the bead on the wire is proportional to $\cos^3\psi$. (N.)

$$x^2 = 4ay$$

$$\Rightarrow \frac{dy}{dx} = \frac{x}{2a}; \quad \frac{d^2y}{dx^2} = \frac{1}{2a}.$$

Therefore at $(2ap, ap^2)$,

$$\varrho = \left|\frac{\left\{1 + \left(\frac{dy}{dx}\right)^2\right\}^{3/2}}{\frac{d^2y}{dx^2}}\right| = \frac{(1 + p^2)^{3/2}}{\frac{1}{2a}} = 2a(1 + p^2)^{3/2}.$$

Figure 9.5 shows the forces acting on the bead when it is at P, where the gradient of the parabola is $\tan\psi$ and the direction of motion has turned through an angle ψ.

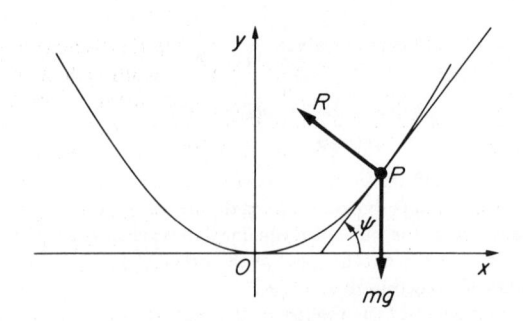

FIG. 9.5.

The equations of motion for the particle are,

(1) $$R - mg \cos \psi = \frac{mv^2}{\varrho},$$

(2) $$v^2 = U^2 - 2gap^2.$$

But $$\tan \psi = \frac{dy}{dx} = p \;\Rightarrow\; \cos \psi = \frac{1}{\sqrt{(1+\tan^2 \psi)}} = \frac{1}{\sqrt{(1+p^2)}}$$
$$\Rightarrow \varrho = 2a \sec^3 \psi.$$

Then, from equation (1),

$$R = \frac{m(U^2 - 2ga \tan^2 \psi)\cos^3 \psi}{2a} + mg \cos \psi$$

$$= m\left(\frac{U^2}{2a}\cos^3 \psi - g \sin^2 \psi \cos \psi + g \cos \psi\right) = m\left(\frac{U^2}{2a} + g\right)\cos^3 \psi.$$

The force exerted by the bead on the wire is therefore proportional to $\cos^3 \psi$.

Exercise 9.3

1. A small bead of mass m slides on a smooth wire bent in the form of a parabola with its axis vertical and its vertex at the lowest point. The bead is released from rest at one end of the latus rectum. Find the reaction of the wire on the bead when the bead is at the vertex of the parabola.

2. A small bead moves on a smooth wire in the shape of the curve $y = \cos x$ from $x = 0$ to $x = \pi/2$, the wire being held in a vertical coordinate plane in which the axis of y is vertical. Show that, if the bead is displaced slightly from rest at $x = 0$, the reaction of the wire on the bead is outward (i.e. away from the origin) throughout the motion from $x = 0$ to $x = \pi/2$. Calculate the speed of the bead when it reaches the point $(\pi/2, 0)$ and the reaction of the wire on the bead there.

3. A smooth wire in the shape of the curve $y = \ln \sec x$ from $x = 0$ to $x = \pi/3$ is held in a vertical coordinate plane in which the axis of y is vertical. A small bead can move on the wire. The bead starts from rest at the point $x = \pi/3$ and slides down the wire.
   (i) Calculate the reaction of the wire on the bead at the point $x = \alpha$, $(\pi/3 > \alpha > 0)$.
   (ii) Show that if the bead reaches the origin with speed v, and if the reaction of the wire on the bead there is R, then
   $$R = mv^2 + mg.$$

4. A particle of mass m slides on a smooth cycloid $s = 4a \sin \psi$ which is held fixed with axis vertical and vertex downwards. If the particle is released from rest at a point whose distance from the vertex, measured along the arc, is c, prove that it passes through the vertex with speed $c\sqrt{(g/4a)}$ and that the thrust on the curve is then $mg(1 + c^2/16a^2)$. (L.)

5. A smooth curve having the form of the catenary $y = c \cosh(x/c)$ is fixed with the positive direction of the y-axis pointing vertically downwards. A particle of mass m, initially resting on the curve at its highest point, is projected along the curve with speed $\sqrt{(gc/2)}$. Show that the speed v of the particle when it is moving on the curve in a direction inclined at an angle ψ to the horizontal is given by $2v^2 = gc(4 \sec \psi - 3)$. Find an expression for the reaction between the particle and the curve in terms of ψ, and show that the particle leaves the curve when $\cos \psi = 2/3$. (N.)

6. A small bead P of mass m is threaded on a smooth circular wire, of radius a and centre O, fixed in a vertical plane. The bead is initially at the lowest point A of the circle and is projected along the wire with a speed which is just sufficient to carry it to the highest point. Denoting the angle POA by θ find the reaction between the wire and the bead in terms of θ.

Deduce from the energy equation by means of integration that the time taken for the bead to reach the position $\theta = \pi/2$ is $\sqrt{(a/g)} \ln(1 + \sqrt{2})$. (N.)

7. A smooth wire in the form of the curve $y = \ln \sec x$, $0 \leqslant x < \pi/2$, is fixed with the y-axis vertically downwards. A small bead of mass m can slide on the wire and is displaced gently from rest at the origin.

Show that the magnitude of the reaction of the wire on the bead at the point where $x = \pi/3$ is $mg(\ln 2 - \frac{1}{2})$, and indicate, in a diagram, the sense of this reaction.

Show that, when the normal reaction is instantaneously zero, the tangential acceleration of the bead is $g\sqrt{\{(e-1)/e\}}$. (N.)

8. A smooth wire in the form of the curve whose parametric equations are $x = a\cos^3 \phi$, $y = a\sin^3 \phi$, $(0 \leqslant \phi \leqslant \pi/2)$, is fixed in a vertical plane with the y-axis vertically upwards. Prove that the tangent to the curve at the point with parameter ϕ makes an angle $\pi - \phi$ with the positive direction of the x-axis. Prove also that at the point A, where $\phi = \pi/4$, the radius of curvature is $3a/2$.

A bead of mass m is released from rest at the point where $\phi = \pi/2$ and slides down the wire. Find the magnitude of the reaction between the bead and the wire when the bead is at A. (N.)

9. A smooth wire of length $8a$, bent in the form of an arch of a cycloid, is fixed in a vertical plane with its mid-point O as its highest point. The intrinsic equation of the curve is

$$s = 4a \sin \psi, \quad -\pi/2 \leqslant \psi \leqslant \pi/2,$$

where ψ is the angle between the tangent at O and the tangent at the point P such that the arc OP is of length s. A small bead of mass m is threaded on the wire. Show that the equation of motion of the bead is

$$d^2s/dt^2 = gs/4a.$$

Given that the speed of the bead at O is $\sqrt{(2ag)}$, show that its speed v at P is given by

$$v^2 = 2ga(1 + 2\sin^2 \psi).$$

Prove that the magnitude of the reaction between the bead and the wire at P is $|mg(4\cos \psi - 3\sec \psi)/2|$. (L.)

9:4 MOTION UNDER A CENTRAL FORCE—POLAR COORDINATES

When the resultant force acting on a particle is always directed towards, or away from, a fixed point O, the force is said to be a *central force*, the path a *central orbit*, and the point O the *centre of force*. Polar coordinates, with the pole at the centre O, are particularly appropriate for the solution of the dynamical problems associated with central orbits, and the general theorems given below are an essential guide to the solution, whatever the law of force may be. In particular, the angular momentum, introduced in equation (9.17), is an important and fundamental concept.

1. A central orbit is a plane curve

At any point C of its motion, where the velocity is directed along CD, the acceleration is along CO (see Fig. 9.6). Hence the velocity at the next instant remains in the plane OCD, and the particle continues subsequently to move in this plane.

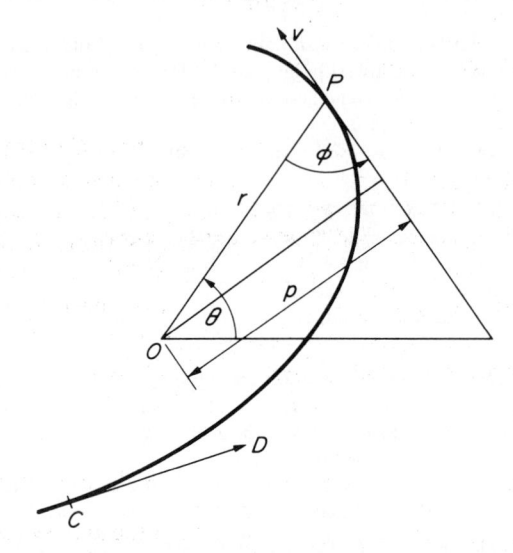

FIG. 9.6.

Suppose that $\mathbf{r}$ is the position vector of the particle P referred to O and that $\mathbf{F}$ is the resultant force on P acting along $\overrightarrow{OP}$. The equation of motion is

$$m\ddot{\mathbf{r}} = \mathbf{F}.$$

Since $\dfrac{d}{dt}(\mathbf{r} \times m\dot{\mathbf{r}}) = \dot{\mathbf{r}} \times m\dot{\mathbf{r}} + \mathbf{r} \times m\ddot{\mathbf{r}} = \mathbf{r} \times m\ddot{\mathbf{r}}$, the equation of motion, when multiplied vectorially by $\mathbf{r}$, gives

$$\frac{d}{dt}(\mathbf{r} \times m\dot{\mathbf{r}}) = \mathbf{r} \times \mathbf{F}. \tag{9.16}$$

The vector $\mathbf{r} \times m\dot{\mathbf{r}}$ is *defined* to be the *moment of momentum* of P about the point O (cf. the definition of the moment of a force about a point) and equation (9.16) shows that its rate of change is equal to the moment about O of the force acting on the particle.

If $\mathbf{F}$ acts along OP and so is parallel to $\mathbf{r}$, then $\mathbf{r} \times \mathbf{F} = \mathbf{0}$ and

$$\mathbf{r} \times m\dot{\mathbf{r}} = \mathbf{h}, \tag{9.17}$$

where $\mathbf{h}$ is a *constant (vector)*, the moment of momentum of P about O. Taking the scalar product of equation (9.17) with $\mathbf{r}$ gives $\mathbf{h} \cdot \mathbf{r} = 0$ so that $\mathbf{r}(= \overrightarrow{OP})$ always lies in the plane through O perpendicular to the fixed vector $\mathbf{h}$.

2. The angular momentum integral

If the force on P is $mf(r,\theta)$ in the direction PO, so that $f(r,\theta)$ is the force per unit mass, then the equation of motion resolved along and perpendicular to the radius vector OP gives respectively

$$m(\ddot{r} - r\dot{\theta}^2) = -mf(r, \theta), \tag{9.18}$$

$$\frac{m}{r}\frac{d}{dt}(r^2\dot{\theta}) = 0. \tag{9.19}$$

Equation (9.19) integrates to give

$$mr^2\dot{\theta} = mh, \tag{9.20}$$

where h is constant. But $mr^2\dot{\theta} = r \cdot mr\dot{\theta}$ is the moment about O of the momentum of the particle localised at the particle and is the (scalar) *moment of momentum* or *angular momentum* of P about O. (The vector $\mathbf{h}$ of equation (9.17) has magnitude $|\mathbf{h}| = mh$.) Hence when a particle moves in a central field of force its moment of momentum, h per unit mass, about the centre of force remains constant. If v is the speed of P, directed along the tangent to the path, and p is the length of the perpendicular from O to its tangent, Fig. 9.6, then the moment of momentum of P about O is mpv, i.e. $h = pv$. A counter-clockwise moment is usually given a positive sign.

3. The theorem of areas

The area of the sector bounded by $r = f(\theta)$, $\theta = \theta_1$ and $\theta = \theta_2$, $(\theta_2 > \theta_1)$, is $\int_{\theta_1}^{\theta_2} \frac{1}{2}r^2 \, d\theta$. When a particle is moving in a central field the area swept out by the radius vector joining the centre of force to the particle, from time t_1 to time t_2, is

$$\int_{t_1}^{t_2} \frac{1}{2}r^2 \frac{d\theta}{dt} \, dt = \int_{t_1}^{t_2} \frac{1}{2}r^2\dot{\theta} \, dt = \frac{1}{2}r^2\dot{\theta}(t_2 - t_1),$$

since $\frac{1}{2}r^2\dot{\theta}$ is a constant for this field,

Therefore *for a particle moving in a central field* the rate of sweeping out of area by the radius vector is constant.

This law, in relation to the motion of the planets around the sun, was discovered by Kepler and stated as his Second Law in a book published in 1609. Kepler's discovery formed a substantial part of the foundation upon which Newton developed his Laws of Universal Gravitation published in the *Principia* in 1687.

Example 1. A particle of mass m on a smooth horizontal table is connected to a fixed point of the table by an elastic spring of natural length a and modulus λ. The particle is projected horizontally with velocity $2a\omega$ in a direction perpendicular to the spring when the length of the latter is $2a$. Show that when the length of the spring is r in the subsequent motion, its angular speed will be $4a^2\omega/r^2$. Write down the energy equation and show that, if the maximum length of the spring is $4a$, then $\lambda = 3ma\omega^2/8$. (N.)

After the particle is projected the only force acting upon it is the tension in the elastic spring and that is directed towards O (and is a function of the distance of the particle from O). Therefore the angular momentum about O is conserved. Hence, in the subsequent motion, when the length of the spring is r, if the angular speed of the particle is then Ω,

$$mr^2\Omega = m(2a)^2\omega$$

$$\Leftrightarrow \Omega = \frac{4a^2\omega}{r^2}.$$

Also, since, *after the particle is projected*, the system of forces acting on the particle is a conservative system, the energy of the system is conserved
$$\Rightarrow \text{(Elastic P.E. of spring + K.E. of particle) is constant}$$
$$\Rightarrow \frac{\lambda(r-a)^2}{2a} + \frac{1}{2}m(r^2\Omega^2 + \dot{r}^2) = \frac{1}{2}m(2a\omega)^2 + \frac{\lambda a^2}{2a},$$

where $\dot{r}$ is the speed of the particle in the direction of the spring.

When the spring reaches its maximum length $r = 4a$ and $\dot{r} = 0$
$$\Rightarrow \frac{\lambda(4a-a)^2}{2a} + \frac{1}{2}m(4a)^2 \frac{16a^4\omega^2}{(4a)^4} = 2ma^2\omega^2 + \frac{\lambda a^2}{2a}.$$
$$\Rightarrow \lambda = 3ma\omega^2/8.$$

Example 2. A particle P, of mass m, lies on a smooth table and is attached by a string of length l passing through a small hole O in the table and carries an equal particle Q hanging vertically. The former particle is projected along the table at right angles to the string with speed $\sqrt{(2gh)}$ when at a distance a from the hole. If r is the distance from the hole at time t, prove the following results:

(α)
$$\dot{r}^2 = gh\left(1 - \frac{a^2}{r^2}\right) + g(a-r),$$

(β) the lower particle will be pulled up to the hole if
$$l \leqslant \frac{h}{2} + \sqrt{\left(ah + \frac{1}{4}h^2\right)},$$

(γ) the tension in the string is
$$T = \frac{mg}{2}\left(1 + \frac{2a^2h}{r^3}\right). \tag{L.}$$

After projection, the only force acting on the particle P is towards the hole and the angular momentum of this particle is therefore conserved (Fig. 9.7). Also, the energy of the whole system is conserved. At time t, the speed of the hanging particle Q is $\dot{r}$ upwards, where $OP = r$.

Conservation of angular momentum for P gives
$$ma\sqrt{(2gh)} = mr^2\dot{\theta}. \tag{1}$$

Conservation of energy gives
$$\tfrac{1}{2}m(r^2\dot{\theta}^2 + \dot{r}^2) + \tfrac{1}{2}m\dot{r}^2 - mg(a-r) = mgh$$
$$\Rightarrow m\dot{r}^2 + \frac{1}{2}mr^2 \frac{m^2a^22gh}{m^2r^4} - mg(a-r) = mgh$$
$$\Leftrightarrow \dot{r}^2 = gh\left(1 - \frac{a^2}{r^2}\right) + g(a-r)$$
$$\Leftrightarrow \dot{r}^2 = \frac{g}{r^2}(r-a)\{h(r+a) - r^2\}$$
$$\Leftrightarrow \dot{r}^2 = -\frac{g}{r^2}(r-a)(r-\lambda_1)(r-\lambda_2), \tag{2}$$

where λ_1, λ_2 are the roots of $r^2 - hr - ah = 0$ and $\lambda_1 > 0 > \lambda_2$. Since $\dot{r}^2 \geqslant 0$, it follows from (2) that $(r-a)(r-\lambda_1) \leqslant 0$, i.e. that r lies in the range a to λ_1 inclusive. Note that steady motion of P in a circle of radius a is only possible when $m(2gh)/a = mg$, i.e. when $h = \tfrac{1}{2}a$. If this is not the case r will vary from a to λ_1 and Q will be pulled up to O if $l \leqslant \lambda_1$. But $\lambda_1 = \tfrac{1}{2}h + \sqrt{(ah + \tfrac{1}{4}h^2)}$ and therefore result (β) follows.

The equation of motion for the hanging particle Q in the direction of the string is
$$mg - T = m\frac{d^2}{dt^2}(l-r) = -m\ddot{r},$$

and the equation of motion in the radial direction for the particle P on the table is

$$-T = m(\ddot{r} - r\dot{\theta}^2)$$
$$\Rightarrow mg - 2T = -m\ddot{r} + m\ddot{r} - mr\dot{\theta}^2$$
$$\Rightarrow 2T = m(g + r\dot{\theta}^2).$$

But, from (1),

$$\dot{\theta}^2 = 2a^2 gh/r^4$$
$$\Rightarrow T = \frac{mg}{2}\left(1 + \frac{2ha^2}{r^3}\right).$$

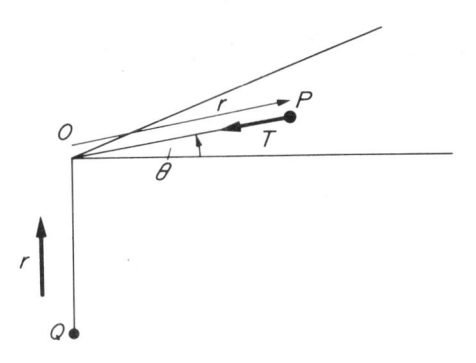

FIG. 9.7.

Exercise 9.4

1. A particle P of mass m lies at rest in a smooth straight tube AB which is free to rotate in a horizontal plane about a fixed vertical axis through A. The particle is attached to A by an elastic string, of modulus mg, which is initially at its natural length l. If the tube is now set rotating with constant angular speed $\omega = \sqrt{\{g/(2l)\}}$, show that after time t, $AP = l(2 - \cos \omega t)$ and find the magnitude of the action of the particle on the tube then.

2. A particle of mass m is held on a smooth table. A string attached to this particle passes through a hole in the table and supports a mass $3m$. Motion is started by the particle on the table being projected with speed v at right angles to the string. If a is the original length of the string on the table, show that, when the hanging mass has descended a distance $\frac{1}{2}a$ (assuming this to be possible) its speed will be $\frac{1}{2}\{3(ga - v^2)\}^{1/2}$.   (L.)

3. Two particles each of mass M are attached to the ends of a taut string of length $2a$ which lies at rest on a smooth horizontal table, passing through a small ring attached to the table; the ring is at the middle point of the string. An impulse I is applied to one of the particles in a direction perpendicular to the string. Prove that, when the other particle reaches the ring, the speeds of the particles are in the ratio $\sqrt{5}:\sqrt{3}$.   (O.C.)

4. A particle of mass m moves so that its position vector, referred to an origin O, is given at time t by

$$\mathbf{r} = \mathbf{a}\cos\omega t + \mathbf{b}\sin\omega t,$$

where $\mathbf{a}$, $\mathbf{b}$ are constant vectors and ω is constant. Calculate the velocity vector $d\mathbf{r}/dt$ and the acceleration vector $d^2\mathbf{r}/dt^2$ of the particle. Deduce that the angular momentum of the particle about O is constant and find in terms of m, ω and $\mathbf{r}$ the force acting on the particle.   (L.)

5. An elastic string of modulus mg and natural length a is attached at one end to a point O on a smooth horizontal table and at the other to a particle of mass m on the table. The particle is projected with speed v in a horizontal direction perpendicular to the string when the length of the string is $3a$. If the maximum length of the string in the subsequent motion is $5a$, show that $v^2 = 75ga/4$.   (N.)

6. The acceleration of a particle P moving in the plane of the coordinate axes Ox, Oy is $-n^2(x\mathbf{i} + y\mathbf{j})$, where n is constant and (x, y) are the coordinates of P. At time $t = 0$ the particle is projected with velocity $na\mathbf{j}$ from the point A, whose position vector is $2a\mathbf{i}$. Show that P describes the curve

$$\frac{x^2}{4a^2} + \frac{y^2}{a^2} = 1$$

and first returns to A after time $2\pi/n$.

Show also that the moment of momentum of P, about an axis passing through O and perpendicular to the plane of the motion, is constant. (L.)

7. A particle of unit mass moves in a plane under the action of a force μ/r^3 towards a fixed point O, where r is the distance from O. Show that it can describe a circle about O of radius a with speed V where $V^2 = \mu/a^2$.

From a point A of this circle the particle is projected along the tangent at A with speed U ($U > V$). Prove that, after time t

$$r^2 = a^2 + (U^2 - V^2)t^2.$$

Prove also that the particle moves off to infinity in a direction making an angle

$$\frac{U}{\sqrt{(U^2 - V^2)}} \cdot \tfrac{1}{2}\pi \text{ with } OA.$$ (O.C.)

8. A particle P describes an orbit about a point O under a central attractive force $f(r)$ per unit mass, where $r = OP$. Explain why the motion is confined to a plane. By considering the transverse component of acceleration, or otherwise, show that if θ is the angle between OP and a line fixed in the plane, then $r^2\dot\theta$ has a constant value, denoted by h. By writing down the equation of motion in the radial direction and then eliminating $\dot\theta$ obtain the equation

$$\ddot r = \frac{h^2}{r^3} - f(r)$$ (N.)

9. It is observed that for the motion of a planet P about the sun S,
1. the path lies in a plane through S,
2. the polar equation of the path is $r = b/(1 + k\cos\theta)$ where S is taken as the pole, the initial line is suitably chosen and b and k are constants,
3. the moment of the velocity about S, $r^2\dot\theta$, has a constant value a.

(i) Show that $\dot r = \dfrac{ka}{b}\sin\theta$ and write down an expression for $r\dot\theta$ in terms of a, b, k and θ. Hence show that the velocity of P is the resultant of two velocities of constant magnitude, one perpendicular to SP and the other perpendicular to the initial line.

(ii) Prove that the acceleration of P is directed along PS and is inversely proportional to SP^2. (N.)

10. A smooth straight tube rotates, with uniform angular speed ω, in a horizontal plane about a point O of its length. Initially, a particle P of mass m, inside the tube, is at rest relative to the tube at a distance a from O. Show that

$$r = a\cosh\omega t,$$

where r is the distance of P from O at time t.

Find the magnitude of the force exerted by the tube on the particle at time t. (L.)

11. A smooth thin straight rod, of length $2a$, rotates in a horizontal plane about one end A with constant angular speed ω. At time $t = 0$, a bead, of mass m, threaded on the rod is at rest relative to the rod and at a distance a from A. Show that the bead leaves the rod after a time $T = [\ln(2 + \sqrt{3})]/\omega$.

Show also that the magnitude of the force exerted by the rod on the bead when the bead is at a distance $3a/2$ from A is $m\sqrt{(5a^2\omega^4 + g^2)}$. (L.)

12. The attractive force exerted on a space probe by the Sun is μ/r^2 per unit mass directed towards the Sun, where r is the distance from the Sun and μ is a constant. Show that the angular momentum of the space probe about the Sun is conserved. Show also that $\tfrac{1}{2}v^2 - \mu/r$ is a constant of the motion, where v is the speed of the probe.

A space probe is projected in a direction tangential to the Earth's orbit and in the same sense as the Earth's motion. It meets the planet Venus in a direction tangential to the motion of the latter. The orbits of the Earth and Venus are to be considered as coplanar circles with the Sun as centre and with radii a_1 and a_2 respectively, where $a_2 < a_1$. Neglecting the gravitational attractions of the planets in comparison with that of the Sun, show that the initial speed of the probe must be less than that of the Earth by an amount

$$\left(\frac{\mu}{a_1}\right)^{1/2} - \left(\frac{2\mu a_2}{a_1(a_1+a_2)}\right)^{1/2}.$$ (L.)

9:5 THE MOTION OF PROJECTILES

In §1:6 and §1:7 we considered some elementary examples on the motion of projectiles under gravity. We now consider some further problems.

1. The range on an inclined plane

A particle is projected from a point O in a vertical plane containing the line of greatest slope of a plane through O which is inclined at an angle θ to the horizontal. The particle is projected at an angle α with the horizontal and it strikes the inclined plane at a point P (Fig. 9.8). The components of the initial velocity of the particle parallel (up) and perpendicular to the plane respectively are $u \cos(\alpha - \theta)$, $u \sin(\alpha - \theta)$. The components of acceleration of the particle parallel (up) and perpendicular to the plane are respectively $-g \sin \theta$, $-g \cos \theta$.

For the component of the motion of the particle perpendicular to the inclined plane,

$$s = ut \sin(\alpha - \theta) - \frac{1}{2}gt^2 \cos \theta.$$

The particle strikes the plane again when $s = 0$, so that the time of flight from O to P is

$$\frac{2u \sin(\alpha - \theta)}{g \cos \theta}.$$ (9.21)

For this period of time the particle has been moving with uniform horizontal velocity $u \cos \alpha$, so that the horizontal distance from O to P is

$$ON = \frac{u \cos \alpha \times 2u \sin(\alpha - \theta)}{g \cos \theta} = \frac{2u^2 \cos \alpha \sin(\alpha - \theta)}{g \cos \theta}.$$

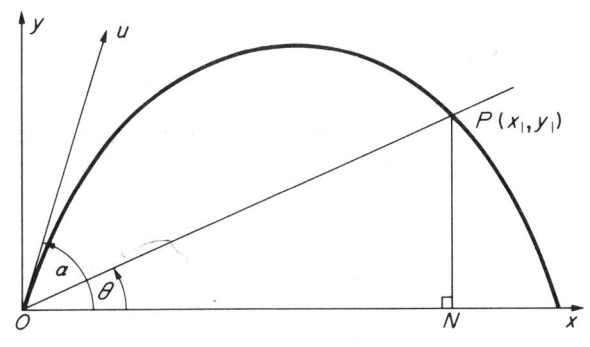

FIG. 9.8.

The range on the inclined plane is

$$OP = \frac{2u^2 \cos \alpha \sin(\alpha - \theta)}{g \cos^2 \theta}. \qquad (9.22)$$

This range is a maximum, as α varies, when $\cos \alpha \sin(\alpha - \theta)$ is a maximum, i.e., when $\tfrac{1}{2}\{\sin(2\alpha - \theta) - \sin \theta\}$ is a maximum,

$$\Rightarrow 2\alpha - \theta = \tfrac{1}{2}\pi,$$
$$\Leftrightarrow \alpha = \tfrac{1}{4}\pi + \tfrac{1}{2}\theta, \qquad (9.23)$$

i.e. when the direction of projection bisects the angle between the inclined plane and the vertical.

The maximum range on the inclined plane is

$$\frac{2u^2 \cdot \tfrac{1}{2}(1 - \sin \theta)}{g \cos^2 \theta} = \frac{u^2}{g(1 + \sin \theta)}. \qquad (9.24)$$

Example 1. A particle is projected with a given speed in a vertical plane from a point on an inclined plane so as to have the greatest possible range up the plane. Prove that, neglecting air resistance, the direction of motion of the particle as it strikes the plane is perpendicular to the direction of projection. (L.)

If the particle is projected with velocity u at an angle α with the horizontal, the equation of the path referred to horizontal and vertical axes through the point of projection is

$$y = x \tan \alpha - \frac{gx^2}{2u^2}(1 + \tan^2 \alpha)$$

and, if the particle passes through the point (x_1, y_1) in the plane,

$$gx_1^2 \tan^2 \alpha - 2u^2 x_1 \tan \alpha + 2u^2 y_1 + gx_1^2 = 0.$$

We showed in §1:7 that (x_1, y_1) is a point of limiting range if the roots of this equation in $\tan \alpha$ are equal, i.e. if

$$\tan \alpha = \frac{u^2}{gx_1}.$$

The direction of motion of the particle at any point is given by the gradient of its curve at that point, i.e. by dy/dx

$$\Rightarrow \tan \psi = \left(\frac{dy}{dx}\right)_{(x_1, y_1)} = \tan \alpha - \frac{gx_1}{u^2}(1 + \tan^2 \alpha)$$

where ψ is the angle between the direction of motion and Ox. If, therefore, the particle has the greatest possible range up the plane,

$$\tan \psi = \tan \alpha - \frac{1}{\tan \alpha}(1 + \tan^2 \alpha) = -\cot \alpha$$
$$\Rightarrow \psi = \alpha + 90°.$$

Therefore the direction of motion of the particle as it strikes the plane is perpendicular to the direction of projection.

Example 2. A particle is projected from a point O with speed u at an angle α with the horizontal, so as to strike at right angles at P a plane inclined at an angle β to the horizontal. Prove that, if OP makes an angle θ with the horizontal,

$$\tan \alpha = \cot \beta + 2 \tan \theta.$$

The initial component of velocity of the particle in a direction parallel to the inclined plane is $u\cos(\alpha-\beta)$, and its acceleration in this direction is $-g\sin\beta$ (Fig. 9.9). When the particle strikes the inclined plane at right angles its velocity in a direction parallel to that plane is zero

$$\Rightarrow u\cos(\alpha-\beta) - gT\sin\beta = 0,$$

where T is the time of flight

$$\Leftrightarrow T = \frac{u\cos(\alpha-\beta)}{g\sin\beta}. \qquad (1)$$

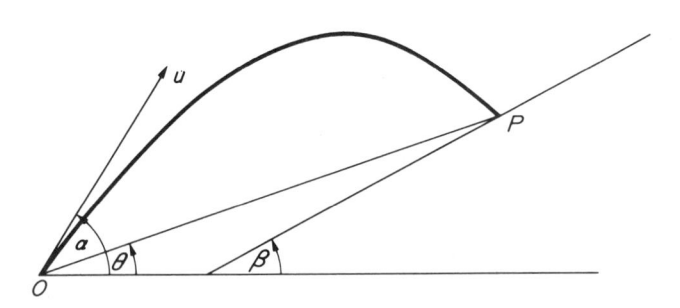

FIG. 9.9.

At time T from the instant of projection, the displacement of the particle in the direction perpendicular to OP vanishes

$$\Rightarrow uT\sin(\alpha-\theta) - \tfrac{1}{2}gT^2\cos\theta = 0$$

$$\Leftrightarrow T = \frac{2u\sin(\alpha-\theta)}{g\cos\theta}. \qquad (2)$$

Equating the values of T given by (1) and (2) gives

$$2\sin(\alpha-\theta)\sin\beta - \cos\theta\cos(\alpha-\beta) = 0$$
$$\Leftrightarrow 2\sin\alpha\cos\theta\sin\beta - 2\cos\alpha\sin\theta\sin\beta - \cos\theta\cos\alpha\cos\beta - \cos\theta\sin\alpha\sin\beta = 0$$
$$\Leftrightarrow \sin\alpha\cos\theta\sin\beta = \cos\theta\cos\alpha\cos\beta + 2\cos\alpha\sin\theta\sin\beta$$
$$\Leftrightarrow \tan\alpha = \cot\beta + 2\tan\theta.$$

Exercise 9.5(a)

In each of questions 1–5 a particle is projected from a point O in a plane inclined at an angle θ to the horizontal. The particle is projected with speed u at an inclination α to the horizontal, and in a plane containing a line of greatest slope of the inclined plane. Give answers, where necessary, correct to 2 significant figures.

1. $u = 70$ ms^{-1}, $\alpha = 60°$ and $\theta = 30°$. Calculate the range on the inclined plane.
2. $u = 40$ ms^{-1}, $\alpha = 45°$ and $\theta = 30°$. Calculate the magnitude and direction of the velocity with which the particle strikes the plane.
3. $u = 80$ ms^{-1}, $\alpha = 60°$, $\theta = 45°$. Calculate the time of flight of the particle.
4. $u = 21$ ms^{-1}, $\theta = 30°$. Calculate the maximum range of the particle (i) up the plane, (ii) down the plane.
5. Calculate in terms of u, α, θ the greatest perpendicular distance from the inclined plane reached by the particle.
6. A particle is projected with speed u from a point on a plane of slope β. Prove that the maximum range down the plane is $u^2/[g(1-\sin\beta)]$. (L.)

7. A particle is projected with speed u from the foot of a plane of inclination β, the direction of projection lying in the vertical plane through the line of greatest slope and making an angle α with the horizontal. If the particle strikes the plane at right angles, prove that

$$1 + 2\tan^2\beta = \tan\alpha\tan\beta.$$ (O.C.)

8. A particle is projected under gravity with speed V from a point on a plane inclined at an angle α to the horizontal. Find its maximum range (i) up the plane, (ii) down the plane.

If r_1 and r_2 are these maximum ranges, prove that

$$\frac{1}{r_1} + \frac{1}{r_2}$$

is independent of the inclination of the plane. (O.C.)

9. A, B are two points on a line of greatest slope of a plane of inclination β, A being above B. A particle projected from A with speed u can just reach B. Prove that the least speed with which a particle can be projected from B so as to reach A is

$$u\tan\left(\frac{\pi}{4} + \frac{\beta}{2}\right).$$

and that the times of flight in the two cases are equal. (O.C.)

10. A particle is projected under gravity with speed u at elevation α. Find the time when it is farthest from a plane inclined at an angle β to the horizontal through the point of projection, the vertical plane through the trajectory cutting the inclined plane in the line of greatest slope.

Find this greatest distance and show that the component of the velocity parallel to the plane is then $u\cos\alpha\sec\beta$. (L.)

11. A heavy particle is projected from a point in an inclined plane, inclined at 2β to the *vertical*, and moves towards the upper part of the plane in the vertical plane through a line of greatest slope of the inclined plane; the initial speed of the particle is $u\cos\beta$ and its initial direction of motion is inclined at β to the *vertical*. Prove that the time of flight of the particle is u/g, its range on the plane is $u^2/2g$, the speed with which it strikes the plane is $u\sin\beta$, and its direction of motion has then turned through a right angle. (O.C.)

12. A particle is projected with initial velocity $\mathbf{u}$ and moves freely under gravity. If $\mathbf{r}$ is the position vector of the particle after time t show that

(i) $\dot{\mathbf{r}} = \mathbf{u} + \mathbf{g}t,$ (ii) $\mathbf{r} = \mathbf{u}t + \tfrac{1}{2}\mathbf{g}t^2.$

If the particle is moving at right angles to its original direction of motion after time t_1, and strikes the horizontal plane through the point of projection after time t_2 (both times being measured from the instant of projection), and if $t_1 < t_2 < 2t_1$, show that the speed, u, of projection is given by

$$u^2 = \tfrac{1}{2}g^2 t_1 t_2,$$

and the range, R, on the horizontal plane by

$$R = \tfrac{1}{2}gt_2(2t_1 t_2 - t_2^2)^{1/2}.$$ (N.)

2. The bounding parabola

In Chapter 1, Part 1, we derived the equation of the path of a particle projected with initial speed u at an angle α with the horizontal and referred to horizontal and vertical axes through the point of projection, in the form

$$y = x\tan\alpha - \frac{gx^2}{2u^2}(1 + \tan^2\alpha).$$ (9.25)

The position of the particle at time t from the instant of projection is given by the

equations
$$x = ut\cos\alpha,$$
$$y = ut\sin\alpha - \tfrac{1}{2}gt^2.$$

Eliminating α from these equations we obtain
$$x^2 + (y + \tfrac{1}{2}gt^2)^2 = u^2 t^2$$
$$\Leftrightarrow g^2 t^4 - 4(u^2 - gy)t^2 + 4(x^2 + y^2) = 0. \qquad (9.26)$$

From equation (9.25) the possible angles of projection for a particle projected from the origin O with speed u to pass through the point $P(h, k)$ are given by the quadratic equation in $\tan\alpha$,
$$k = h\tan\alpha - \frac{gh^2}{2u^2}(1 + \tan^2\alpha).$$

According as this equation has real distinct roots, equal roots, or complex roots, there are two, one, or no possible paths for the particle from O to P. When the roots of the equation are equal, the point P is the point of maximum range for the particle on the line OP.

We consider now the geometric problem of finding the possible trajectories for a particle projected from O with speed u to pass through P. In either case the path of the particle is a parabola with directrix $y = u^2/(2g)$.

Since O is a point on the parabola, the focus of the parabola must lie on the circle with centre O and radius $u^2/(2g)$, Fig. 9.10 (i). The focus must also lie on the circle with centre P and radius PN, where PN is the perpendicular from P to the directrix. The focus must therefore be either at S or S', where S and S' are the points of intersection of the circles, if these points exist. If there are two such points, the possible trajectories of the particles are:

1. The parabola through O and P with focus S and directrix $y = u^2/(2g)$.
2. The parabola through O and P with focus S' and directrix $y = u^2/(2g)$.

If the two circles do not interesect, it is not possible to project the particle from O through P with initial speed u.

If the circles touch [Fig. 9.10 (ii)] there is only one such path for the particle from O to P and in this case the point P is a point of maximum range on the line OP.

It is a proposition of geometry that the tangent to a parabola at a point K on the parabola bisects the angle between the perpendicular from K to the directrix and the line joining K to the focus. The possible directions of projection of the particle from O to P therefore bisect the angles between the y-axis and OS and between the y-axis and OS' respectively [Fig. 9.10 (i)]. The possible directions of motion of the particle at P bisect the angles between PN and PS and between PN and PS' respectively.

In Fig. 9.10 (ii), BR is the line $y = u^2/g$ and PR is perpendicular to BR so that $PO = PR$. It follows that the locus of points P which are such that OP is a maximum range on the line OP for a projectile fired from O with speed u is a *parabola* with focus O and directrix BR. The equation of this parabola *referred to AO as axis of x and AN as axis of y* is
$$y^2 = 4.OA.x,$$
$$\Rightarrow y^2 = \frac{2u^2}{g}x. \qquad (9.27)$$

(i)

(ii)

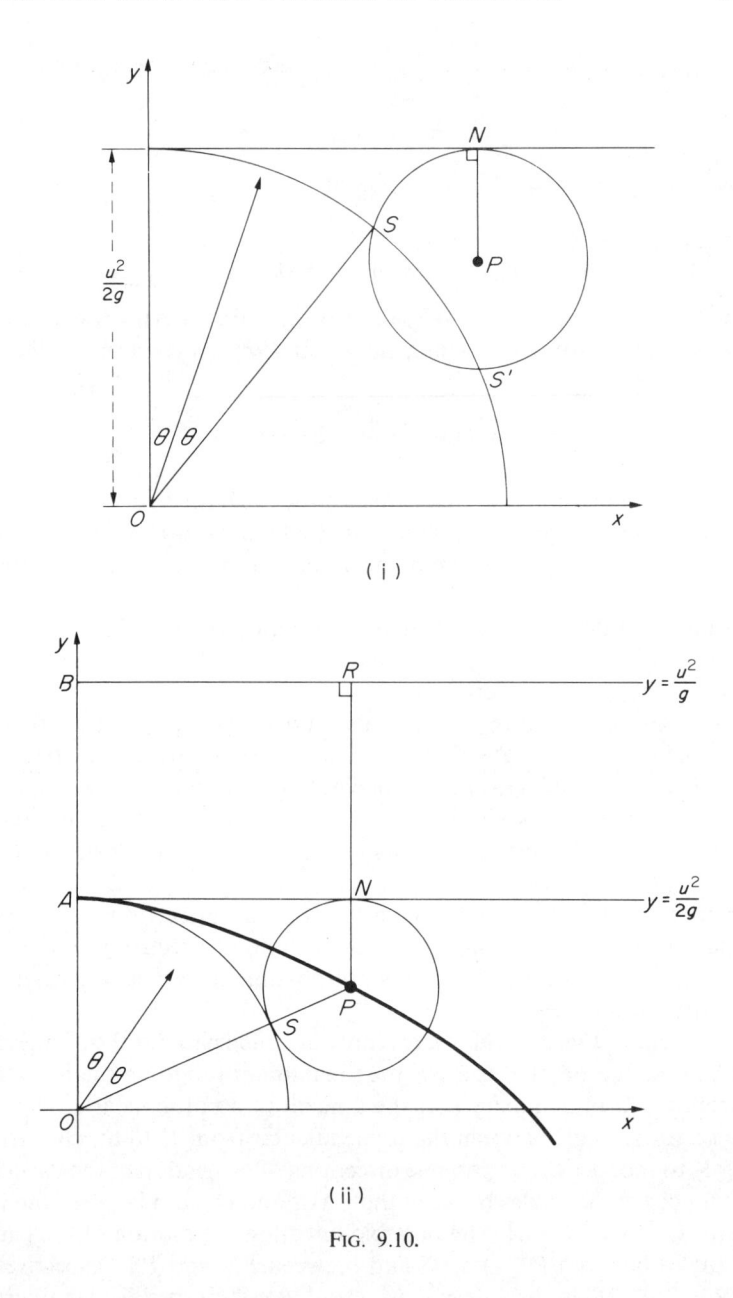

Fig. 9.10.

This parabola is called the *bounding parabola* for particles projected in the vertical plane from O with speed u. A particle cannot be projected from O with speed u to pass through a point outside the bounding parabola and all points on the bounding parabola are points which can just be reached by a particle projected from O with speed u.

Note on envelopes. The *envelope* of a family of curves was defined in Volume 2, Part I, § 20:13 and it was shown that for the family of curves $f(x, y, \alpha) = 0$, where α

varies, the equation of the envelope is given by the eliminant of α from the equations

$$f(x, y, \alpha) = 0, \qquad \frac{\partial}{\partial \alpha}\{f(x, y, \alpha)\} = 0$$

We consider in this way the envelope of the trajectories of all particles projected from O with speed u, i.e. the envelope of the curves

$$f(x, y, \alpha) = x \tan \alpha - \frac{gx^2}{2u^2}(1 + \tan^2 \alpha) - y = 0$$

as α varies. We have

$$\frac{\partial}{\partial \alpha}\{f(x, y, \alpha)\} = x \sec^2 \alpha - \frac{gx^2}{u^2} \tan \alpha \sec^2 \alpha,$$

and the required envelope is therefore given by eliminating α between $f(x, y, \alpha) = 0$ and $\tan \alpha = u^2/(gx)$. The eliminant is

$$\frac{xu^2}{gx} - \frac{gx^2}{2u^2}\left(1 + \frac{u^4}{g^2 x^2}\right) - y = 0$$

$$\Leftrightarrow x^2 = -\frac{2u^2}{g}\left(y - \frac{u^2}{2g}\right). \tag{9.28}$$

Referred to horizontal and vertical axes of x and y this is the equation of a parabola with latus rectum $(2u^2)/g$ and vertex $\{0, u^2/(2g)\}$, i.e. the equation of the bounding parabola we have defined above.

The bounding parabola is sometimes called the *enveloping parabola* or the *parabola of safety*.

Equations (9.25), (9.26) and the enveloping parabola between them afford a choice of methods for the solution of certain problems concerning projectiles. These methods are illustrated in the examples which follow.

Example 1. A shot is fired with speed V from a point O which is distance d vertically above a plane inclined to the horizontal at an angle β. Find the maximum range down the plane.

In Fig. 9.11, P is the point $V^2/(2g)$ vertically above O. The equation of the bounding parabola referred to vertically downwards and horizontal axes of x and y through P is

$$y^2 = \frac{2V^2}{g} x.$$

The section of the inclined plane in the plane of projection is the straight line through $\{d + V^2/(2g), 0\}$ inclined at $(\frac{1}{2}\pi - \beta)$ to the x-axis and the equation of this line is

$$y = \left(x - d - \frac{V^2}{2g}\right) \cot \beta.$$

The line meets the bounding parabola where

$$y^2 = \frac{2V^2}{g}\left[\frac{y}{\cot \beta} + d + \frac{V^2}{2g}\right]$$

$$\Leftrightarrow y^2 - \frac{2V^2 y}{g \cot \beta} - \frac{V^4}{g^2} - \frac{2V^2 d}{g} = 0.$$

326   PARTICLE WITH TWO DEGREES OF FREEDOM   Ch. 9 §9:5

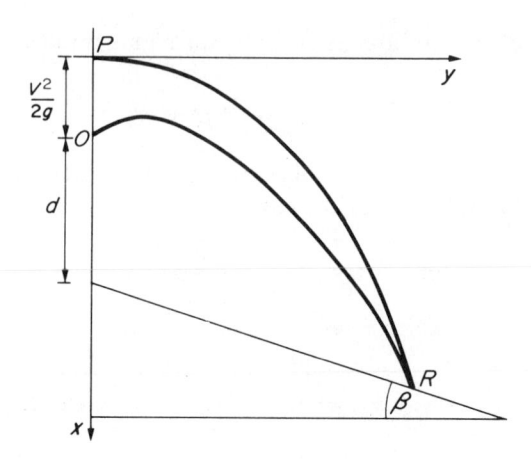

FIG. 9.11.

The maximum range of the particle on the inclined plane is therefore $y_1 \sec \beta$, where y_1 is the positive root of this equation in y. This maximum range is therefore

$$\left\{ \frac{V^2}{g \cot \beta} + \sqrt{\left(\frac{V^4}{g^2 \cot^2 \beta} + \frac{V^4}{g^2} + \frac{2V^2 d}{g} \right)} \right\} \sec \beta$$

$$= \frac{V^2}{g \cos \beta} \left\{ \tan \beta + \sqrt{\left(\sec^2 \beta + \frac{2dg}{V^2} \right)} \right\}.$$

Example 2. A particle is to be projected with speed v from a point O to strike an object at a vertical distance h below O and at a horizontal distance d from O. Find the least value of v for which this is possible and show that for larger values of v there are two possible directions of projection. Show also that the values of v for which the two possible directions of projection are at right angles is $d\sqrt{(g/h)}$.

Let the angle of projection be α. Then the equation of the path of the projectile is

$$y = x \tan \alpha - \frac{gx^2}{2v^2}(1 + \tan^2 \alpha)$$

referred to horizontal and upward vertical axes Ox and Oy through O. Therefore, since the projectile passes through $(d, -h)$,

$$-h = d \tan \alpha - \frac{gd^2}{2v^2}(1 + \tan^2 \alpha)$$

$$\Leftrightarrow gd^2 \tan^2 \alpha - 2v^2 d \tan \alpha + (gd^2 - 2v^2 h) = 0.$$

If the point $(d, -h)$ is a point of limiting range for the particle, this equation in $\tan \alpha$ has equal roots. Hence, in this case,

$$v^4 d^2 = gd^2(gd^2 - 2v^2 h)$$

$$\Leftrightarrow v^4 + 2ghv^2 - g^2 d^2 = 0.$$

The least value of v^2 for which it is possible to reach $(d, -h)$ is therefore the positive root of this equation in v^2, i.e. $-gh + g\sqrt{(h^2 + d^2)}$. The least value of v for which it is possible to reach $(d, -h)$ is thus $+\sqrt{\{-gh + g\sqrt{(h^2 + d^2)}\}}$.

The equation in $\tan \alpha$ is a quadratic and gives two values of $\tan \alpha$ for values of v above the minimum value. There are therefore two possible directions of projection for such values of v. If the directions of projection (say α_1 and α_2 with the horizontal) are at right angles, one direction is above the horizontal and one is below the horizontal so that $\alpha_1 - \alpha_2 = 90°$. Therefore $\tan\alpha_1 \tan\alpha_2 = -1$ and from the equation in $\tan\alpha$,

$$\tan\alpha_1 \tan\alpha_2 = 1 - \frac{2v^2 h}{gd^2}$$

$$\Rightarrow -1 = 1 - \frac{2v^2 h}{gd^2}$$

$$\Leftrightarrow v = d\sqrt{(g/h)}.$$

Exercise 9.5(b)

1. From a point on a plane, inclined at an angle α to the horizon, a particle is projected with speed V at an angle β to the plane in the vertical plane through the line of greatest slope. Find for what angle β the range up the plane is greatest and the length of this range.

For this trajectory show that the direction of projection and the direction of impact upon the inclined plane are perpendicular. (N.)

2. A particle is projected from a given point so as just to pass over a wall of height h which is in such a position that the *top* of the wall is at a distance r from the point of projection. Show that the least speed of projection necessary for this purpose is $\sqrt{\{g(h+r)\}}$, and that with this speed the particle will have reached its greatest height before grazing the wall.

Find also the angle of projection. (N.)

3. A ball thrown with initial speed $\sqrt{(2gh)}$ strikes a vertical wall which stands at a distance d from the point of projection. Show that the point on the wall that is hit by the ball cannot be at a height greater than $(4h^2 - d^2)/4h$ above the point of projection.

Show also that the area of the wall that is within range of the ball is bounded by a parabola. (L.)

4. A particle is projected from a point A with speed u. Prove that, whatever the angle of projection, the particle describes a parabola whose directrix is at a height $u^2/2g$ above A.

If the particle is to pass through a given point B, prove that the number of possible angles of projection is 0, 1, or 2 according as AB is greater than, equal to, or less than $(u^2 - gh)/g$, where h is the height of B above the level of A. (O.C.)

5. A ball is kicked from a point A on level ground so that it leaves A with a speed u for which the maximum horizontal range is R. If the ball clears a wall of height h at a horizontal distance $\frac{1}{2}R$ from A, find the greatest possible value of h. Find also the direction in which the ball leaves A to clear the highest possible wall. (L.)

6. If a particle is projected horizontally with a speed V from a point A at height $V^2/2g$ above another point O, show that it will describe the enveloping parabola of the family of trajectories obtained by projecting particles from O with speed V in varying directions in the same vertical plane.

A particle is projected horizontally from A with speed V, and another is projected from O with the same speed at any angle to the horizontal and in the vertical plane containing the first trajectory. Prove that the difference of the squares of their speeds at the common point of their paths is V^2. (N.)

7. A gun fires a shell with velocity V at an angle of elevation θ to hit an aeroplane which flies overhead at a constant height h ($h < V^2/2g$) above the ground. Find the quadratic equation in x whose two solutions for x are the horizontal distances from the gun at which the aeroplane can be hit. (Neglect air resistance.)

Letting θ now be variable, and regarding the quadratic equation as giving x implicitly as a function of θ (or otherwise), show that the maximum horizontal distance from the gun at which the aeroplane is at risk of being hit is

$$\frac{V}{g}\sqrt{(V^2 - 2gh)}.$$

Find the angle of elevation at which the gun must be fired in order to achieve this distance. (L.)

8. A projectile is fired at an angle of elevation α from the top of a vertical tower of height H. The projectile

hits the horizontal plane through the base of the tower at a distance D from the foot of the tower. Show that the maximum height of the projectile above the plane is

$$H + \frac{D^2 \tan^2 \alpha}{4(H + D \tan \alpha)},$$

and that the total time of flight of the projectile is

$$\sqrt{[2(H + D \tan \alpha)/g]}. \quad \text{(L.)}$$

9. A particle is projected from a point O with initial speed $2\sqrt{(ga)}$ at an angle θ to the horizontal. It strikes a target which is at a horizontal distance a from O and at vertical distance $3a/4$ above the level of O. Find the possible values of $\tan \theta$. Find also the least speed of projection of a particle from O if it is to strike the target.
(L.)

10. A particle is at rest at the highest point of the surface of a fixed smooth sphere of radius a. The particle is given a horizontal velocity u. Show that if $u^2 < ag$ the particle will begin to slide down the surface of the sphere and find the distance below the highest point of the sphere at which the particle will leave the surface.
If $a = 5$ m and $u = 2\sqrt{5}$ ms^{-1}, show that the particle will strike the horizontal plane through the lowest point of the sphere $(9\sqrt{3} - 3\sqrt{2})/(5\sqrt{5})$ seconds after leaving the surface of the sphere. (Take g as 10 ms^{-2}.)
(L.)

11. An aircraft is climbing with uniform speed u in a straight line that makes an angle β with the horizontal. At the instant when the aircraft is at a height h vertically above an anti-aircraft gun on the ground, the gun fires a shell at an angle of elevation α. Show that, whatever the initial velocity of the shell, the shell cannot hit the aircraft unless

$$\tan \alpha \geqslant k \sec \beta + \tan \beta \text{ where } ku = \sqrt{(2gh)}.$$

Show that, for the shell to hit the aircraft, the initial speed of the shell must be at least $u\sqrt{(1 + k^2 + 2k \sin \beta)}$.
(L.)

9:6 OBLIQUE IMPACT OF ELASTIC BODIES

In the general collision problem we suppose that the momenta $m_1 \mathbf{u}_1$ and $m_2 \mathbf{u}_2$ of the two colliding bodies before impact and the coefficient of restitution e are known. We require to find the velocities $\mathbf{v}_1$ and $\mathbf{v}_2$ after impact.

Conservation of linear momentum gives

$$m_1 \mathbf{u}_1 + m_2 \mathbf{u}_2 = m_1 \mathbf{v}_1 + m_2 \mathbf{v}_2, \quad (9.29)$$

which in two dimensions is equivalent to two scalar equations.

Newton's law of restitution gives

$$(\mathbf{v}_1 - \mathbf{v}_2) \cdot \hat{\mathbf{n}} = -e(\mathbf{u}_1 - \mathbf{u}_2) \cdot \hat{\mathbf{n}}, \quad (9.30)$$

where $\hat{\mathbf{n}}$ is a unit vector along the common normal at the point of impact. Thus we have only three equations available to find the four components of $\mathbf{v}_1$ and $\mathbf{v}_2$, and we must use the additional fact that (assuming the surfaces at the point of contact to be smooth) the momentum of *each* body is unchanged along the common tangent at the point of contact. The following examples illustrate the procedure.

Example 1. A smooth sphere, of centre A and mass $2m$, moving with velocity $v(\mathbf{i} + \sqrt{3}\mathbf{j})$ collides with a smooth sphere, of centre B and mass m, moving with velocity $v\mathbf{j}$, where $\mathbf{i}$ is a unit vector in the direction $\overrightarrow{AB}$. The coefficient of restitution between the spheres is $\frac{1}{2}$. Calculate the velocities of the spheres after impact and the loss of kinetic energy sustained by the system during the impact.

Since the spheres are smooth, the components of velocity perpendicular to AB are unaltered by the collision. Let the velocities after impact of the spheres with centres A and B be $v(a\mathbf{i}+\sqrt{3}\mathbf{j})$, $v(b\mathbf{i}+\mathbf{j})$ respectively. Linear momentum is conserved

$$\Rightarrow 2mv(\mathbf{i}+\sqrt{3}\mathbf{j}) + mv\mathbf{j} = 2mv(a\mathbf{i}+\sqrt{3}\mathbf{j}) + mv(b\mathbf{i}+\mathbf{j})$$
$$\Rightarrow 2 = 2a + b.$$

The restitution equation gives
$$b - a = \tfrac{1}{2} \Rightarrow a = \tfrac{1}{2}, b = 1$$

$\Rightarrow$ velocities after impact are $v(\tfrac{1}{2}\mathbf{i}+\sqrt{3}\mathbf{j})$, $v(\mathbf{i}+\mathbf{j})$
$\Rightarrow$ the loss of kinetic energy is

$$\left\{\tfrac{1}{2}(2m)4v^2 + \tfrac{1}{2}mv^2\right\} - \left\{\tfrac{1}{2}(2m)\frac{13v^2}{4} + \tfrac{1}{2}m.2v^2\right\} = \tfrac{1}{4}mv^2.$$

Example 2. A red ball is stationary on a rectangular billiard table $OABC$. It is then struck by a white ball of equal mass and equal radius with velocity

$$u(-2\mathbf{i}+11\mathbf{j}),$$

where $\mathbf{i}$ and $\mathbf{j}$ are unit vectors along OA and OC respectively. After impact the red and white balls have velocities parallel to the vectors $-3\mathbf{i}+4\mathbf{j}$, $2\mathbf{i}+4\mathbf{j}$ respectively. Prove that the coefficient of restitution between the two balls is $\tfrac{1}{2}$. (L.)

Since the impulse on each ball acts along the line of centres of the balls, the red ball must move along this line after impact. Hence $\tfrac{1}{5}(-3\mathbf{i}+4\mathbf{j})$ is a unit vector in this direction.
Let the velocities of the red and white balls after impact be $v_1(-3\mathbf{i}+4\mathbf{j})$ and $u_1(2\mathbf{i}+4\mathbf{j})$ respectively.
Linear momentum is conserved

$$\Rightarrow u(-2\mathbf{i}+11\mathbf{j}) = u_1(2\mathbf{i}+4\mathbf{j}) + v_1(-3\mathbf{i}+4\mathbf{j}) \quad \text{(masses are equal)}$$
$$\Rightarrow -2u = 2u_1 - 3v_1,$$
$$11u = 4u_1 + 4v_1$$
$$\Rightarrow u_1 = \tfrac{5}{4}u, \; v_1 = \tfrac{3}{2}u.$$

Restitution:
$$v_1|-3\mathbf{i}+4\mathbf{j}| - u_1(2\mathbf{i}+4\mathbf{j}).\tfrac{1}{5}(-3\mathbf{i}+4\mathbf{j}) = eu(-2\mathbf{i}+11\mathbf{j})\cdot\tfrac{1}{5}(-3\mathbf{i}+4\mathbf{j})$$
$$\Rightarrow 5v_1 - 2u_1 =.10\, eu \Rightarrow 10\, eu = 5u$$
$$\Rightarrow e = \tfrac{1}{2}.$$

Example 3. An elastic sphere impinges on an equal sphere at rest. Show that, whatever the coefficient of restitution, the deviation in the direction of motion of the first sphere cannot exceed a right angle.

Figure 9.12 shows the velocities of the spheres (with centres A, B) immediately before and immediately after impact.
Momentum resolute along line of centres $ma + mb = mu\cos\theta$, where m is the mass of each sphere,
Restitution $a - b = -eu\cos\theta$

$$\Rightarrow a = \tfrac{1}{2}u(1-e)\cos\theta, \qquad b = \tfrac{1}{2}u(1+e)\cos\theta.$$

The sphere with centre A therefore moves after impact with a velocity which makes an angle ϕ with AB, where

$$\tan\phi = \frac{u\sin\theta}{\tfrac{1}{2}u(1-e)\cos\theta} = \frac{2\tan\theta}{1-e}.$$

330   PARTICLE WITH TWO DEGREES OF FREEDOM   Ch. 9 §9:6

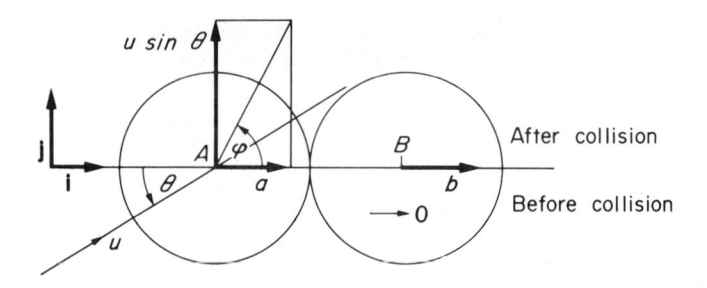

FIG. 9.12.

The deviation in the direction of motion of this sphere is $(\phi - \theta)$, and

$$\tan(\phi - \theta) = \frac{\dfrac{2\tan\theta}{1-e} - \tan\theta}{1 + \dfrac{2\tan^2\theta}{1-e}} = \frac{(1+e)\tan\theta}{2\tan^2\theta + 1 - e}.$$

Since the greatest value of θ which makes the collision geometrically possible is 90°, $\tan\theta$ is positive; since also $e \leqslant 1$, $\tan(\phi - \theta)$ is positive for all possible values of θ and e. The deviation $\phi - \theta$ cannot therefore exceed 90°.

Example 4. A ball is projected with speed V at an inclination β to the horizontal from a point K distant a from a smooth vertical wall. After striking the wall the ball returns to K. If e is the coefficient of restitution between the ball and the wall, show that

$$eV^2 \sin 2\beta = ga(1+e).$$

Show also that the height above K of the point of impact with the wall is $(a\tan\beta)/(1+e)$.   (N.)

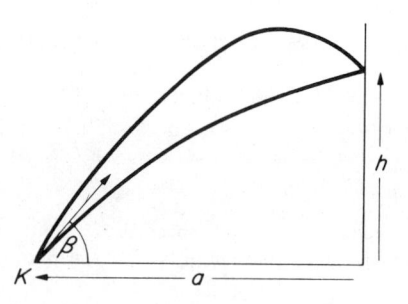

FIG. 9.13.

When the ball strikes the wall, the horizontal component of its velocity becomes $eV\cos\beta$ in magnitude and the vertical component of its velocity is unchanged (Fig. 9.13). Therefore the time of flight from K to the wall and back, deduced from the horizontal component of the motion, is

$$\frac{a}{V\cos\beta}\left(1 + \frac{1}{e}\right),$$

Ch. 9 §9:6 **OBLIQUE IMPACT OF ELASTIC BODIES**

and this time deduced from the vertical component of the motion is $\dfrac{2V \sin \beta}{g}$

$$\Rightarrow \frac{2V \sin \beta}{g} = \frac{a(1+e)}{eV \cos \beta}$$

$$\Rightarrow eV^2 \sin 2\beta = ga(1+e).$$

If h is the height above K of the point of impact with the wall,

$$h = Vt \sin \beta - \tfrac{1}{2}gt^2,$$

where t is the time of flight from K to the wall,

$$\Rightarrow h = \frac{Va \sin \beta}{V \cos \beta} - \frac{ga^2}{2V^2 \cos^2 \beta}$$

$$= a \tan \beta \left(1 - \frac{ga}{V^2 \sin 2\beta}\right)$$

$$= a \tan \beta \left(1 - \frac{e}{1+e}\right) = \frac{a \tan \beta}{1+e}.$$

Example 5. A small smooth sphere falls freely from rest through a distance h and then strikes a fixed smooth plane which is inclined at an angle α to the horizontal. If the coefficient of restitution between the sphere and the plane is $\tfrac{1}{2}$, prove that the distance between the first and fourth points of impact is $\tfrac{105}{16} h \sin \alpha$.   (L.)

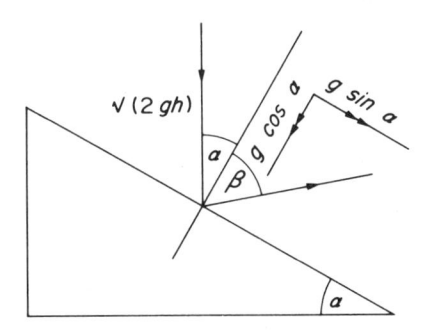

FIG. 9.14.

The sphere strikes the plane with a velocity $\sqrt{(2gh)}$ vertically (Fig 9.14), i.e. at an angle α with the normal to the plane. The sphere rebounds with a component of velocity parallel to the plane unchanged at $\sqrt{(2gh)} \sin \alpha$ and a component of velocity perpendicular to the plane of $\tfrac{1}{2}\sqrt{(2gh)} \cos \alpha$.

Then, *for the first bounce*, the time of flight t_1 is given by

$$0 = \tfrac{1}{2}\sqrt{(2gh)}\, t_1 \cos \alpha - \tfrac{1}{2}gt_1^2 \cos \alpha$$

$$\Rightarrow t_1 = \sqrt{\left(\frac{2h}{g}\right)}.$$

For the second bounce, the initial component of velocity perpendicular to the plane is $\tfrac{1}{4}\sqrt{(2gh)} \cos \alpha$ and the time of flight is

$$t_2 = \frac{1}{2}\sqrt{\left(\frac{2h}{g}\right)}.$$

For the third bounce, the initial component of velocity perpendicular to the plane is $\tfrac{1}{8}\sqrt{(2gh)} \cos \alpha$ and the time of flight is

$$t_3 = \frac{1}{4}\sqrt{\left(\frac{2h}{g}\right)}.$$

Therefore the total time for three bounces is

$$\frac{7}{4}\sqrt{\left(\frac{2h}{g}\right)}.$$

The equation of motion in the direction parallel to the plane, which gives the total distance R between the first and fourth points of impact is

$$R = \sqrt{(2gh)}\sin\alpha \cdot \frac{7}{4}\sqrt{\left(\frac{2h}{g}\right)} + \frac{1}{2}g\sin\alpha \cdot \frac{49}{16}\frac{2h}{g}$$

$$\Rightarrow R = \tfrac{105}{16} h \sin\alpha.$$

Note that the total time which elapses before the sphere ceases to bounce on the plane is the sum, T, of the (infinite) geometrical progression

$$\left(1 + \frac{1}{2} + \frac{1}{2^2} + \frac{1}{2^3} \cdots\right)\sqrt{\left(\frac{2h}{g}\right)} = 2\sqrt{\left(\frac{2h}{g}\right)}.$$

At the moment when bouncing ceases the ball is distant

$$\sqrt{(2gh)}\sin\alpha\, T + \tfrac{1}{2}g\sin\alpha\, T^2 = 8h\sin\alpha$$

from the point of first impact.

Exercise 9.6

In questions 1–6 a smooth sphere of centre A and mass m_1 collides with a smooth sphere of centre B and mass m_2. The coefficient of restitution between the spheres is e. Immediately before impact A is moving with speed u at an anticlockwise angle α with the positive direction AB and B is moving with velocity of magnitude v at an anticlockwise angle β with the positive direction AB. Immediately after the collision A is moving with velocity of magnitude u_1 at an angle α_1 with AB and B is moving with velocity of magnitude v_1 at an angle β_1 with AB.

1. $m_1 = 2\text{kg}$, $m_2 = 4\text{kg}$, $u = 4\text{ms}^{-1}$, $v = 2\text{ms}^{-1}$, $\alpha = 45°$, $\beta = 135°$, $e = \tfrac{1}{2}$; calculate $u_1, \alpha_1, v_1, \beta_1$.
2. $m_1 = m_2 = m$, $\alpha = 45°$, $v = 0$, $e = \tfrac{1}{2}$; calculate $u_1, v_1, \alpha_1, \beta_1$.
3. $m_1 = m_2 = m$, $\alpha = 90°$, $\beta = 180°$, $u = v = V$, $e = \tfrac{1}{2}$; calculate $u_1, v_1, \alpha_1, \beta_1$.
4. $m_1 = m$, $m_2 = 2m$, $\alpha = 30°$, $\beta = 150°$, $u = v = V$, $e = \tfrac{2}{3}$; calculate $u_1, v_1, \alpha_1, \beta_1$.
5. $m_1 = m$, $m_2 = nm$, $\alpha = 60°$, $u = V$, $v = 0$, $\alpha_1 = 90°$; calculate e, u_1, v_1, β_1.
6. $m_1 = m_2 = m$, $v = 0$, $\alpha = 45°$, $e = \tfrac{3}{4}$; calculate α_1.

7. A smooth sphere collides with an equal smooth sphere which is at rest and the line of centres makes an acute angle α with the direction of motion just before impact. If the coefficient of restitution is e, find the angle through which the first sphere is deflected by the impact and show that this angle is greatest if $2\tan^2\alpha = 1 - e$. (L.)

8. A ball B is at rest. It is struck by another ball A, of equal mass and volume, which is moving so that its centre is travelling along a tangent to the ball B. It is noticed that the direction of A is turned through an angle of $30°$ by the impact. Find the coefficient of restitution. (L.)

9. A smooth sphere of mass m impinges obliquely on a heavier sphere of the same radius, but of mass km, at rest on a horizontal table. If, after impact, the directions of motion of the two spheres are perpendicular, prove that the coefficient of restitution must be $1/k$.

If, when $k = 2$, the kinetic energy lost due to the impact is one-quarter of the original kinetic energy, determine the inclination of the initial direction of motion to the line of centres. (L.)

10. A smooth sphere rests, but is free to move, on a horizontal inelastic plane. Another identical sphere falls vertically with speed V, striking the first sphere so that at the instant of impact the line of centres of the spheres makes an angle α with the vertical. If the coefficient of restitution between the spheres is e, prove that the speed of the first sphere after impact is

$$\frac{V(1+e)\sin 2\alpha}{3 - \cos 2\alpha}.$$

Prove that when $\alpha = \tfrac{1}{4}\pi$ the fractional loss of kinetic energy due to the collision is $\tfrac{1}{3}(1-e^2)$. (L.)

11. A smooth sphere moving with velocity $2u\mathbf{i} + u\mathbf{j}$ strikes an equal sphere which is at rest. If the coefficient of restitution between the spheres is e and if the line of centres on impact is parallel to the vector $\mathbf{i}$, find the velocity of each sphere immediately afterwards.

Find the direction in which the two spheres have equal momenta after impact, and find also the value of e for which their final kinetic energies are equal. (L.)

12. Two smooth spheres A and B have masses $2m$ and m respectively, and velocity vectors $3u\mathbf{i} + 4u\mathbf{j}$ and $-4u\mathbf{i} + 3u\mathbf{j}$ respectively, when they collide with their line of centres parallel to the unit vector $\mathbf{i}$. If the impact causes a loss of energy equal to the original kinetic energy of the sphere B, prove that the coefficient of restitution between the spheres is $\sqrt{(23/98)}$. (L.)

13. A smooth sphere A is at rest on a smooth horizontal plane when it is struck by an identical sphere B moving in a direction making an acute angle α with the line of centres at the instant of impact. As a result of the collision the sphere B loses one-quarter of its kinetic energy. Prove that $\sqrt{2} \leqslant \tan\alpha \leqslant \sqrt{3}$ and that the acute angle made with the line of centres by the direction of motion of B immediately after impact is not less than $\tan^{-1}(2\sqrt{2})$. (N.)

14. A particle projected from a point O on a smooth inclined plane strikes the plane normally at the second impact. At the sixth impact it is again at O. If e is the coefficient of restitution, show that

$$e^4 + e^2 - 1 = 0.$$ (N.)

15. An elastic ball strikes a fixed smooth plane obliquely. Show that if α, β are the inclinations to the plane of the directions of motion just before and just after impact, then $\tan\beta = e\tan\alpha$ where e is the coefficient of restitution.

A horizontal circular tray has a vertical rim round its edge. Show that, if a small smooth sphere is projected along the tray from a point in the edge in a direction making an angle α with the radius to the point and after two impacts on the rim, returns to its starting point, then $\tan^2\alpha = e^3/(1 + e + e^2)$. (O.C.)

16. A ball is projected from a point O at the foot of a smooth plane inclined at an angle α to the horizontal. Its components of velocity are U parallel to the line of greatest slope and V perpendicular to the plane. The ball bounces twice on the plane before returning to O at the third bounce. Prove that

$$(1 + e + e^2)V\tan\alpha = U,$$

where e is the coefficient of restitution.

Prove also that, at the first bounce, the ball is moving up or down the plane according as e is greater than or less than $\tfrac{1}{2}(\sqrt{5}-1)$. (O.C.)

17. A ball is projected from a point on a smooth fixed plane inclined at an angle α to the horizontal. The initial direction of projection is up the plane, makes an angle $\alpha + \beta$ with the horizontal, and is in the vertical plane containing the line of greatest slope through the point of projection. Show that the condition that the ball will have just ceased *bouncing* when it returns to its point of projection is

$$\tan\alpha\tan\beta = 1 - e,$$

where e is the coefficient of restitution for the ball and the plane.

18. A smooth sphere of mass m and centre O rests on a smooth horizontal table. Two smooth spheres, each equal to the first sphere and with centres A, B, moving towards each other with the same speed, strike the first sphere simultaneously, without striking each other. The coefficient of restitution is e. At the moment of impact AO makes an angle α with AB. Prove that, for variable α, the subsequent speed of O is greatest when $\alpha = \tfrac{1}{6}\pi$.

Prove that the direction of motion of A turns through a right angle if $\tan^2\alpha = \tfrac{1}{3}e$.

19. A particle of mass m is sliding with speed u on a smooth horizontal table when it encounters a smooth fixed wedge. The surface of the wedge is inclined upwards from the table at an angle α to the horizontal and the vertical plane through the direction of motion of the particle intersects the surface of the wedge in a line of greatest slope. The coefficient of restitution between the particle and the wedge is e. Show that the speed of the particle immediately after its impact with the wedge is $u(\cos^2\alpha + e^2\sin^2\alpha)^{1/2}$ and find the impulse exerted on the particle at the impact.

Prove that the particle will fall on the wedge after the impact if $e < \cot^2\alpha$. Prove also that, if $e < \cos^2\alpha$, then the particle ceases to bounce on the wedge at time

$$\frac{2\,eu\tan\alpha}{g(1-e)}$$

after it first strikes the wedge. (L.)

20. A smooth sphere S of mass 4 kg, moving with velocity $9\mathbf{j}$ ms^{-1}, is struck so that it moves with velocity $(8\mathbf{i}-6\mathbf{j})$ ms^{-1}. Calculate the impulse given to S.

The sphere S then collides with a second smooth sphere T of equal radius and mass 2 kg. Before the collision T is moving with velocity $(2\mathbf{i}+4\mathbf{j})$ ms^{-1} and, at the instant of collision, the line of centres of S and T is parallel to $\mathbf{j}$. The coefficient of restitution between S and T is $\frac{1}{4}$. Calculate the velocity vectors of S and T after the collision. (L.)

21. Show that the vectors $\mathbf{p}_1 = 5\mathbf{i}+12\mathbf{j}$ and $\mathbf{p}_2 = 12\mathbf{i}-5\mathbf{j}$ are at right angles.

If $\mathbf{n}$ and $\mathbf{t}$ are unit vectors in the directions $\mathbf{p}_1$ and $\mathbf{p}_2$ respectively, show that the vectors $13\mathbf{i}+13\mathbf{j}$ and $26\mathbf{i}$ can be expressed as $17\mathbf{n}+7\mathbf{t}$ and $10\mathbf{n}+24\mathbf{t}$ respectively.

Two perfectly elastic smooth spheres, S_1 and S_2, of equal masses and equal radii, collide. At the instant of collision S_1 and S_2 have velocity vectors $\mathbf{v}_1 = 13\mathbf{i}+13\mathbf{j}$ and $\mathbf{v}_2 = 26\mathbf{i}$ respectively and $\mathbf{n}$ is the unit vector in the direction of the line of centres. Find the velocity vectors of the spheres after impact in terms of $\mathbf{n}$ and $\mathbf{t}$ and hence find the velocity vectors after impact in terms of $\mathbf{i}$ and $\mathbf{j}$. (L.)

22. A smooth sphere A of mass 4 units and velocity $(\mathbf{i}-\mathbf{j}+2\mathbf{k})$ units collides with a smooth sphere B of mass 2 units and velocity $(\mathbf{i}+5\mathbf{j}-\mathbf{k})$. If the subsequent velocity of A is $(2\mathbf{i}+\mathbf{j}+\mathbf{k})$ find the subsequent velocity of B.

If α is the angle through which the direction of motion of A is turned by the collision, and β the corresponding angle for B, show that $2\cos\alpha = 3\cos\beta$.

Find the impulse exerted by B on A and deduce the direction of the line of centres at the moment of impact. Hence show that the components of relative velocity in the direction of the line of centres before and after impact are in the ratio $5:-1$. (N.)

9:7 THE GALILEAN TRANSFORMATION

Suppose that, referred to a fixed (Newtonian) origin O, the positions of two observers, who may be moving, are O_1 and O_2. The position vectors of a particle P, of mass m, relative to O_1, O_2 are $\overrightarrow{O_1P}$, $\overrightarrow{O_2P}$ respectively, Fig. 9.15.

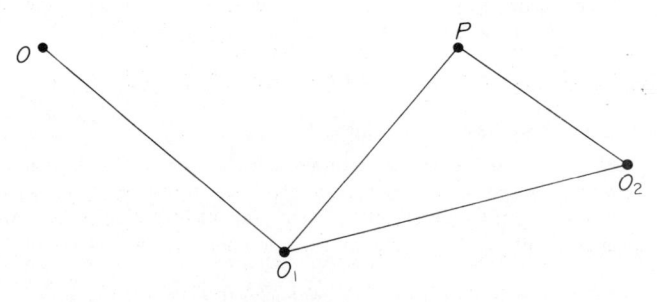

FIG. 9.15.

Relative to O_1, the acceleration of P is

$$\frac{d^2}{dt^2}(\overrightarrow{O_1P})$$

and so observer O_1, who measures forces, velocities, etc., as if he were at rest, *defines* the force $\mathbf{F}_1$, which he considers is acting on P, by the equation

$$\mathbf{F}_1 = m \frac{d^2}{dt^2}(\overrightarrow{O_1 P}). \tag{9.31}$$

Similarly, observer O_2 defines the force $\mathbf{F}_2$ which he considers is acting on P by the equation

$$\mathbf{F}_2 = m \frac{d^2}{dt^2}(\overrightarrow{O_2 P}). \tag{9.32}$$

Subtraction of equation (9.32) from (9.31)

$$\Rightarrow \mathbf{F}_1 - \mathbf{F}_2 = m \frac{d^2}{dt^2}(\overrightarrow{O_1 P} - \overrightarrow{O_2 P})$$

$$\Rightarrow \mathbf{F}_1 - \mathbf{F}_2 = m \frac{d^2}{dt^2}(\overrightarrow{O_1 O_2}). \tag{9.33}$$

This equation can be used to relate the forces which the separate observers define to be acting on P. Two special cases are considered below.

Since an observer must use a set of coordinate axes (or a coordinate frame) to measure physical quantities such as forces and velocities, we shall henceforth refer to our observers' *frames of reference* which move with them. Relative to his frame an observer considers himself to be at rest. [It is not possible to define an absolute rest-frame; only relative motion can be determined by observation.] Further, in this book only non-rotating frames will be considered.

1. Frames with uniform relative velocity

When O_2 is moving with uniform velocity $\mathbf{v}$ relative to O_1 and if, at time $t = 0$, $\overrightarrow{OO_1} = \mathbf{a}_1$, $\overrightarrow{OO_2} = \mathbf{a}_2$, then

$$\overrightarrow{O_1 O_2} = \mathbf{a}_2 - \mathbf{a}_1 + \mathbf{v}t$$

$$\Rightarrow \frac{d^2}{dt^2}(\overrightarrow{O_1 O_2}) = \mathbf{0}$$

$$\Rightarrow \mathbf{F}_1 = \mathbf{F}_2.$$

Thus forces measured or defined by observers who are in motion, but with constant relative velocity, are identical. For example, the forces (e.g. weights) measured in a train or a lift moving with constant velocity are the same as if the train or lift was at rest. This result has many important experimental applications. Thus, the forces experienced by a model aircraft fixed in a wind-tunnel, with air flowing down the tunnel with (undisturbed) velocity $\mathbf{v}$, are identical with those which would be experienced by the model if it was (could be!) projected with velocity $-\mathbf{v}$ through the tunnel containing stationary air.

2. Frames with uniform relative acceleration

Suppose now that the acceleration of O_2 relative to O_1 is $\mathbf{f}$. [Here we consider only the case where **f** is constant.] Then, in this case,

$$\frac{d^2}{dt^2}(\overrightarrow{O_1O_2}) = \mathbf{f}$$

and equation (9.33) gives

$$\mathbf{F}_2 = \mathbf{F}_1 - m\mathbf{f}. \tag{9.34}$$

This equation implies that observer O_2 would regard particle P to be under the influence of a uniform gravitational field (giving rise to an acceleration $-\mathbf{f}$) in excess of any field which O_1 may measure.

Example. If O_1 is stationary and O_2 is in a lift moving vertically with acceleration $\mathbf{f}$, $\mathbf{F}_1 = m\mathbf{g}$

$$\Rightarrow \mathbf{F}_2 = m(\mathbf{g}-\mathbf{f}).$$

Special Cases

(i) If $\mathbf{f} = \mathbf{g}$ so that the lift falls freely under gravity with acceleration $\mathbf{g}$, O_2, in the absence of any information from the outside world, could consider himself in a "gravity-free" situation. [A more complicated example, where **f** is not constant, is the "gravity-free" experience of cosmonauts when their space-ship is in free-flight.]

(ii) If **f** is in the sense of $-\mathbf{g}$, i.e. upwards, O_2 would claim that the weight of P exceeds the weight assigned by O_1. Suppose, for example, that $\mathbf{f} = -\mathbf{g}/10$, [an acceleration of 1 ms^{-2} upwards]. Then $\mathbf{F}_2 = 11m\mathbf{g}/10$, i.e. O_2 in the lift asserts that the weight of P has increased by 10%.

The results of this paragraph will be used in §9:9 to simplify many otherwise complicated problems.

The transformation between two frames moving with constant relative velocity is called a *Galilean transformation*.

9:8 THE MOTION OF A SYSTEM OF PARTICLES—GENERAL THEOREMS

We now consider some very general theorems concerning the motion of a system of interacting particles. Special cases of these theorems will be used in Chapter 10 when we discuss the motion of rigid bodies. Note that our results are established in the three-dimensional case, although in this book we consider only two-dimensional applications.

1. Notation

The system consists of n particles $P_1, \ldots, P_i, \ldots, P_n$ having respective masses $m_1, \ldots, m_i, \ldots, m_n$. Particle P_i has position vector $\mathbf{r}_i$ referred to a Newtonian origin O and is acted upon by a force $\mathbf{F}_i$ from external causes, e.g. gravity. In addition, there is a force $\mathbf{F}'_i$ acting on P_i which is the result of "internal actions" from other particles of the system. In general $\mathbf{F}_i$ is given but $\mathbf{F}'_i$ is unknown. The only fact we know concerning $\mathbf{F}'_i$ comes from the law of action and reaction. Suppose that one element of force contributing to $\mathbf{F}'_i$ is due to the action of particle P_j; then the law tells us that there is an equal and opposite element of force acting on P_j which contributes to $\mathbf{F}'_j$. Hence, if we consider the whole set of forces $\mathbf{F}'_1, \ldots, \mathbf{F}'_i, \ldots, \mathbf{F}'_n$, they must be in equilibrium

$$\Rightarrow \Sigma \mathbf{F}'_i = 0, \tag{9.35a}$$

$$\Sigma (\mathbf{r}_i - \mathbf{r}_A) \times \mathbf{F}'_i = 0, \tag{9.35b}$$

where $\mathbf{r}_A$ is the position vector of an arbitrary point A. Here and in the following Σ implies summation over the system of n particles.

The position vector, $\bar{\mathbf{r}}$, of the centre of mass G is given by

$$M\bar{\mathbf{r}} = \Sigma m_i \mathbf{r}_i, \quad M = \Sigma m_i. \tag{9.36}$$

By differentiation, the velocity, $\dot{\bar{\mathbf{r}}}$, and acceleration, $\ddot{\bar{\mathbf{r}}}$, of G are given by

$$M\dot{\bar{\mathbf{r}}} = \Sigma m_i \dot{\mathbf{r}}_i, \quad M\ddot{\bar{\mathbf{r}}} = \Sigma m_i \ddot{\mathbf{r}}_i. \tag{9.37}$$

We also introduce the position vector, $\mathbf{R}_i$, of P_i referred to G, and its derivatives by

$$\mathbf{r}_i = \bar{\mathbf{r}} + \mathbf{R}_i, \quad \dot{\mathbf{r}}_i = \dot{\bar{\mathbf{r}}} + \dot{\mathbf{R}}_i, \quad \ddot{\mathbf{r}}_i = \ddot{\bar{\mathbf{r}}} + \ddot{\mathbf{R}}_i. \tag{9.38}$$

The importance of the vector $\mathbf{R}_i$ and its derivatives lies in the relations

$$\Sigma m_i \mathbf{R}_i = 0, \quad \Sigma m_i \dot{\mathbf{R}}_i = 0, \quad \Sigma m_i \ddot{\mathbf{R}}_i = 0. \tag{9.39}$$

The essential feature of systems of particles is that the momentum vector $m_i \dot{\mathbf{r}}_i$ is *localised* at the point P_i. In Chapter 7 we introduced the idea of the moment of a force to take account of localisation. We adopt a similar procedure below with momentum and introduce the idea of *moment of momentum* or *angular momentum*. The moment of momentum of the system about the arbitrary point A is defined to be

$$\mathbf{h}(A) = \Sigma (\mathbf{r}_i - \mathbf{r}_A) \times m_i \dot{\mathbf{r}}_i. \tag{9.40}$$

2. *Linear momentum*

The linear momentum of the system is

$$\mathbf{p} = \Sigma m_i \dot{\mathbf{r}}_i$$

$$\Rightarrow \mathbf{p} = M\dot{\bar{\mathbf{r}}}, \tag{9.41}$$

on using equation (9.37), i.e.
the linear momentum of the system is the same as that of a particle of mass M (the total mass of the system) moving with the velocity of the centre of mass G.

3. *Kinetic energy*

This is defined to be T the sum of the kinetic energies of the constituent particles of the system, so that

$$T = \tfrac{1}{2} \Sigma m_i \dot{\mathbf{r}}_i^2$$
$$\Leftrightarrow T = \tfrac{1}{2} \Sigma m_i (\dot{\bar{\mathbf{r}}} + \dot{\mathbf{R}}_i)^2$$
$$\Leftrightarrow T = \tfrac{1}{2} \Sigma m_i \dot{\bar{\mathbf{r}}}^2 + \Sigma m_i \dot{\bar{\mathbf{r}}} \cdot \dot{\mathbf{R}}_i + \tfrac{1}{2} \Sigma m_i \dot{\mathbf{R}}_i^2$$
$$\Leftrightarrow T = \tfrac{1}{2} (\Sigma m_i) \dot{\bar{\mathbf{r}}}^2 + \dot{\bar{\mathbf{r}}} \cdot \Sigma m_i \dot{\mathbf{R}}_i + \tfrac{1}{2} \Sigma m_i \dot{\mathbf{R}}_i^2$$
$$\Leftrightarrow T = \tfrac{1}{2} M\dot{\bar{\mathbf{r}}}^2 + \tfrac{1}{2} \Sigma m_i \dot{\mathbf{R}}_i^2, \tag{9.42}$$

on using equations (9.36), (9.39). Thus
the kinetic energy is made up of one term depending on the motion of G only (the kinetic energy of the whole mass moving with the velocity of the centre of mass) and of another term which involves only motion relative to G.

4. Moment of momentum

The moment of momentum of the system about the arbitrary point A, as defined by equation (9.40),

$$\Rightarrow \mathbf{h}(A) = \Sigma\, (\bar{\mathbf{r}} - \mathbf{r}_A) \times m_i \dot{\mathbf{r}}_i + \Sigma\, \mathbf{R}_i \times m_i \dot{\mathbf{r}}_i$$
$$\Rightarrow \mathbf{h}(A) = (\bar{\mathbf{r}} - \mathbf{r}_A) \times M\dot{\bar{\mathbf{r}}} + \Sigma\, \mathbf{R}_i \times m_i (\dot{\bar{\mathbf{r}}} + \dot{\mathbf{R}}_i)$$
$$\Rightarrow \mathbf{h}(A) = (\bar{\mathbf{r}} - \mathbf{r}_A) \times M\dot{\bar{\mathbf{r}}} + \Sigma\, \mathbf{R}_i \times m_i \dot{\mathbf{R}}_i \qquad (9.43)$$

since $\Sigma\, \mathbf{R}_i \times m_i \dot{\bar{\mathbf{r}}} = (\Sigma\, m_i \mathbf{R}_i) \times \dot{\bar{\mathbf{r}}} = \mathbf{0}$ by equations (9.39). The first term on the right-hand side of equation (9.43) is the moment of momentum about A of the whole mass located at G and moving with the velocity of G. The other term is the *relative moment of momentum of the system about G* (relative because the velocity of G has been subtracted from the velocity of P_i).

5. The motion of the centre of mass

The equation of motion of P_i is

$$m_i \ddot{\mathbf{r}}_i = \mathbf{F}_i + \mathbf{F}'_i. \qquad (9.44)$$

Summing over the system

$$\Rightarrow \Sigma\, m_i \ddot{\mathbf{r}}_i = \Sigma\, \mathbf{F}_i + \Sigma\, \mathbf{F}'_i$$
$$\Rightarrow \Sigma\, m_i \ddot{\bar{\mathbf{r}}} + \Sigma\, m_i \ddot{\mathbf{R}}_i = \Sigma\, \mathbf{F}_i + \Sigma\, \mathbf{F}'_i$$
$$\Rightarrow M\ddot{\bar{\mathbf{r}}} = \Sigma\, \mathbf{F}_i \qquad (9.45)$$

on using equations (9.36) and (9.39). Therefore
the motion of the centre of mass of the system is the same as if the whole mass and all the external forces were concentrated there.
This result applies, for example, to the motion of the centre of mass of a spacecraft or of the centre of mass of the earth in its motion around the sun.

Note that, if $\Sigma\, \mathbf{F}_i = \mathbf{0}$, then

$$\Sigma\, m_i \ddot{\mathbf{r}}_i = \mathbf{0} = M\ddot{\bar{\mathbf{r}}}$$
$$\Rightarrow \Sigma\, m_i \dot{\mathbf{r}}_i = M\dot{\bar{\mathbf{r}}} = \mathbf{P}_o, \qquad (9.46)$$

a constant vector. Here we have established the theorem of *conservation of linear momentum*, i.e.
the linear momentum of a system unacted upon by external forces remains constant (i.e. the velocity of G remains constant).

6. Motion about the centre of mass

Taking vector product of equation (9.44) with $\mathbf{r}_i$ and summing for the system

$$\Rightarrow \Sigma\, \mathbf{r}_i \times m_i \ddot{\mathbf{r}}_i = \Sigma\, \mathbf{r}_i \times \mathbf{F}_i + \Sigma\, \mathbf{r}_i \times \mathbf{F}'_i \qquad (9.47)$$
$$\Rightarrow \Sigma\, (\bar{\mathbf{r}} + \mathbf{R}_i) \times m_i (\ddot{\bar{\mathbf{r}}} + \ddot{\mathbf{R}}_i) = \Sigma\, (\bar{\mathbf{r}} + \mathbf{R}_i) \times \mathbf{F}_i$$

[on using (9.35b) with $r_A = 0$]

$$\Rightarrow \bar{r} \times M\ddot{\bar{r}} + (\Sigma m_i R_i) \times \ddot{\bar{r}} + \bar{r} \times \Sigma m_i \ddot{R}_i + \Sigma R_i \times m_i \ddot{R}_i = \bar{r} \times \Sigma F_i + \Sigma R_i \times F_i.$$

The second and third terms of the left-hand side vanish because of equations (9.39), whereas the first terms on each side of the equation are equal by equation (9.45). Therefore

$$\Sigma R_i \times m_i \ddot{R}_i = \Sigma R_i \times F_i. \tag{9.48}$$

This is of the same form as equation (9.47) with, of course, $\Sigma r_i \times F'_i = 0$, and implies that

the motion of the system about its centre of mass is the same as if the centre of mass were fixed and the same (external) forces acted.

7. *Motions generated by simultaneously applied impulses*

Suppose that a set of impulses is applied to our system over the infinitesimally short period from $t = t_0$ to $t = t_1$ so that our typical external impulse

$$J_i = \int_{t_0}^{t_1} F_i \, dt.$$

Then integration of equation (9.45) over this interval gives

$$\left[M\dot{\bar{r}} \right]_{t = t_0}^{t = t_1} = \Sigma J_i$$

$$\Rightarrow M(v_1 - v_0) = \Sigma J_i, \tag{9.49}$$

where v_0, v_1 are the velocities of G just before and just after the application of the impulses. Therefore
the change in the velocity of G is the same as the change in velocity of a particle of mass M at G under the action of all the external impulses transferred to act at G.
Similarly integration of equation (9.47) gives

$$\Sigma r_i \times m_i (v_{i1} - v_{i0}) = \Sigma r_i \times J_i. \tag{9.50}$$

Here v_{i0} and v_{i1} are the velocities of our typical particle just before and just after application of the impulses.
Therefore
the change in the moment of momentum (angular momentum) about ANY POINT equals the moment of the external impulses about that point.
[Note that we have assumed that r_i does not vary in the interval t_0 to t_1. All impulsive motion problems assume no change of position during the impulse but a discontinuous change in velocity.]

9:9 THE MOTION OF CONNECTED PARTICLES

In this section we give some examples of the motion of particles which are usually connected together by means of strings. The work is greatly simplified by use of the results of §9:7 and 9:8.

Example 1. *The rectilinear motion in a horizontal plane of two particles which are connected by an elastic spring*

Figure 9.16 shows two particles of masses m_1 at A, m_2 at B on a smooth horizontal plane connected by an unstressed elastic spring AB of modulus λ and natural length $a_1 + a_2$. The centre of mass of the particles is G, where $AG = a_1$, $GB = a_2$.

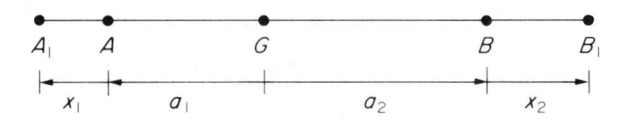

FIG. 9.16.

If the spring is stretched or compressed and then released, in the subsequent motion the only horizontal forces acting on the particles are the equal and opposite tensions in the spring. The resultant of the forces acting on the system of two particles is therefore zero and the centre of mass of the particles will remain at rest during the motion. If, at time t from the beginning of the motion, the particle of mass m_1 is at A_1, where $AA_1 = x_1$, the particle of mass m_2 is at B_1 where $BB_1 = x_2$, and the tension in the spring is T_1, then

$$\frac{a_1}{a_2} = \frac{a_1 + x_1}{a_2 + x_2} = \frac{m_2}{m_1} \Rightarrow \frac{x_1}{x_2} = \frac{m_2}{m_1}.$$

But

$$T = \frac{\lambda(x_1 + x_2)}{(a_1 + a_2)}$$

$$\Rightarrow T = \frac{\lambda}{a_1 + a_2}\left(x_1 + \frac{m_1 x_1}{m_2}\right) = \frac{\lambda(m_1 + m_2)x_1}{m_2(a_1 + a_2)}.$$

Therefore the equation of motion for the particle m_1 is

$$m_1 \ddot{x}_1 = -\frac{\lambda(m_1 + m_2)}{m_2 l} x_1, \quad \text{where } l = a_1 + a_2,$$

$$\Leftrightarrow \ddot{x}_1 = -\frac{\lambda(m_1 + m_2)}{m_1 m_2 l} x_1.$$

Particle m_1 therefore moves in SHM of period

$$2\pi \sqrt{\left[\frac{m_1 m_2 l}{\lambda(m_1 + m_2)}\right]}.$$

The symmetry of this result shows this also to be the period of the SHM in which the particle m_2 moves.

Example 2. Two particles A and B, each of mass m, are connected by a light elastic string of natural length a. When A is held at rest and B executes simple harmonic vibrations under gravity in the vertical line through A, the period is T. The particles are now placed on a smooth horizontal table and released from rest at a distance $a + b$ apart. Show that they will collide after a time

$$\left(\frac{1}{4} + \frac{a}{2\pi b}\right)\frac{T}{\sqrt{2}}. \tag{N.}$$

The centre of mass of the particles is the mid-point of the straight line joining them.

(a) When A is held at rest and B executes SHM under gravity in a vertical line through A, the equation of motion of B is

$$m\ddot{x} = -\frac{\lambda x}{a},$$

THE MOTION OF CONNECTED PARTICLES

where λ is the modulus of elasticity of the string,

$$\Rightarrow T = 2\pi \sqrt{\left(\frac{am}{\lambda}\right)}.$$

(b) When the particles are released from rest on the table and when the string is of length $a + 2x$, then the tension in the string is $\lambda 2x/a$ and the equation of motion for each particle is

$$m\ddot{x} = -\frac{2\lambda x}{a},$$

where, for each particle, $\tfrac{1}{2}a + x$ is its distance from the fixed centre of mass of the two particles. Each particle therefore executes SHM of period $2\pi\sqrt{(am/2\lambda)} = T/\sqrt{2}$ and amplitude $\tfrac{1}{2}b$ from the point of release to the centre of its motion in time $T/(4\sqrt{2})$ and when each particle reaches the centre of its motion the string becomes slack. Each particle now has a speed

$$\tfrac{1}{2}b\sqrt{\left(\frac{2\lambda}{ma}\right)} = \frac{b\pi\sqrt{2}}{T}.$$

Therefore, the time taken by the particles to cover a relative distance a is

$$a \bigg/ \left(\frac{2b\pi\sqrt{2}}{T}\right) = \frac{aT}{2b\pi\sqrt{2}}$$

and the particles collide after a total time

$$\frac{T}{\sqrt{2}}\left(\frac{1}{4} + \frac{a}{2\pi b}\right)$$

from the beginning of the motion.

Example 3. A bead A, of mass m, slides on a smooth horizontal rail, and a particle B, also of mass m, is attached to the bead by a light inelastic string of length $2a$. The system is let go from rest with the string taut and in the vertical plane through the rail, and with AB making an acute angle α with the downward vertical. Prove that, if the inclination of the string to the vertical at time t is θ, then

$$\tfrac{1}{2}\dot{\theta}^2 = \frac{g}{a}\left(\frac{\cos\theta - \cos\alpha}{2 - \cos^2\theta}\right).$$

By considering the acceleration of the bead, or otherwise, prove that the tension in the string at any time during the motion is

$$\frac{\cos\theta\,(6 - \cos^2\theta) - 4\cos\alpha}{(2 - \cos^2\theta)^2} mg.$$

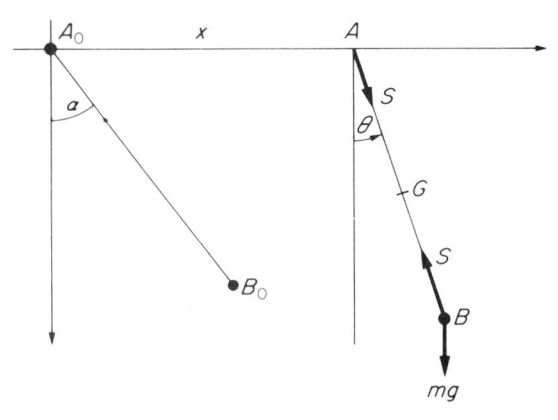

FIG. 9.17.

If $A_0 B_0$ (Fig. 9.17) is the initial position and AB is the position at time t, the coordinates of G are then $(x + a \sin \theta, a \cos \theta)$. The velocity components of G are $(\dot{x} + a\dot\theta \cos \theta, -a\dot\theta \sin \theta)$. Since the rail is smooth there is no horizontal force on the system and G therefore has no horizontal velocity (The horizontal component of linear momentum of the system is zero.)

$$\Rightarrow \dot{x} + a\dot\theta \cos \theta = 0. \quad (1)$$

The velocity components of A are $(\dot{x}, 0)$ and those of B are $(\dot{x} + 2a\dot\theta \cos \theta, -2a\dot\theta \sin \theta)$; hence the kinetic energy

$$T = \tfrac{1}{2} m \dot{x}^2 + \tfrac{1}{2} m \{(\dot{x} + 2a\dot\theta \cos \theta)^2 + 4a^2 \dot\theta^2 \sin^2 \theta\}$$
$$= \tfrac{1}{2} m a^2 \dot\theta^2 \cos^2 \theta + \tfrac{1}{2} m \{a^2 \dot\theta^2 \cos^2 \theta + 4a^2 \dot\theta^2 \sin^2 \theta\}$$

on using (1). The potential energy is $V = -2mga \cos \theta$. Hence the energy equation is

$$ma^2 \dot\theta^2 (2 - \cos^2 \theta) - 2mga \cos \theta = E.$$

The initial condition, $\dot\theta = 0$ when $\theta = \alpha$,

$$\Rightarrow \tfrac{1}{2} a \dot\theta^2 (2 - \cos^2 \theta) - g \cos \theta = -g \cos \alpha. \quad (2)$$

This leads to the required expression for $\dot\theta^2$, from which $\ddot\theta$ may be obtained by differentiation.

To find the tension S we consider the motion of A only.

$$S \sin \theta = m \ddot{x} = m \frac{d}{dt}(-a\dot\theta \cos \theta) = -ma\ddot\theta \cos \theta + ma\dot\theta^2 \sin \theta.$$

Substitution of the expressions for $\dot\theta^2$ and $\ddot\theta$ obtained from (2) leads to the required value of S.

Example 4. A light inextensible string, of length $2a$, has equal particles, each of mass m, attached to its ends and a third particle of mass M attached to its mid-point. The particles lie in a straight line on a smooth horizontal table with the string just taut and M is projected along the table with speed V perpendicular to the string. Show that, if the two particles at the ends collide after a time T when the displacement of M from its initial position is x, then

$$(M + 2m) x = M V T + 2ma.$$

Show also that the tension in the string just before the collision is

$$mM^2 V^2 / \{(M + 2m)^2 a\}.$$

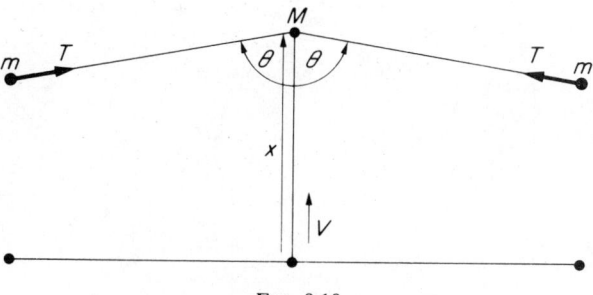

FIG. 9.18.

Suppose the position at time t is as shown in Fig. 9.18. Then the equation of conservation of linear momentum along the original direction of motion of M is

$$M \dot{x} + 2m \frac{d}{dt}(x - a \cos \theta) = MV. \quad (1)$$

Integration gives

$$(M+2m)x = MVt + 2ma\cos\theta,$$

which gives the required result when $\theta = 0$.

The energy equation is

$$\tfrac{1}{2}M\dot{x}^2 + m\left[\left(\frac{d}{dt}(x-a\cos\theta)\right)^2 + (a\dot\theta\sin\theta)^2\right] = \tfrac{1}{2}MV^2. \tag{2}$$

Also the equation of motion of M is

$$M\ddot{x} = -2T\cos\theta.$$

But by differentiating (1)

$$(M+2m)\ddot{x} = -2ma(\dot\theta^2\cos\theta + \ddot\theta\sin\theta).$$

Therefore just before the collision $\theta = 0$, $T = T_0$, where

$$T_0 = Mma\dot\theta_0^2/(M+2m).$$

But from (1) and (2) $a^2\dot\theta_0^2 = MV^2/(M+2m)$ and the required result follows.

Example 5. A light rod of length $2a$ is free to rotate in a horizontal plane about its mid-point O which is fixed. Two small smooth equal rings, P, Q, are free to slide on the rod, and are initially equidistant from, but on opposite sides of, O. The rod is given an angular speed Ω when P, Q, are distant $3a/5$ from O and at rest relative to the rod. If the system is now left to itself, show that at the instant when the rings leave the rod, the rings are moving with speed $3a\Omega/5$, and the angular speed of the rod is $9\Omega/25$.

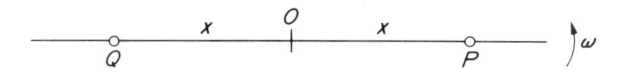

FIG. 9.19.

In the position shown (Fig. 9.19) the angular speed of the rod is ω. Since the rod rotates freely about O, the angular momentum about O remains constant. Since the action of the rod on a ring is perpendicular to the rod this internal force does no work when the ring slides on the rod, and the work done on the particles in the rotation by this force is exactly cancelled by the work done on the rod by the reactions. Hence, the kinetic energy remains constant.

In the position shown the (scalar) angular momentum is

$$h(O) = 2mx^2\omega = 2m\frac{9a^2}{25}\Omega. \tag{1}$$

The kinetic energy is

$$T = m(\dot{x}^2 + \omega^2 x^2) = m\frac{9a\Omega^2}{25}. \tag{2}$$

From (1), $\omega = 9\Omega a^2/(25x^2)$. Hence using (2)

$$\dot{x}^2 + \frac{81\Omega^2 a^4}{625 x^2} = \frac{9a^2\Omega^2}{25}.$$

When $x = a$ the rings leave the rod and at this instant

$$\dot{x}^2 = \frac{9a^2\Omega^2}{25}\left(1 - \frac{9}{25}\right) = \frac{9.16a^2\Omega^2}{625}$$

$\Rightarrow \dot{x} = 12a\Omega/25$. Also, when $x = a$, $\omega = 9\Omega/25$. The speed v of a ring is given by $v^2 = \dot{x}^2 + \omega^2 x^2$ and so, when $x = a$,

$$v^2 = \left(\frac{144}{625} + \frac{81}{625}\right)a^2\Omega^2 = \frac{225}{625}a^2\Omega^2$$

$$\Rightarrow v = \tfrac{3}{5}a\Omega.$$

Examples 6. A light inextensible string AB, of length $2a$ and carrying equal particles of mass m at its ends, lies at rest on a smooth horizontal table with the string just taut. A horizontal impulse of magnitude J is applied to A in a direction making an acute angle β with BA produced. Discuss the subsequent motion.

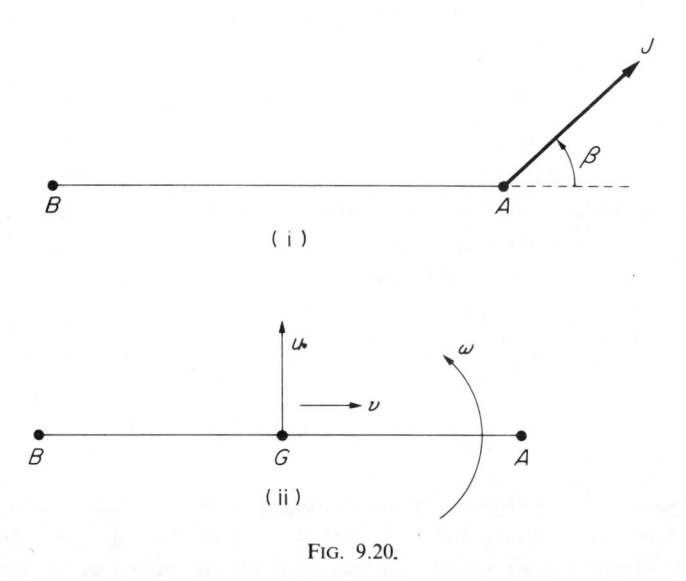

FIG. 9.20.

Figure 9.20 (i) shows the impulse on the system. Suppose that, immediately after the impulse is applied, G has velocity components u, v respectively perpendicular and parallel to BA and the angular speed of the string is ω as shown in Fig. 9.20(ii). Then in this case the resolutes of equation (9.49) [for the change in velocity of G] are

$$2mu = J\sin\beta, \quad 2mv = J\cos\beta$$
$$\Rightarrow u = (J\sin\beta)/(2m), \quad v = (J\cos\beta)/(2m) \tag{1}$$

Since no horizontal forces act on the system, G continues to move with this (constant) velocity.

The moment of the impulse about G is $a \cdot J\sin\beta$ and the moment of momentum equation gives

$$2ma \cdot a\omega = aJ\sin\beta$$
$$\Rightarrow \omega = (J\sin\beta)/(2ma). \tag{2}$$

Ch. 9 §9:9 THE MOTION OF CONNECTED PARTICLES 345

Because no external forces have a moment about G, the string continues to rotate with this constant angular speed.

Further, since G moves with constant velocity the force exerted by the string on each particle is the same as if G were fixed and the particle moved with uniform angular speed ω. The tension in the string is therefore constant and of magnitude $ma\omega^2$.

Example 7. Two particles A and B of masses m and $2m$ respectively are connected by a light inextensible string and placed on a smooth horizontal table with the string just taut. The particle B is given an impulse $\mathbf{I}$ that sets both particles in motion. If $\hat{\mathbf{n}}$ is the unit vector in the initial direction of AB, and if $\mathbf{u}$ and $\mathbf{v}$ are the initial velocities of A and B respectively, show that

$$\mathbf{u} = \frac{1}{3m}(\mathbf{I}.\hat{\mathbf{n}})\hat{\mathbf{n}},$$

$$\mathbf{v} = \frac{1}{2m}\left(\mathbf{I} - \frac{1}{3}(\mathbf{I}.\hat{\mathbf{n}})\hat{\mathbf{n}}\right).$$

Show that the impulsive tension in the string is of magnitude $\frac{1}{3}\mathbf{I}.\hat{\mathbf{n}}$.
Verify that the kinetic energy of the system generated by the impulsie is $\frac{1}{2}\mathbf{I}.\mathbf{v}$.

FIG. 9.21.

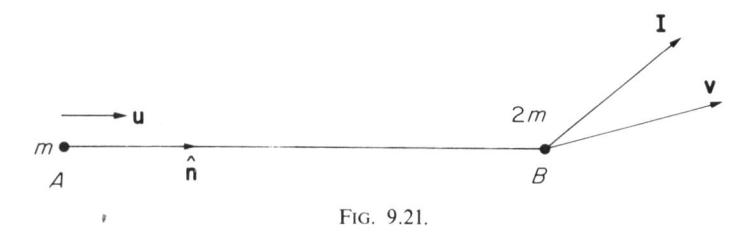

With reference to Fig. 9.21, particle A must move along the direction of $\hat{\mathbf{n}}$. Further, continuity of the components of velocity along AB implies

$$\mathbf{v}.\hat{\mathbf{n}} = \mathbf{u}.\hat{\mathbf{n}}$$
$$\Rightarrow (\mathbf{v}.\hat{\mathbf{n}})\hat{\mathbf{n}} = \mathbf{u}.$$

Linear momentum along AB gives

$$m\mathbf{u} + 2m(\mathbf{v}.\hat{\mathbf{n}})\hat{\mathbf{n}} = (\mathbf{I}.\hat{\mathbf{n}})\hat{\mathbf{n}}$$
$$\Rightarrow \mathbf{u} = \frac{1}{3m}(\mathbf{I}.\hat{\mathbf{n}})\hat{\mathbf{n}}.$$

Then the general linear momentum equation

$$\mathbf{I} = m\mathbf{u} + 2m\mathbf{v}$$
$$\Rightarrow \mathbf{v} = \frac{1}{2m}\mathbf{I} - \frac{1}{2}\mathbf{u}$$
$$\Rightarrow \mathbf{v} = \frac{1}{2m}\left[\mathbf{I} - \frac{1}{3}(\mathbf{I}.\hat{\mathbf{n}})\hat{\mathbf{n}}\right]. \tag{1}$$

The impulse $\mathbf{T}$ exerted by the string on A is given by

$$\mathbf{T} = m\mathbf{u}$$
$$\Rightarrow \mathbf{T} = \frac{1}{3}(\mathbf{I}.\hat{\mathbf{n}})\hat{\mathbf{n}}$$

so that the impulsive tension in the string is of magnitude $\frac{1}{3}(\mathbf{I}.\hat{\mathbf{n}})$.

The kinetic energy generated by the impulse is K where

$$K = \tfrac{1}{2}(m\mathbf{u}^2 + 2m\mathbf{v}^2)$$

$$= \frac{1}{2m}\left[\frac{1}{9}(\mathbf{I}\cdot\hat{\mathbf{n}})^2 + \frac{1}{2}\left(\mathbf{I} - \frac{1}{3}(\mathbf{I}\cdot\hat{\mathbf{n}})\hat{\mathbf{n}}\right)^2\right]$$

$$= \frac{1}{4m}\left[\mathbf{I}^2 - \frac{1}{3}(\mathbf{I}\cdot\hat{\mathbf{n}})^2\right]$$

$$= \tfrac{1}{2}\mathbf{I}\cdot\mathbf{v}$$

on using equation (1).

Exercise 9.9

1. Two particles of masses m and $4m$ respectively are connected by a light elastic string of modulus λ and natural length a. They are placed at rest on a smooth horizontal table at a distance a apart and equal impulses I applied simultaneously in opposite directions act upon them along the line of the string so that the string extends. Prove that in the ensuing motion the greatest extension of the string is attained in time $\pi\sqrt{(ma/5\lambda)}$ and find its value. (N.)

2. Two particles A and B, each of mass m, connected by an elastic string of natural length a and modulus $ma\omega^2/2$, are initially at rest on a smooth horizontal table and distant a apart. The particle A is held at rest and B is projected away from A in the direction AB with speed $\sqrt{(3/2)a\omega}$. Show that when the length of the string is $2a$ the speed of B is $a\omega$.

At this instant A is released. Write down the equations of motion of the particles during the subsequent motion in terms of the displacements, x and y, of A and B respectively from their initial positions of rest. Deduce that at time t after the release of A the length of the string is

$$a(1 + \cos\omega t + \sin\omega t).$$

Show that the string will first become slack when $t = 3\pi/4\omega$, and that the particles will collide when

$$t = \frac{3\pi}{4\omega} + \frac{1}{\omega\sqrt{2}}. \qquad \text{(N.)}$$

3. Two perfectly elastic particles of equal masses m attract one another with a force μm times their distance apart and move under their mutual attraction, starting from rest at a distance a apart. Find the time that elapses between successive collisions and the greatest speed acquired by the particles. (L.)

4. Two particles A and B, of masses m_1 and m_2 respectively, are joined together by an elastic string of natural length a and modulus λ and are initially at rest on a smooth horizontal table at a distance a apart. An impulse m_2V is given to the particle B in the direction AB. Show that at time t after the impulse the length of the string will be $(na + V\sin nt)/n$, where

$$n^2 = \lambda(m_1 + m_2)/am_1m_2,$$

provided that $t < \pi/n$.

Find the velocities of the particles, and the distance through which A has moved from its initial position, at the moment when the string first becomes slack. (N.)

5. A light rod of length a can turn freely about the end A, which is fixed, and to the other end B is attached a particle of mass m. The end B is joined by a string of length a to a small ring C of mass m, which can slide on a smooth horizontal wire passing through A. The system is released from rest when AC is $2a$. Show that, when the angle CAB is $60°$, the angular acceleration of the rod is $-g/(16a)$, and find the tension in the string at this instant.

6. A ring A, of mass m, can move freely along a smooth horizontal wire, which is fixed just above a smooth horizontal table. The ring is attached by an inextensible string of length a to a particle B, of mass m, resting on the table. Initially the string AB is taut and alongside the wire, and then B is given a horizontal velocity u perpendicular to the wire. Find the angular speed of AB when it makes an angle θ with the wire and show that the tension of the spring is then

$$\frac{2mu^2}{a(1+\cos^2\theta)^2}.$$

7. A particle moves in a plane under a force directed towards a fixed point S whose coordinates referred to rectangular axes are $(a, 0)$. The particle passes through the origin O and through the point $P(a, 2a)$. The velocity at O is $u\mathbf{j}$; the velocity at P is in the direction of the vector $\mathbf{i}+\mathbf{j}$. Use the principle of conservation of moment of momentum to find the speed at P. (L.)

8. Two particles A, B, each of mass m, are connected by a light straight rigid rod of length $2a$ which is free to move on a smooth horizontal table. Initially the particles are at rest when particle A is given a horizontal impulse of magnitude J making an angle $\theta (<\pi/2)$ with $\overrightarrow{AB}$. Find the impulse in the rod. Find also the kinetic energy generated by the impulse. Describe the subsequent motion of the rod. (L.)

9. A small ring of mass m is free to slide on a smooth horizontal wire, and is connected by a light rod of length l to a particle of mass m. The rod is released from rest when parallel to the wire. If the angular speed of the rod is ω when the rod makes an angle θ with the horizontal, prove that

$$\omega^2 = \frac{4g\sin\theta}{l(1+\cos^2\theta)}.$$ (L.)

10. A shell of mass $4m$ is fired from a gun. At the instant when its velocity is $u\mathbf{i}+v\mathbf{j}$ it explodes into two fragments of masses $3m$ and m. If, immediately after the explosion the velocity of the fragment of mass m relative to the fragment of mass $3m$ is $-2u\mathbf{i}+2v\mathbf{j}$, calculate the velocity of each fragment at this instant. Show that their directions of motion are perpendicular to each other if $5v^2 = 3u^2$. (N.)

11. Show that the kinetic energy of two particles, one of mass m_1 moving with velocity $\mathbf{u}_1$ and the other of mass m_2 moving with velocity $\mathbf{u}_2$, may be expressed in the form

$$\tfrac{1}{2}\mu\mathbf{v}^2 + \tfrac{1}{2}\mathbf{P}^2/M,$$

where $M = m_1+m_2$, $\mu = m_1 m_2/M$, $\mathbf{v} = \mathbf{u}_1 - \mathbf{u}_2$ and $\mathbf{P}$ is the total momentum.

A firework of mass 0.4 kg explodes during flight into two parts A and B, where A is of mass 0.3 kg and B is of mass 0.1 kg. If the speed of A relative to B is 200 ms^{-1} find the increase in kinetic energy as a result of the explosion, and find the speeds of A and B relative to the centre of mass of the system.

If, furthermore, the velocity of the firework before the explosion is 50 m s^{-1} at an angle of $30°$ to the direction of motion of A just after the explosion, show that the speed of A after the explosion is $50\sqrt{3}$ m s^{-1}, and find the speed of B. (N.)

12. Two particles, each of mass m, are fixed to the ends of a light inextensible string of length $2a$, and a particle of mass $2m$ is fixed to the mid-point. The string is laid in a straight line on a smooth horizontal table. The middle particle is then struck so that it begins to move along the table in a direction perpendicular to the string with speed u. By considering the momentum and energy of the system just before the end particles collide, or otherwise, prove that the speed of each of these particles, perpendicular to the string, at this instant is $u/\sqrt{2}$.

If the coefficient of restitution between these particles is e, prove that the energy of the system after the impact is $\tfrac{1}{2}mu^2(1+e^2)$. Hence, or otherwise, find the speeds of the middle particle and each end particle when the string is next in a straight line. (O.C.)

9:10 THE TWO-DIMENSIONAL MOTION OF A PROJECTILE IN A RESISTING MEDIUM

A particle of mass m is projected from O (Fig. 9.22) in a vertical plane with speed u at an angle α with the horizontal in a medium whose resistance $\mathbf{R}$ to the motion of the particle is $mk\mathbf{v}$, where $\mathbf{v}$ is the velocity of the particle and k is constant. [In this book we consider only problems where the resistance varies as the velocity.] It is assumed that at every instant $\mathbf{R}$ is acting in the direction opposite to the velocity of the particle at that

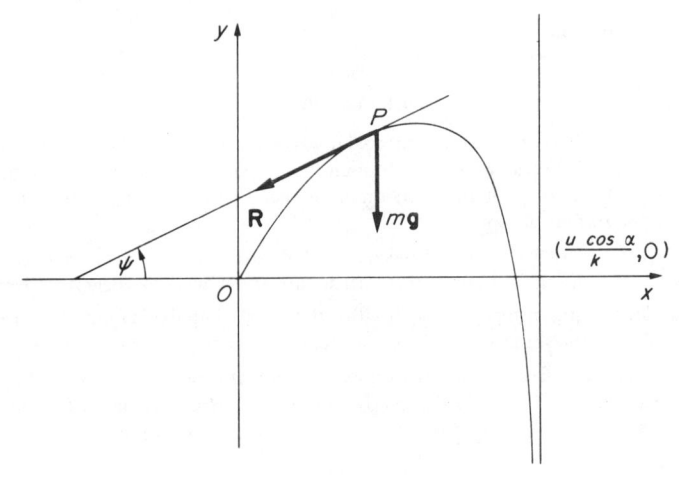

FIG. 9.22.

instant. The figure shows the forces acting on the particle at an instant of its flight, giving as the equation of motion

$$m\mathbf{g} + \mathbf{R} = m\frac{d\mathbf{v}}{dt} = m\mathbf{g} - mk\mathbf{v}. \tag{9.51}$$

We solve the vector equation (9.51) by equating like components, but Example 2 on p. 350 shows how the solution can be obtained directly by vector methods.

When the coordinates of the particle referred to horizontal and vertical axes through the point of projection are (x, y) the equations of the resolved parts of the motion, parallel to the x- and y-axes respectively, are

$$m\ddot{x} = -mkv\cos\psi, \tag{9.52}$$
$$m\ddot{y} = -mkv\sin\psi - mg, \tag{9.53}$$

where ψ is the angle which the direction of motion of the particle makes with Ox.

If s is the length of the arc of the trajectory from O to P, (x, y), then $v = ds/dt$, $\cos\psi = dx/ds$ and $\sin\psi = dy/ds$.

Hence equations (9.52) and (9.53) can be written

$$\ddot{x} + k\frac{ds}{dt}\frac{dx}{ds} = 0$$
$$\Rightarrow \ddot{x} + k\dot{x} = 0, \tag{9.54}$$

and
$$\ddot{y} + k\frac{ds}{dt}\frac{dy}{ds} + g = 0$$
$$\Rightarrow \ddot{y} + k\dot{y} = -g. \tag{9.55}$$

The solutions of these equations are

$$x = Ae^{-kt} + B,$$

$$y = Ce^{-kt} + D - \frac{gt}{k},$$

where A, B, C and D are constants.

The initial conditions $x = 0$, $dx/dt = u \cos \alpha$ when $t = 0$ give

$$A + B = 0 \quad \text{and} \quad u \cos \alpha = -kA$$
$$\Rightarrow A = -(u \cos \alpha)/k, \quad B = (u \cos \alpha)/k$$
$$\Rightarrow x = \frac{u \cos \alpha}{k}(1 - e^{-kt}). \tag{9.56}$$

The initial conditions $y = 0$, $dy/dt = u \sin \alpha$ when $t = 0$ give

$$C + D = 0, \quad -kC - \frac{g}{k} = u \sin \alpha,$$
$$\Rightarrow C = -\frac{u \sin \alpha}{k} - \frac{g}{k^2}, \quad D = \frac{u \sin \alpha}{k} + \frac{g}{k^2}$$
$$\Rightarrow y = \left(\frac{u \sin \alpha}{k} + \frac{g}{k^2}\right)(1 - e^{-kt}) - \frac{gt}{k}. \tag{9.57}$$

The equation of the path of the projectile obtained by eliminating t from (9.56) and (9.57) is

$$y = x\left(\tan \alpha + \frac{g}{ku \cos \alpha}\right) + \frac{g}{k^2} \ln\left(1 - \frac{kx}{u \cos \alpha}\right). \tag{9.58}$$

Equation (9.56) shows that

$$\text{as} \quad t \to \infty, \quad x \to \frac{u \cos \alpha}{k},$$

and it follows that the path has a vertical asymptote.

Example 1. Show that the effect of the resistance, if k is sufficiently small, is to reduce the horizontal range, R, of the particle by $(4kVR \sin \alpha)/3g$, approximately.

Expanding the logarithm term in equation (9.58) gives

$$y = x \tan \alpha - \frac{gx^2}{2V^2 \cos^2 \alpha} - \frac{gkx^3}{3V^3 \cos^3 \alpha} + O(k^2).$$

Neglecting terms of $O(k^2)$ we find that when $y = 0$, x satisfies the equation

$$x = \frac{2V^2 \sin \alpha \cos \alpha}{g} - \frac{2kx^2}{3V \cos \alpha}. \tag{1}$$

Successive approximations to the positive root of this equation are

$$x = \frac{2V^2 \sin \alpha \cos \alpha}{g} + O(k),$$

$$x = \frac{2V^2 \sin \alpha \cos \alpha}{g} - \frac{8kV^3 \sin^2 \alpha \cos \alpha}{3g^2} + O(k^2),$$

on putting $x^2 = (4V^4 \sin^2 \alpha \cos^2 \alpha)/g^2$ in the second term on the right-hand side of equation (1). This second approximation gives the required result.

To check the dimensions of this result, $R = mkv$, so k has dimensions T^{-1},

$$\frac{V^2}{g} = L^2T^{-2} \times \frac{1}{LT^{-2}} = L, \quad \frac{kV^3}{g^2} = T^{-1} \times L^3T^{-3} \times \frac{1}{L^2T^{-4}} = L.$$

Example 2. We discuss the solution of equation (9.51) by vector methods.

The equation of motion can be written

$$\mathbf{f} = \frac{d\mathbf{v}}{dt} = -g\hat{\mathbf{z}} - k\mathbf{v}, \tag{1}$$

where $\hat{\mathbf{z}}$ is a unit vector directed vertically upwards and $\mathbf{f}$ is the acceleration. Differentiation with respect to t gives

$$\frac{d\mathbf{f}}{dt} = -k\mathbf{f} \tag{2}$$

$$\Rightarrow \mathbf{f} = \mathbf{f}_0 e^{-kt}, \tag{3}$$

where $\mathbf{f}_0$ is a constant vector. Hence the acceleration is always parallel to the (fixed) direction of the initial acceleration $\mathbf{f}_0$ and tends to zero as $t \to \infty$. Hence, from (1), as $t \to \infty$

$$\mathbf{v} \to -\frac{g}{k}\hat{\mathbf{z}}, \tag{4}$$

and this gives the limiting or terminal (vertically downwards) velocity. Since $\mathbf{v} = d\mathbf{r}/dt$, where $\mathbf{r}$ is the position vector referred to the point of projection as origin, (1) integrates to give

$$\mathbf{v} = -gt\hat{\mathbf{z}} - k\mathbf{r} + \mathbf{v}_0, \tag{5}$$

where $\mathbf{v}_0$ is the initial velocity.

Let $\mathbf{i}$ be a unit horizontal vector. Then the horizontal displacement

$$x = \mathbf{i}.\mathbf{r} = \mathbf{i}.(\mathbf{v}_0 - \mathbf{v})/k \tag{6}$$

from (5). Hence as $t \to \infty$, we find from (4) and (6) the limiting horizontal range given by

$$\lim_{t \to \infty} \mathbf{i}.\mathbf{r} = \mathbf{i}.\mathbf{v}_0/k. \tag{7}$$

Equation (5) can be written

$$\frac{d}{dt}(\mathbf{r}e^{kt}) = -gt\,e^{kt}\hat{\mathbf{z}} + \mathbf{v}_0 e^{kt}$$

which integrates to give

$$\mathbf{r} = \frac{\mathbf{v}_0}{k}(1 - e^{-kt}) + \frac{g}{k^2}\hat{\mathbf{z}}(1 - kt - e^{-kt}) \tag{8}$$

$$\Rightarrow \mathbf{v} = \frac{d\mathbf{r}}{dt} = \mathbf{v}_0 e^{-kt} - \frac{g}{k}\hat{\mathbf{z}}(1 - e^{-kt}). \tag{9}$$

The time of flight T is given by

$$\mathbf{v}.\mathbf{z} = \mathbf{v}_0.\hat{\mathbf{z}}e^{-kT} - \frac{g}{k}\hat{\mathbf{z}}^2(1 - e^{-kT}) = 0$$

$$\Rightarrow T = \frac{1}{k}\ln\left[1 + \frac{k}{g}(\mathbf{v}_0.\hat{\mathbf{z}})\right]. \tag{10}$$

Exercise 9.10

1. A projectile moves under gravity in a medium whose resistance per unit mass is $k\mathbf{v}$, where $\mathbf{v}$ is the velocity and k a constant. If the velocity of projection has horizontal and vertical components U and V respectively, prove that the highest point of the trajectory is at a horizontal distance $UV/(g + kV)$ from the point of projection.

2. A particle is projected under gravity with speed V at an inclination α to the horizontal in a medium

whose resistance per unit mass is k times the speed. Show that the particle is moving in a direction at right angles to the direction of projection after a time

$$\frac{1}{k}\ln\left(1+\frac{Vk}{g\sin\alpha}\right).$$

3. A particle moves under gravity g in a medium that produces a retardation equal to k times the velocity. The particle is projected with speed u at angular elevation α and its path is observed to meet the horizontal plane through the point of projection at an angle β after time τ/k. Show that β satisfies the equation

$$\tan\alpha + \tan\beta = \frac{g}{ku}\sec\alpha(e^{\tau}-1).$$

4. A particle of mass m is projected with a velocity of magnitude u inclined to the horizontal at an angle α, in a medium in which the horizontal component of resistance is mkv, where v is the horizontal component of the velocity of the particle, and in which the vertical component of the resistance is negligible compared with gravity. Show that the range of the particle on the horizontal plane through the point of projection is

$$\frac{u\cos\alpha}{k}\left(1-e^{-2ku\sin\alpha/g}\right).$$

5. A particle is projected in a vertical plane in a medium the resistance of which to the motion of the particle varies as the speed of the particle and is equal to n times the weight of the particle when the speed of the particle is V. The initial horizontal and vertical components of the velocity of the particle are U, V respectively. Show that the particle reaches its maximum height after a time

$$\frac{V}{ng}\ln(n+1)$$

from the instant of projection and that the horizontal displacement of the particle from the point of projection is then

$$\frac{UV}{(n+1)g}. \qquad \text{(L.)}$$

6. A particle of mass m is projected under gravity with horizontal and vertical components of velocity U and V respectively. Motion takes place in a medium which only produces a horizontal resistance of magnitude mku, where k is a constant and u is the horizontal component of the velocity of the particle. Show that R, the horizontal range from the point of projection, is

$$\frac{U(1-e^{-kT})}{k},$$

where $T = 2V/g$.

Show that the horizontal distance D travelled by the particle before reaching the highest point of its path is

$$\frac{U(1-e^{-\frac{1}{2}kT})}{k} \quad \text{and that } D > \tfrac{1}{2}R. \qquad \text{(L.)}$$

7. A particle of mass m is projected with speed u at an angle of elevation α and moves under gravity in a medium which offers a resistance of magnitude nmg, where n is a positive constant, in a direction opposite to the velocity. Write down the tangential and normal equations of motion, and deduce that, when the particle is travelling with speed v in a direction inclined at ψ to the horizontal,

$$\frac{dv}{d\psi} = v(n\sec\psi + \tan\psi).$$

Hence, or otherwise, show that

$$\frac{u}{v} = \frac{\cos^{n+1}\psi}{\cos^{n+1}\alpha} \cdot \frac{(1+\sin\alpha)^n}{(1+\sin\psi)^n}. \qquad \text{(L.)}$$

Miscellaneous Exercise 9

1. A particle of mass m moves under the action of a variable force $\mathbf{F}$ so that its positive vector $\mathbf{r}$ at time $t\,(\geqslant 0)$ is given by

$$\mathbf{r} = 2t^3\mathbf{i} - 6t^2\mathbf{j} - 15t\,\mathbf{k}.$$

Find, in terms of t, the component of the velocity of the particle in the direction of the vector $\mathbf{a}$, where

$$\mathbf{a} = \mathbf{i} - 2\mathbf{j} + 2\mathbf{k}.$$

Find also the vector moment of $\mathbf{F}$ about the origin at the instant when this component of the velocity of the particle is zero. (N.)

2. The position vector of a particle of unit mass at time t is given by

$$\mathbf{r} = \mathbf{i}e^{-t}\sin t + \mathbf{j}e^{-t}\cos t.$$

If the particle is moving under the action of a force $\mathbf{F}$ and of a frictional resisting force given by $-2\dot{\mathbf{r}}$, show that $\mathbf{F} = \lambda \mathbf{r}$ where λ is a scalar constant.

Evaluate $\ddot{\mathbf{r}}\cdot\dot{\mathbf{r}}$, and show that the component of acceleration in the direction of motion at any instant is the negative of the speed at that instant. (N.)

3. A particle P of mass m moves in the plane of rectangular axes Ox, Oy under the influence of a force $\mathbf{F}$ whose components when the particle is at the point (x,y) are $-mn^2x$, $-mn^2y$. At time $t = 0$ the particle passes through the point $(a,0)$ with a velocity whose components are $0, u$. At the instant, after $t=0$, when P first crosses the y-axis it is struck by a particle of equal mass moving with a velocity whose components are $0, v$. The particles coalesce and the composite particle continues to move under the influence of a force $2\mathbf{F}$. Prove that the equation of the path of the composite particle is

$$4u^2x^2 + a^2n^2\left(y + \frac{vx}{an}\right)^2 = a^2u^2.$$ (N.)

4. Prove that, when a particle moves in a plane, the radius of curvature, ρ, of its path is given by

$$\rho = |\mathbf{v}|^3/|\mathbf{v}\times\mathbf{f}|,$$

where $\mathbf{f}$, $\mathbf{v}$ are the acceleration and velocity vectors respectively.

5. Show that, for motion in a circle with constant speed about the point $\mathbf{r} = \mathbf{a}$,

$$\ddot{\mathbf{r}} = c(\mathbf{a}-\mathbf{r}),$$

where c is a scalar constant.

A heavy particle P of mass m is attached to the end of an elastic string of natural length a and modulus mg. This string passes through a fixed smooth ring at O and its other end is fixed to a point A at a distance a from O. The particle is free to move in a vertical plane. If $\mathbf{r}$ is the vector OP, show that $\ddot{\mathbf{r}} + k\mathbf{r} = \mathbf{g}$, where k is a scalar constant to be determined.

Show that it is possible for the particle to take part in circular motion with constant speed. Find the position of the centre of the circle and the initial velocity of P for such motion if it starts from O. (O.C.M.E.I.)

6. A particle of mass m has position vector $\mathbf{r}$ relative to a fixed point O. It is acted upon by a force $\lambda\mathbf{k}\times\dot{\mathbf{r}} - \mu\mathbf{r}$, where λ and μ are positive constants, and $\mathbf{k}$ is one of a right-handed set, $\mathbf{i}$, $\mathbf{j}$, $\mathbf{k}$, of mutually orthogonal unit vectors. Write down the vector equation of motion of the particle.

Show that if $\mathbf{r} = x\mathbf{i} + y\mathbf{j} + z\mathbf{k}$ then z satisfies the equation $m\ddot{z} + \mu z = 0$. Deduce that if $\mathbf{r}$ and $\dot{\mathbf{r}}$ are initially perpendicular to $\mathbf{k}$ the particle moves in a plane.

Show also that the vector equation of motion is satisfied by $\mathbf{r} = a(\mathbf{i}\cos\omega t + \mathbf{j}\sin\omega t)$ where a is an arbitrary constant and ω is either of the two real roots of a certain quadratic equation. Deduce that the particle can describe a circle about O under this force with either of two constant angular speeds. (N.)

7. The position vector $\mathbf{r}$ of a particle of mass m is given, at time t, by

$$\mathbf{r} = (a\cos nt)\mathbf{i} + (b\sin nt)\mathbf{j} + (ct)\mathbf{k},$$

where n, a, b and c are constants, $(a\neq b)$. Find the force on the particle at time t and the rate at which this force is then doing work. Deduce that the work done by the force in the period $t = 0$ to $t = \pi/2n$ is $\frac{1}{2}(a^2 - b^2)mn^2$. (N.)

8. At time t a particle is in motion with velocity **v** and is being acted upon by a variable force **F**. Write down expressions for (i) the power at time t, (ii) the work done by **F** during the time interval $0 \leqslant t \leqslant T$.

The particle, of mass m, moves in a plane where **i** and **j** are perpendicular unit vectors so that its position vector at time t is given by

$$\mathbf{r} = 2a \cos 2t\, \mathbf{i} + a \sin 2t\, \mathbf{j},$$

where a is a positive constant. Derive expressions for the velocity **v** and the force **F** at time t.

Obtain an expression in terms of t for the power at time t and show that the work done by **F** during the interval $0 \leqslant t \leqslant T$ is $3ma^2(1 - \cos 4T)$.

If T varies, find the maximum value of the work done by **F** and determine also the smallest value of T for which this maximum value is reached. (N.)

9. The position vector at time t of a particle of mass m moving under the action of a variable force **F** is given by

$$\mathbf{r} = a \cos nt\, \mathbf{i} + a \sin nt\, \mathbf{j} + bt\, \mathbf{k}$$

where a, b, n are positive constants. Find the angle between the velocity vector and the acceleration vector at time t.

Show that if P is the position of the particle and Q is the point whose position vector is $bt\mathbf{k}$, then the force **F** on the particle at time t is proportional to $\overrightarrow{PQ}$, the constant of proportionality being independent of t.

Show that the vector $\overrightarrow{PQ}$ is of constant length and rotates parallel to the plane of **i** and **j** with a constant angular speed. (N.)

10. A particle of mass m is projected from a point O with speed u at an angle α above the horizontal in a medium whose resistance to the motion of the particle is of magnitude mk times the speed, where k is a positive constant.

At time t after projection the particle has coordinates (x, y) referred to axes Ox, horizontal, and Oy, vertically upwards. Obtain equations of motion in the form

$$\frac{d^2x}{dt^2} + k\frac{dx}{dt} = 0,$$

$$\frac{d^2y}{dt^2} + k\frac{dy}{dt} = -g.$$

Find x and y at time t and show that the particle is at its highest point when

$$x = u^2 \sin \alpha \cos \alpha / (ku \sin \alpha + g). \tag{N.}$$

11. A particle resting on a smooth horizontal table is connected to a fixed point of the table by an elastic string of unstretched length a. The particle is projected horizontally with velocity $a\omega$ in a direction perpendicular to the string when the string is just taut. If in the subsequent motion the distance of P from O attains the value $2a$, find the angular speed of the string when $OP = 2a$. (L.)

12. A particle is projected with speed u from a point A on the top of a cliff of height h above sea level. The particle strikes the sea at a point B which is at a horizontal distance s from A. Show that, for s to have its largest possible value, the angle of elevation α at which the particle should be projected is given by $\tan \alpha = u^2/(gs)$. Show also that the maximum value of s is

$$\frac{u\sqrt{(u^2 + 2gh)}}{g}. \tag{L.}$$

13. At time t the position vector of a particle is given by

$$\mathbf{r} = a \cos(t^2)\mathbf{i} + a \sin(t^2)\mathbf{j},$$

where a is a constant. Show that the rate of change with time of the speed of the particle is constant.

Show also that the scalar product of the velocity vector and the acceleration vector equals $4a^2 t$, and express the tangent of the angle between these vectors in terms of t. (L.)

14. Two identical smooth spheres are moving on a horizontal table with velocity vectors $3\mathbf{i} + 4\mathbf{j}$ and $-\mathbf{i} + \mathbf{j}$

and collide when the line joining their centres is parallel to the vector **i**. If the coefficient of restitution between the spheres is $\frac{1}{2}$, find the velocity vectors of the spheres after impact. Find also the ratio of the magnitudes of the velocities, before and after impact, of the spheres relative to each other. If, at this instant of impact, the centres of the spheres are 2 units of distance apart, find the distance between their centres 1 unit of time later.

(L.)

15. A small smooth sphere S moving on a horizontal table with speed V impinges on an identical sphere T which is at rest on the table. Before impact the direction of motion of S makes an angle β with the line of centres at the moment of impact. The coefficient of restitution between the spheres is e. Find the components after impact of the velocities of the spheres along and perpendicular to the line of centres at the moment of impact. Show that the direction of motion of S deviates through an angle θ where

$$\tan\theta = \frac{(1+e)\tan\beta}{2\tan^2\beta + 1 - e}.$$

(L.)

16. A, B, C, D are the corners of a smooth horizontal rectangular table, bounded along all four edges by a smooth vertical rim. From the mid-point of AB a particle is projected along the surface of the table in a direction making an angle α with AB, to strike in turn the rims BC, CD, DA. If the sides AB, BC are of length a, b, respectively, and if e is the coefficient of restitution between the particle and the rim, show that the particle will return to its starting point if

$$\tan\alpha = \frac{2eb}{a(1+e)},$$

and that if this condition is satisfied the particle will always pursue the same path.

17. O is a point on a smooth plane inclined at an angle α to the horizontal. A ball is projected from O with speed V in a direction inclined at an angle β to the plane and $\alpha + \beta$ to the horizontal. The ball arrives at O again on the nth bounce. Prove, by treating separately the motions along and perpendicular to the plane, that

$$\cot\alpha\cot\beta = \frac{1-e^n}{1-e},$$

where e is the coefficient of restitution.

(O.C.S.M.P)

18. Two equal particles A, B are attached to the ends of a light inextensible string of length $3a$ which passes through a small smooth ring fixed at the point O on a smooth horizontal table. When AO, OB are straight and AOB is a right angle with $OA = a, OB = 2a$, A is given a velocity V parallel to $\overrightarrow{BO}$. Show that B reaches O with speed $2V/3$ after a time $4a/V$.

(L.)

19. A small ring A of mass m is free to slide on a fixed smooth horizontal wire. A light inextensible string of length $2a$ has one end attached to A and the other end to a particle B of mass $2m$. Initially B is supported close to the wire at a distance $a\sqrt{3}$ from A. If B is allowed to fall freely, find the impulsive tension in the string when it becomes taut

(i) if A is fixed,

(ii) if A is free to slide along the wire.

(L.)

20. A particle is projected with initial velocity $u\mathbf{i} + v\mathbf{j}$, $\mathbf{i}$ and $\mathbf{j}$ being unit vectors in the horizontal and upward vertical directions respectively, and moves with constant gravitational acceleration of magnitude g. Establish the formulae

$$\dot{\mathbf{r}} = u\mathbf{i} + (v - gt)\mathbf{j}$$

and

$$\mathbf{r} = ut\mathbf{i} + (vt - \tfrac{1}{2}gt^2)\mathbf{j}$$

for the velocity and position vectors of the particle at time t.

The particle is projected from the origin with initial velocity $28\mathbf{i} + 100\mathbf{j}$ towards an inclined plane whose line of greatest slope has the equation $\mathbf{r} = 480\mathbf{i} + \lambda(2\mathbf{i} + \mathbf{j})$ where λ is a parameter. Show that the particle strikes the plane after 20 seconds and determine the distance along the line of greatest slope from the point where $\lambda = 0$ to the point of impact.

Show also that α, the acute angle between the direction of motion at impact and the inclined plane, is given by $5\sqrt{5}\cos\alpha = 2$.

[Velocity components are measured in m s^{-1} and displacements in m. Take g as 9·8 m s^{-2}.] (N.)

21. Two particles A_1, A_2 of masses m_1, m_2 respectively can move on a smooth horizontal table, and they attract each other with a mutual attraction acting along the line joining the particles. If $\mathbf{R}$ is the force exerted by A_1 on A_2, and $\mathbf{r}$ is the position vector of A_2 relative to A_1, show that

$$\mu\ddot{\mathbf{r}} = \mathbf{R} \quad \text{where} \quad \mu = m_1 m_2/(m_1 + m_2).$$

The magnitude of the attractive force is $\gamma m_1 m_2/x^2$ when the particles are at a distance x apart, where γ is a constant. Initially the particles are at a distance a apart and particle A_2 is moving away from A_1 with relative velocity v along the line $A_1 A_2$. If

$$v^2 < 2\gamma(m_1 + m_2)/a$$

find the distance between the particles when both of them are moving with the same velocity. If A_1 is initially at rest find this common velocity. (N.)

22. A system of n particles is confined to a plane Oxy, where Ox, Oy, Oz form a right-handed system of mutually perpendicular axes with origin O. The ith particle has mass m_i, position vector $\mathbf{r}_i$ (magnitude r_i) and is acted on by an external force $\mathbf{F}_i$ acting in the plane Oxy ($i = 1, 2, \ldots, n$). In addition there are internal interactions between the particles, the force of the jth particle on the ith being $\mathbf{F}_{ij}$ ($j = 1, 2, \ldots, n$), and $\mathbf{F}_{ii}$ is taken to be zero.

(a) Show that the sum of the moments of the two forces $\mathbf{F}_{ij}$ and $\mathbf{F}_{ji}$ about O is zero.
(b) State what can be deduced about the sum of the moments of all the internal forces about O.
(c) If $\mathbf{L}$ is the sum of the moments of all the external forces about O, show that

$$\mathbf{L} = \frac{d}{dt}\left[\sum_i m_i \mathbf{r}_i \times \dot{\mathbf{r}}_i\right].$$

Consider the particular case for which every particle is moving in a circle, centre O, with the same, variable, angular speed ω in the sense Ox to Oy. Show that in this case
(d) at any instant $|\dot{\mathbf{r}}_i| = \omega r_i$,
(e) $|\mathbf{r}_i \times \dot{\mathbf{r}}_i| = r_i^2 \omega$,
(f) $\mathbf{L} = I\dot{\omega}\mathbf{k}$ where I is the moment of inertia of the system about Oz and $\mathbf{k}$ is a unit vector along Oz.
(N.)

23. A particle P rests at the highest point of a smooth fixed sphere of radius a and centre O. The particle P is given a horizontal velocity of magnitude u, where $u = \sqrt{(ag/2)}$, and slides on the outer surface of the sphere. Show that P leaves the sphere when OP makes an angle θ with the upward vertical, where $\cos\theta = 5/6$. Find the speed of P at this instant. (L.)

24. A particle describes the curve $r = 3e^\theta$ so that the radial velocity of the particle when it is at a distance r from the pole O is $2/r$. Show that the acceleration of the particle is $8/r^3$ directed towards O. (L.)

25. A particle moving on a smooth horizontal plane describes the curve $r = 2a/(2 - \sin\theta)$ under the action of a force directed towards the point $r = 0$. When $r = 2a$, the speed of the particle is V. Show that the value of $r^2 d\theta/dt$ is constant and equal to $2aV$.

Show also that
(a) $dr/dt = V\cos\theta$,
(b) $d^2r/dt^2 = -(2aV^2\sin\theta)/r^2$.
Find the acceleration of the particle when its speed is a maximum. (L.)

26. (i) At time t seconds, the position vectors $\mathbf{r}_1$, $\mathbf{r}_2$ of two particles A_1, A_2 are given by

$$\mathbf{r}_1 = 2t\mathbf{i} + (3t^2 - 4t)\mathbf{j} - t^3\mathbf{k},$$
$$\mathbf{r}_2 = t^3\mathbf{i} - 2t\mathbf{j} + (2t^2 - 1)\mathbf{k}$$

respectively. Find the velocity and acceleration of A_2 relative to A_1 when $t = 2$.

(ii) The acceleration $\mathbf{a}$ of a particle P at time t seconds is given by

$$\mathbf{a} = 2e^{-t}\mathbf{i} + (5\cos t)\mathbf{j} + (3\sin t)\mathbf{k}.$$

Given that, at time $t = 0$, P is at the point with position vector $(\mathbf{i}+\mathbf{j}+\mathbf{k})$ and has velocity $(\mathbf{i}+3\mathbf{j}+\mathbf{k})$, find the velocity and position vector of P at time t seconds. (L.)

27. A small bead of mass m slides on a smooth wire in such a way that its position vector at time t relative to a fixed origin O is

$$\mathbf{r} = at^2\mathbf{i} + (at^2 - ut)\mathbf{j} + (at^2 + ut)\mathbf{k}$$

where a and u are constant scalars.

In addition to the reaction of the wire the motion is maintained by a constant force $ma(4\mathbf{i}+\mathbf{j}+\mathbf{k})$ acting on the bead.

Verify that the reaction of the wire on the bead is perpendicular to the direction of motion of the bead.
(N.)

28. A particle describes the curve $r(3 + \cos\theta) = 2a$ in such a way that the transverse component of its acceleration is always zero. Given that its maximum speed is u, find its minimum speed.

Show that the point (r, θ) its radial velocity is $(u\sin\theta)/4$ and its radial acceleration is inversely proportional to r^2. (L.)

29. (i) At time t the velocity $\mathbf{v}$ of a particle P satisfies the equation $d\mathbf{v}/dt + 2\mathbf{v} = \mathbf{0}$. At time $t = 0$ the position vector and velocity of P are $(\mathbf{i}+\mathbf{j})$ and $4(\mathbf{j}-\mathbf{i})$ respectively.

Show that at time t the position vector of P is

$$\mathbf{r} = (-\mathbf{i}+3\mathbf{j}) + 2(\mathbf{i}-\mathbf{j})e^{-2t}.$$

(ii) A particle moves in the $x - y$ plane such that its position vector $\mathbf{r}$ satisfies the differential equation

$$\frac{d^2\mathbf{r}}{dt^2} + 4\mathbf{r} = \mathbf{0}.$$

Given that $\mathbf{r} = 3\mathbf{i}$ when $t = 0$ and $\mathbf{r} = \mathbf{j}$ when $t = \pi/4$, show that the particle describes the ellipse

$$x^2 + 9y^2 = 9.$$ (L.)

30. (i) A particle moves so that at time t its position vector is given by

$$\mathbf{r} = \mathbf{i}e^{\omega t}\cos(\omega t) + \mathbf{j}e^{\omega t}\sin(\omega t),$$

where ω is a constant.

Find the magnitudes of the velocity and acceleration vectors, and show that the acceleration vector is always perpendicular to the radius vector.

(ii) A particle moving on the locus $r = a(1 - \sin\theta)$ is at the point (r, θ) at time t. If $d\theta/dt = k$, where k is constant, prove that the radial component of the acceleration is

$$ak^2(2\sin\theta - 1).$$

Determine the transverse component of the acceleration when $\theta = 4\pi/3$. (L.)

31. Particles A and B, of masses $11m$ and $5m$ respectively, are joined by a light inextensible string AB. The system is placed on a smooth horizontal table with the string just taut. A horizontal impulse is given to particle B so that B moves with initial speed V in a direction making an angle θ with AB produced, where $\tan\theta = 4/3$. Calculate the magnitude and direction of the impulse applied to B and the total kinetic energy generated by the impulse. (L.)

32. (i) Solve the differential equation $d\mathbf{r}/dt = 4\mathbf{r}$, given that, when $t = 0$, $\mathbf{r}.\mathbf{i} = 1$ and $\mathbf{r} \times \mathbf{i} = \mathbf{j} + \mathbf{k}$.

(ii) Find the solution of the differential equation

$$d^2\mathbf{r}/dt^2 + 2 d\mathbf{r}/dt + 2\mathbf{r} = \mathbf{0}$$

such that $\mathbf{r} = \mathbf{i} + \mathbf{k}$ when $t = 0$, and $d\mathbf{r}/dt = \mathbf{0}$ when $t = \pi/2$.

(iii) At time t the position vector $\mathbf{r}$ of the point P satisfies the differential equation

$$d^2\mathbf{r}/dt^2 + 4\mathbf{r} = 3\mathbf{i}\sin t.$$

When $t = 0$, P passes through the point $\mathbf{r} = \mathbf{j}$ with velocity $\mathbf{i}$. Find the equation of the locus of P. (L.)

33. A particle of mass m is attached to one end of an inelastic string of length a. The other end of the string is attached to a fixed point A. The particle hangs in equilibrium under gravity. A horizontal impulse J is applied to the particle. Prove that, if the particle just comes momentarily to rest when the string is horizontal, then $J = m\sqrt{(2ga)}$.

If, instead, $J = m\sqrt{(3ga)}$, find the tension in the string when it is horizontal. Find also the height above the level of A to which the particle rises before the string first becomes slack. (L.)

34. The position vector $\mathbf{r}$ of a particle satisfies the equation

$$\ddot{\mathbf{r}} = \mathbf{E} + \dot{\mathbf{r}} \times \mathbf{H},$$

with $\mathbf{r} = \mathbf{0}$, $\dot{\mathbf{r}} = \lambda\mathbf{H}$ at $t = 0$, where $\mathbf{E}$ and $\mathbf{H}$ are constant vectors and λ is a constant scalar. Obtain

(i) an expression for $\dot{\mathbf{r}}^2$ in terms of λ, $\mathbf{E}$, $\mathbf{H}$ and $\mathbf{r}$,

(ii) an expression for $\dot{\mathbf{r}}$ in terms of λ, $\mathbf{E}$, $\mathbf{H}$, $\mathbf{r}$ and t.

Further, if $\mathbf{E}.\mathbf{H} = 0$, show that $\mathbf{r}.\mathbf{H} = \lambda H^2 t$ and deduce that

$$\dot{\mathbf{r}} = \mathbf{r} \times \mathbf{H} + \frac{(\mathbf{r}.\mathbf{H})}{\lambda H^2}\mathbf{E} + \lambda\mathbf{H}.$$ (L.)

10 An Introduction to the Dynamics of a Rigid Body

10:1 ROTATION OF A LAMINA ABOUT A FIXED AXIS

Consider a rigid lamina free to rotate in its own plane about a fixed axis perpendicular to this plane, Fig. 10.1. A particle P_i of the lamina has mass m_i and

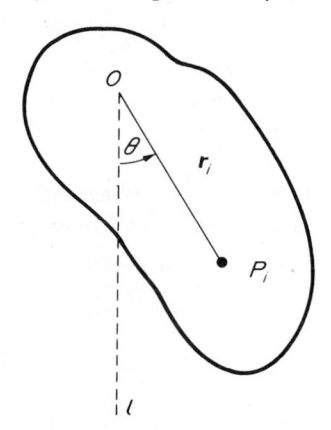

FIG. 10.1.

position vector $\mathbf{r}_i$ relative to the point O where the axis intersects the plane of the lamina. As before, §9:8, let $\mathbf{F}_i$, $\mathbf{F}'_i$ respectively be the resultant external and internal forces acting on P_i. Then, as in §9:8, the equation of motion of P_i is

$$\mathbf{F}_i + \mathbf{F}'_i = m_i \ddot{\mathbf{r}}_i$$
$$\Rightarrow \sum \mathbf{r}_i \times \mathbf{F}_i + \sum \mathbf{r}_i \times \mathbf{F}'_i = \sum \mathbf{r}_i \times m_i \ddot{\mathbf{r}}_i.$$

Now $\sum \mathbf{r}_i \times \mathbf{F}_i$ is the sum of moments about O of the external forces acting on the lamina and reduces to a couple, $L\hat{\mathbf{n}}$ say, where $\hat{\mathbf{n}}$ is a unit vector directed along the axis of rotation. Further, since P_i moves in a circle centre O and radius $|\mathbf{r}_i|$, $\mathbf{r}_i \times \ddot{\mathbf{r}}_i$ is perpendicular to $\mathbf{r}_i$ and is of magnitude $|\mathbf{r}_i|\ddot{\theta}$. Hence

$$\sum \mathbf{r}_i \times m_i \ddot{\mathbf{r}}_i = \left(\sum m_i r_i^2\right) \ddot{\theta} \hat{\mathbf{n}} = I_0 \ddot{\theta} \hat{\mathbf{n}},$$

where $r_i = |\mathbf{r}_i|$ and I_0 is the moment of inertia of the lamina about the axis. [Note that, in

Fig. 10.1, Ol is a fixed line in space.] We therefore obtain an equation of angular motion

$$L = I_0 \ddot{\theta} \qquad (10.1)$$

corresponding to the equation $P = mf$ for linear motion.

Example 1. A wheel and axle can rotate freely about a fixed horizontal axis. A light cord, wound round the axle which is of radius r, carries a particle A of mass M hanging at its free end. If, when the system is allowed to move from rest, A moves through a distance H in a time t_o, show that the moment of inertia of the wheel and axle about its axis is

$$\frac{Mr^2}{2H}(gt_0^2 - 2H). \qquad (L.)$$

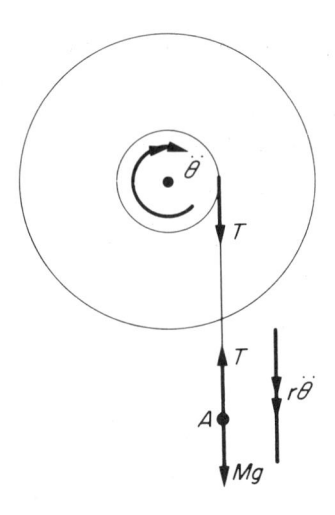

Fig. 10.2.

Figure 10.2 shows the forces acting on the wheel and axle and the forces acting on A. We denote the angular acceleration of the wheel and axle by $\ddot{\theta}$. Since the string does not slip on the axle the linear acceleration of A is equal to the linear acceleration of a point on the circumference of the axle, i.e. $r\ddot{\theta}$.

The equation of motion for A is

$$Mg - T = Mr\ddot{\theta}.$$

The equation of motion for the wheel and axle is

$$Tr = I\ddot{\theta},$$

where I is the M. of I. of the wheel and axle.

From these equations, eliminating T, $\ddot{\theta} = Mgr/(I + Mr^2)$. Since $\ddot{\theta}$ is constant and A moves through a distance H in time t_0,

$$H = \frac{\frac{1}{2}Mgr^2}{I + Mr^2} t_0^2$$

$$\Rightarrow IH = \tfrac{1}{2}Mgr^2 t_0^2 - HMr^2$$

$$\Rightarrow I = \frac{Mr^2}{2H}(gt_0^2 - 2H).$$

Example 2. A uniform circular disc, of radius a and mass m, is free to rotate about a smooth fixed horizontal axis through its centre O and perpendicular to its plane. Over the rim of the disc hangs a light string which carries particles A_1, A_2, of masses m and $2m$ respectively, at its free ends. The system is released from rest with the string taut and the hanging parts vertical. Assuming that the string does not slip on the disc, find the accelerations of the particles and the tensions in the vertical parts of the string during the subsequent motion.

Show that to prevent the string slipping on the disc, the coefficient of friction μ between the string and the disc must not be less than $[\ln(10/9)]/\pi$.

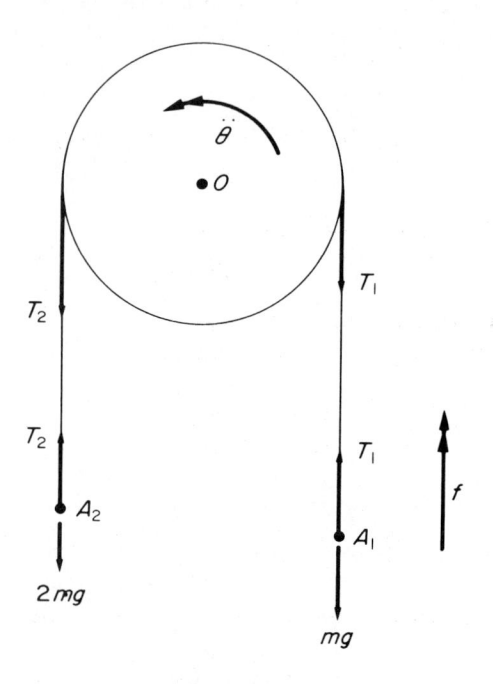

FIG. 10.3.

Let the tensions in the free (vertical) portions of the strings be T_1, T_2 as shown in Fig. 10.3. If the linear acceleration of either particle is of magnitude f, and the angular acceleration of the disc is $\ddot{\theta}$, then $f = a\ddot{\theta}$. The equations of motion for the particles A_1, A_2 and the disc are respectively:

$$T_1 - mg = mf,$$
$$2mg - T_2 = 2mf,$$
$$(T_2 - T_1)a = \tfrac{1}{2}ma^2\ddot{\theta} = \tfrac{1}{2}maf$$
$$\Rightarrow f = 2g/7, \quad T_1 = 9mg/7, \quad T_2 = 10mg/7. \tag{1}$$

Using the result (7.18), the ratio T_2/T_1 cannot exceed $e^{\mu\pi}$, where μ is the coefficient of friction between the string and the disc,

$$\Rightarrow e^{\mu\pi} \geqslant 10/9$$
$$\Rightarrow \mu \geqslant \frac{1}{\pi} \ln(10/9)$$

for no slipping.

10:2 MOMENTUM AND ENERGY EQUATIONS FOR ANGULAR MOTION OF A LAMINA

If, at time t, the angular speed of the lamina is $\dot{\theta} = \omega$, then the equation of angular motion $L = I_0 \ddot{\theta}$ can be written

$$L = I_0 \frac{d\omega}{dt}. \tag{10.2}$$

If $\omega = \omega_0$ when $t = 0$, integration of equation (10.2) gives

$$\int_0^t L\,dt = \int_{\omega_0}^{\omega} I_0\,d\omega = I_0(\omega - \omega_0). \tag{10.3}$$

When L is constant, equation (10.3) becomes

$$Lt = I_0(\omega - \omega_0). \tag{10.3a}$$

The result (10.3a) implies that Lt equals the gain in the *moment of momentum* $I_0 \omega$ of the lamina about O. [This is a special case of the result (9.50).]

Again,

$$L = I_0 \frac{d\omega}{dt} = I_0 \frac{d\omega}{d\theta}\frac{d\theta}{dt} = I_0 \omega \frac{d\omega}{d\theta}$$

$$\Rightarrow \int_{\theta_0}^{\theta} L\,d\theta = \int_{\omega_0}^{\omega} I_0 \omega\,d\omega = \tfrac{1}{2} I_0(\omega^2 - \omega_0^2). \tag{10.4}$$

Note that, since the speed of particle P_i is $r_i \omega$, the kinetic energy of the lamina is

$$T = \sum \tfrac{1}{2} m_i r_i^2 \omega^2 = \tfrac{1}{2} \left(\sum m_i r_i^2 \right) \omega^2 = \tfrac{1}{2} I_0 \omega^2. \tag{10.5}$$

Hence result (10.4) implies that the work done on the lamina is equal to the gain in the (angular) kinetic energy of the lamina.

All the results of this and the preceding section may be extended to the motion of a three-dimensional rigid body about a fixed axis by summation over the sections of the body by planes perpendicular to the fixed axis. Further in § 10:8, these results will be used in conjunction with those of § 9:8 to discuss the general two-dimensional motion of a rigid body.

Example 1. A uniform circular disc of mass m and radius a can turn freely in a vertical plane about a horizontal axis through a point O on its rim. A particle P of mass m is attached to the point of the rim diametrically opposite to O. The system is disturbed from rest with the particle vertically above O.
Show that the angular speed of the disc after it has turned through an angle θ is given by

$$11a\dot{\theta}^2 = 12g(1 - \cos\theta).$$

Find the magnitude of the force exerted by the disc on the particle when $\theta = \pi$. (L.)

The combined M. of I. of the disc and particle P about the axis of rotation is

$$\left(\frac{ma^2}{2} + ma^2\right) + m(2a)^2 = \frac{11ma^2}{2}.$$

The energy equation for the system when it has turned through an angle θ from its initial position is therefore

$$\frac{1}{2}\frac{11ma^2}{2}\dot\theta^2 = 3mga(1-\cos\theta),$$

where potential energy is measured from a horizontal plane through O as origin,

$$\Leftrightarrow 11a\dot\theta^2 = 12g(1-\cos\theta). \tag{1}$$

The particle rotates about O in a circle of radius $2a$ and has, therefore, components of acceleration $2a\dot\theta^2$ towards O and $2a\ddot\theta$ perpendicular to OP. The acceleration of the particle is provided by the resultant of the force which the disc exerts on the particle and the weight of the particle. Differentiation of (1) with respect to t

$$\Rightarrow 11a\ddot\theta = 6g\sin\theta.$$

Therefore, when $\theta = \pi$, $\dot\theta^2 = 24g/(11a)$, $\ddot\theta = 0$ and the force exerted by the disc on the particle is a vertically upward force of magnitude R, where $R - mg = 2ma\dot\theta^2$

$$\Rightarrow R = 2ma\dot\theta^2 + mg = \frac{59mg}{11}.$$

Example 2. A light inextensible string passes round the rim of a rough pulley, which is supported with its axis horizontal, and carries particles A_1, A_2 of mass 5 kg and 4·5 kg respectively at its ends. It is found that A_1 descends a distance of 1·25 m from rest in 2·5 seconds. Assuming that the pulley is a uniform disc of radius 0·1 m and mass 2 kg, find the frictional couple, assumed constant, acting on the pulley at the supports.

If the string and particles are removed and the pulley is made to rotate at the rate of 4 revolutions per second, find how many revolutions it will make before coming to rest, assuming that the frictional couple is now reduced to 1/8 of its value when the particles are in motion. [Take $g = 10$ ms^{-2}.]

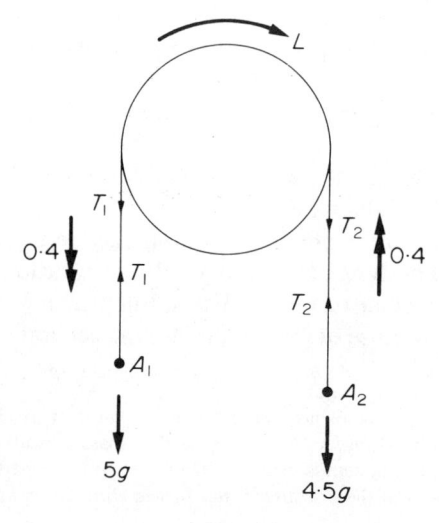

FIG. 10.4.

With reference to Fig. 10.4, let the frictional couple on the pulley be L Nm (or J) and the tensions in the vertical hanging portions be as shown.

The particles move 1·25 m from rest in 2·5 seconds. Hence their acceleration is of magnitude $(1\cdot25)/[\tfrac{1}{2}\times(2\cdot5)^2] = 0\cdot4$ ms^{-2}. The angular acceleration of the disc is therefore $(0\cdot4)/(0\cdot1) = 4$ rad s^{-2}. The

equations of motion are:
$$5g - T_1 = 5 \times 0.4,$$
$$T_2 - 4.5g = 4.5 \times 0.4,$$
$$(T_1 - T_2) \times 0.1 - L = \tfrac{1}{2} \times 2 \times (0.1)^2 \times 4$$
$$\Rightarrow L = 0.08.$$

The frictional couple is 0·08 Nm. When the frictional couple is reduced to 0·01 Nm, bringing the pulley to rest in n revolutions from an angular speed of 4 rev s^{-1} ($= 8\pi$ rad s^{-1}), the energy equation gives
$$0.01 \times 2n\pi = \tfrac{1}{2} \times \tfrac{1}{2} \times 2 \times (0.1)^2 \times (8\pi)^2$$
$$\Rightarrow n = 16\pi.$$

The pulley makes 16π revolutions before coming to rest.

Example 3. A uniform circular disc of mass M and radius a is free to rotate about a horizontal axis through its centre normal to its plane. A particle P of mass m is placed on the rim of the disc at its highest point, the coefficient of friction between P and disc being μ. The system is slightly disturbed from rest, and θ is the angle turned through by the disc at time t. Show that, so long as there is no relative motion between P and the disc,

$$(M + 2m)a(d\theta/dt)^2 = 4mg(1 - \cos\theta)$$
and
$$(M + 2m)a(d^2\theta/dt^2) = 2mg\sin\theta.$$

Find the radial and tangential components of the force exerted on P by the disc. Deduce that P will slip before it loses contact with the disc and show that slipping will occur when

$$M\sin\theta = \mu\{(M + 6m)\cos\theta - 4m\}. \tag{N.}$$

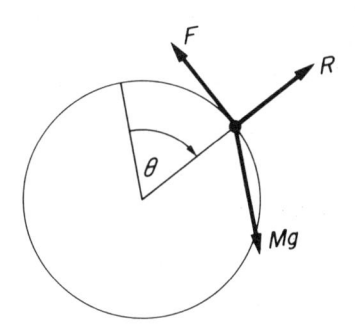

Fig. 10.5.

Figure 10.5 shows the forces acting on the particle when the disc has turned through an angle θ. The energy equation for the system is
$$\tfrac{1}{2}I\dot\theta^2 = mga(1 - \cos\theta),$$
where
$$I = \frac{Ma^2}{2} + ma^2$$
$$\Rightarrow (M + 2m)a\dot\theta^2 = 4mg(1 - \cos\theta). \tag{1}$$

Differentiating with respect to t we have
$$2(M + 2m)a\dot\theta\ddot\theta = 4mg\,\dot\theta\sin\theta$$
$$\Rightarrow (M + 2m)a\ddot\theta = 2mg\sin\theta. \tag{2}$$

The equations of motion for the particle are,
in the radial direction

$$R = mg\cos\theta - ma\dot\theta^2$$
$$= mg\cos\theta - \frac{4m^2g(1-\cos\theta)}{M+2m} \quad \text{[from (1)]},\qquad(3)$$

in the tangential direction

$$F = mg\sin\theta - ma\ddot\theta$$
$$= mg\sin\theta - \frac{2m^2g\sin\theta}{M+2m} \quad \text{[from (2)]},\qquad(4)$$

$$\Rightarrow R = \frac{mg\{(M+6m)\cos\theta - 4m\}}{M+2m},$$

$$F = \frac{Mmg\sin\theta}{M+2m}.$$

Hence F increases and R decreases as θ increases and so the ratio F/R, as given by these values, increases from 0 to ∞ as θ increases from 0 to $\cos^{-1}\{4m/(M+6m)\}$. But for equilibrium $F/R \leq \mu$, and so F/R will equal the finite value μ before R becomes zero and the particle slips before leaving the disc.

When the particle slips $F/R = \mu$

$$\Rightarrow \frac{Mmg\sin\theta}{M+2m} \Big/ \left[\frac{mg\{(M+6m)\cos\theta-4m\}}{M+2m}\right] = \mu$$
$$\Rightarrow M\sin\theta = \mu\{(M+6m)\cos\theta - 4m\}.$$

Exercise 10.2

1. Three uniform rods, each of length l, are joined to form an equilateral triangle ABC. Prove that the radius of gyration of the triangle of rods about an axis through A perpendicular to the plane ABC is $\tfrac{1}{2}l\sqrt{2}$.

The triangle, which can turn freely in a vertical plane about a fixed horizontal axis through A, is released from the position in which BC is vertical and C is above B. Find the speed of B when AB is vertical.

2. Calculate I for a flywheel of mass M in the form of a disc of uniform thickness and radius a.

The flywheel is rotating at the rate of n revolutions per minute, and is braked by a constant force F applied tangentially to the rim. Find the number of revolutions of the wheel before it is brought to rest.

3. Find the moment of inertia of a uniform circular disc of mass M and radius a, about an axis perpendicular to its plane through a given point of its rim.

This axis is fixed horizontally and the disc rotates freely about it in a vertical plane. If the greatest angular speed of the disc is $\sqrt{(3g/a)}$, find its least angular speed. (L.)

4. A uniform circular disc of radius 1 m is pivoted about a horizontal axis through its centre perpendicular to its plane. A particle of twice the mass of the disc is fixed to a point P on the disc 0·5 m from the axis. The disc is held with P almost vertically above the axis, and then released. Find the maximum angular speed of the disc in the subsequent motion.

5. One end of a uniform rod of length l is smoothly hinged to a fixed point. While it is hanging freely, the rod is given an angular speed $\sqrt{(6g/l)}$. Show that it describes a complete circle. (L.)

6. A uniform semicircular lamina, of radius a, can swing freely in a vertical plane about one end of its bounding diameter AB. Show that, if the lamina is released from rest when AB is horizontal and lowermost, its angular speed when AB is again horizontal is

$$\frac{4}{3}\sqrt{\left(\frac{2g}{\pi a}\right)}.\qquad\text{(L.)}$$

7. Two uniform rods AB and CD, of lengths $2a$ and a respectively, are each of the same mass per unit length. The rods are rigidly joined together at right angles so that C coincides with the mid-point of AB. If the system is pivoted at D so that it can move in a vertical plane, and it starts to move from rest with AB vertical, prove that when this rod is horizontal the angular speed of the system is $\sqrt{(5g/3a)}$. (L.)

8. Find the moment of inertia of a uniform circular disc, of mass M and radius a, about an axis perpendicular to its plane through its centre.

This axis is fixed horizontally and the disc can rotate freely about it in a vertical plane. A particle of mass m is attached to the disc at a distance $\tfrac{1}{2}a$ from the centre, and in one revolution the greatest and least angular speeds of the disc are $\sqrt{(g/4a)}$ and $\sqrt{(g/8a)}$, respectively. Show that $m/M = 2/63$. (L.)

9. Prove that the moment of inertia of a uniform circular disc of mass M and radius a about the axis through its centre normal to its plane is $\tfrac{1}{2}Ma^2$.

The disc is mounted so as to be free to revolve without friction about the axis, which is fixed horizontally. A particle P of mass m is attached to the highest point of the disc, and the system is slightly disturbed from rest. Show that the angular momentum at the instant when P is vertically below the centre of the disc is $\sqrt{[2m(M+2m)ga^3]}$. At this instant a constant retarding couple is applied so that the disc comes to rest after rotating through a further $60°$. Find the magnitude of the couple. (N.)

10. A uniform rod, of weight W and of length $2a$, is pivoted at one end. It is released from rest when it is horizontal and the only resistance to motion is a constant friction couple of magnitude $3Wa/5\pi$. Show that it will rise through an angle $60°$ on the other side of the vertical and find its angular speed when it is vertical for the first time. (N.)

11. Prove that the moment of inertia of a uniform circular disc of radius a and mass M, about an axis OQ through the centre O and perpendicular to the plane of the disc, is $\tfrac{1}{2}Ma^2$.

The circular disc can rotate freely about the axis OQ, which is horizontal. A light inextensible string is wound round the rim with one end attached to the rim; to the other end is attached a particle P, of mass m, which hangs vertically. When the system is released from rest prove, by the principle of conservation of energy, that the speed of P, when it has descended through a distance b, is

$$2\sqrt{\left(\frac{bmg}{M+2m}\right)}.$$

Find also the tension in the string. (O.C.)

12. A circular disc of mass M and radius a is free to turn in its own plane (which is vertical) about a horizontal axis through its centre. A light string passes over the disc and masses M and $\tfrac{1}{2}M$ hang at its ends. The free portions of the string are vertical, and there is no slipping of the string over the disc. Show that the acceleration of the masses is $\tfrac{1}{4}g$. (O.C.)

13. A wheel and axle has a moment of inertia I about its axis which is fixed horizontally and about which it can turn freely. Light strings are wrapped round the wheel and the axle and masses m_1, m_2 hang from their respective free ends. If the radius of the wheel is a and that of the axle is b, show that, after the system is released from rest, its angular acceleration is

$$\frac{(m_1 a - m_2 b)g}{m_1 a^2 + m_2 b^2 + I}$$

until one string is fully unwound. (L.)

14. A uniform circular disc, of mass m and radius a, is free to rotate without friction about a horizontal axis OA through its centre O and perpendicular to its plane. One end of a light string is attached to a point on the circumference of the disc and part of the string is wound on the circumference. The other end of the string carries a particle of mass km hanging freely. The system is released from rest. Show that, when the disc has rotated through an angle θ, the speed of the particle will be

$$2\{kga\theta/(1+2k)\}^{1/2}.$$

Find the tension in the vertical part of the string. (N.)

15. A light inextensible string, carrying particles of mass M and m ($< M$), passes over a uniform solid pulley, of mass $2m$, which can rotate freely about a fixed horizontal axis through its centre O and perpendicular to its plane; the groove of the pulley is sufficiently rough to prevent the string from slipping. The system is released from rest when the heavier particle is at a distance b vertically below O. Prove that, when this particle is at a distance x below O,

$$(M+2m)\dot{x}^2 = 2(M-m)g(x-b).$$

Hence find the ratio of the tensions in the parts of the string carrying the particles.

10:3 THE COMPOUND PENDULUM

A rigid body constrained to rotate about a fixed horizontal axis and making small oscillations under gravity about its position of stable equilibrium is called a *compound pendulum*.

The period of small oscillations of a compound pendulum

Figure 10.6 represents a rigid body of mass M rotating about a horizontal axis through P and displaced through an angle θ from its position of stable equilibrium. G is the centre of mass of the body, $PG = h$, and the M. of I. of the body about the axis is I.

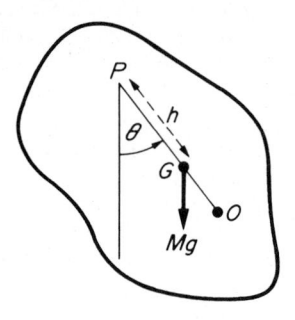

FIG. 10.6.

The equation of angular motion is

$$-Mgh \sin \theta = I\ddot{\theta}$$

$$\Leftrightarrow \ddot{\theta} = -\frac{Mgh \sin \theta}{I}. \tag{10.6}$$

[Note the negative sign, since the couple is tending to *reduce* θ.]

If θ is small and terms involving higher powers of θ than the second are neglected, this equation becomes

$$\ddot{\theta} = -\frac{Mgh}{I}\theta.$$

This is the equation of a SHM of period

$$T = 2\pi \sqrt{\left(\frac{I}{Mgh}\right)}. \tag{10.7}$$

If k is the radius of gyration of the body about an axis parallel to the given axis and through G, then $I = M(k^2 + h^2)$ and the formula for the period of small oscillations becomes

$$T = 2\pi \sqrt{\left(\frac{k^2 + h^2}{gh}\right)}. \tag{10.8}$$

The length l of the equivalent simple pendulum is thus given by the equation

$$l = \frac{k^2 + h^2}{h}. \tag{10.9}$$

The centre of oscillation. If O is the point in PG produced such that $GO = k^2/h$ and the pendulum makes small oscillations about a horizontal axis through O, the length of the E.S.P. is

$$\left(k^2 + \frac{k^4}{h^2}\right) \bigg/ \frac{k^2}{h} = \frac{k^2 + h^2}{h}.$$

The period of small oscillations of the pendulum about the axis through O is therefore the same as the period of small oscillations about the axis through P. The point O is called the *centre of oscillation* corresponding to P. Note that OP has the length of the E.S.P. about either axis.

There are also two points P' and O' in PG and on opposite sides of G from P and O so that $GP' = GP$ and $GO' = GO$ about a horizontal axis through each of which the period of small oscillations is the same as that about the axis through P.

The minimum period of small oscillations as the position of the axis varies

The period of small oscillations $2\pi\sqrt{[(k^2 + h^2)/gh]}$ is a minimum when $(k^2 + h^2)/h$ is a minimum, i.e. if

$$\frac{d}{dh}\left(\frac{k^2}{h} + h\right) \text{ is zero and } \frac{d^2}{dh^2}\left(\frac{k^2}{h} + h\right) > 0.$$

These conditions are satisfied when $h = k$. The period of oscillation is therefore a minimum when the distance between the axis and the centre of gravity of the body is equal to its radius of gyration about a parallel axis through the centre of gravity.

Example 1. A compound pendulum consists of a uniform rod CD, of mass m and length $2a$, rigidly clamped at its centre O to a similar rod AB; the rods are at right angles and $OA = x$. Find the length of the equivalent simple pendulum for free oscillations in the plane of the rods about a horizontal axis through A.

Find the values of x for which the length of the equivalent simple pendulum is unaltered by removal of the rod CD. (L.)

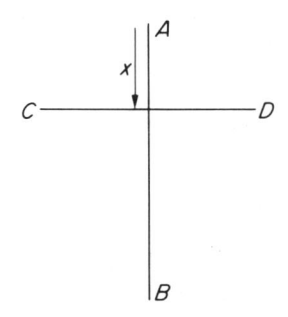

FIG. 10.7.

The M. of I. of AB about the axis is $\frac{4}{3}ma^2$, Fig. 10.7. The M. of I. of CD about the axis is $\frac{1}{3}ma^2 + mx^2$. Therefore the total M. of I. of the pendulum about the axis is

$$\frac{5ma^2}{3} + mx^2.$$

Therefore, in the usual notation,

$$k^2 + h^2 = \left(\frac{5ma^2}{3} + mx^2\right) \bigg/ 2m = \frac{5a^2 + 3x^2}{6}.$$

Also
$$h = (ma + mx)/(2m) = \tfrac{1}{2}(a + x).$$

Therefore the length of the E.S.P. is $(5a^2 + 3x^2)/\{3(a+x)\}$.

When the rod CD is removed the length of the E.S.P. is $4a/3$. The length of the E.S.P. is therefore unaltered by the removal of CD if

$$\frac{5a^2 + 3x^2}{3(a+x)} = \frac{4a}{3} \Leftrightarrow 3x^2 - 4ax + a^2 = 0$$

$$\Leftrightarrow x = a \quad \text{or} \quad \tfrac{1}{3}a.$$

Example 2. A uniform circular disc of mass m and radius a, can turn freely in its own vertical plane about a horizontal axis through a point O on the rim. A particle of mass $2m$ is attached to the disc at P, where P is on the diameter through O, and $OP = x$. Show that the period of small oscillations of the whole system is the same as that of the simple pendulum of length l, given by

$$l = \frac{3a^2 + 4x^2}{2(a + 2x)}.$$

Show also that $a \leqslant l \leqslant 1\cdot 9a$. (N.)

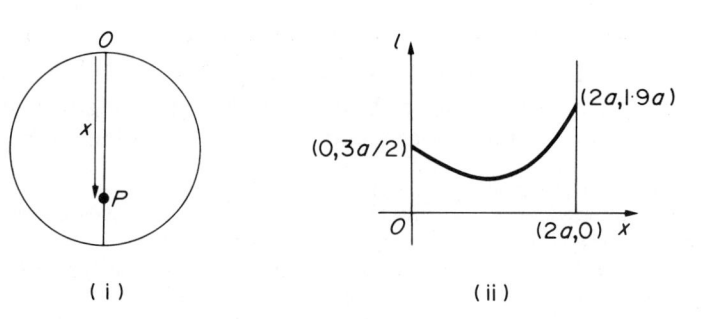

Fig. 10.8.

The M. of I. about the axis is $(3ma^2/2) + 2mx^2$, Fig. 10.8(i). Therefore in the usual notation

$$k^2 + h^2 = (3ma^2 + 4mx^2)/(6m) = (3a^2 + 4x^2)/6,$$
$$h = (ma + 2mx)/(3m) = (a + 2x)/3$$
$$\Rightarrow l = \frac{k^2 + h^2}{h} = \frac{3a^2 + 4x^2}{2(a + 2x)}.$$

The stationary value of l occurs when $dl/dx = 0$ for positive values of x.

$$\frac{dl}{dx} = \frac{(2x + 3a)(2x - a)}{(a + 2x)^2},$$

and therefore l is stationary when $x = \tfrac{1}{2}a, l = a$. This stationary value of l is a minimum value because dl/dx is negative for $-3a/2 < x < \tfrac{1}{2}a$ and positive for $x > \tfrac{1}{2}a$.

Figure 10.8(ii) is a sketch of the graph of l against x for physically possible values of x. This sketch shows that the greatest value of l for values of x in the range $0 \leqslant x \leqslant 2a$ is the value of l when $x = 2a$, i.e. $1\cdot 9a$,

$$\Rightarrow a \leqslant l \leqslant 1\cdot 9a.$$

Example 3. A uniform rod AB of length $2a$ and mass m is freely pivoted to a fixed point at A. One end of a light elastic string of natural length a and modulus mg is attached to B and the other to a fixed point C distant $2a$ vertically above A. Denoting the angle BAC by 2θ show that there is a position of equilibrium in which $\theta = \alpha$ where $3 \sin \alpha = 1$.

Calculate the moment of the forces about A when $\theta = \alpha + \varepsilon$, and by approximating to the first order in ε, where ε is small, show that the period of small oscillations of the rod in the plane BAC is $2\pi(a/2g)^{1/2}$.

(N.)

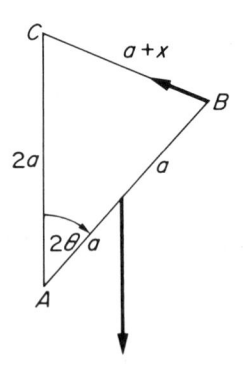

FIG. 10.9.

Figure 10.9 shows the position of the rod and elastic string when $B\hat{A}C = 2\theta$ and the system is in equilibrium with $BC = 4a \sin \theta$.

The forces acting on the rod are the tension in the elastic string, the weight of the rod and the action of the hinge on the rod. The sum of the moments about A of the forces acting on the rod is zero

$$\Rightarrow amg \sin 2\theta = \frac{mg(4a \sin \theta - a)}{a} 2a \cos \theta$$

$$\Leftrightarrow \sin 2\theta = 8 \sin \theta \cos \theta - 2 \cos \theta$$

$$\Leftrightarrow 2 \sin \theta \cos \theta = 8 \sin \theta \cos \theta - 2 \cos \theta$$

$$\Leftrightarrow 2 \cos \theta (3 \sin \theta - 1) = 0.$$

Therefore there is a position of equilibrium in which $\theta = \alpha$ where $3 \sin \alpha = 1$. When $\theta = \alpha + \varepsilon$, the sum of the moments of the forces about A is

$$mga \sin 2(\alpha + \varepsilon) - \frac{mg\{4a \sin (\alpha + \varepsilon) - a\} 2a \cos (\alpha + \varepsilon)}{a}$$

$$= mga \sin 2\alpha \cos 2\varepsilon + mga \cos 2\alpha \sin 2\varepsilon$$
$$- mg(4 \sin \alpha \cos \varepsilon + 4 \cos \alpha \sin \varepsilon - 1)(2a \cos \alpha \cos \varepsilon - 2a \sin \alpha \sin \varepsilon).$$

Therefore, when ε is small so that $\sin \varepsilon \approx \varepsilon$, $\sin 2\varepsilon \approx 2\varepsilon$, $\cos \varepsilon \approx \cos 2\varepsilon \approx 1$, the sum of the moments of the forces about A is approximately

$$mga(\sin 2\alpha + 2\varepsilon \cos 2\alpha) - mga(4 \sin \alpha + 4\varepsilon \cos \alpha - 1)(2 \cos \alpha - 2\varepsilon \sin \alpha).$$

But $\sin \alpha = \frac{1}{3} \Rightarrow \cos \alpha = \frac{2\sqrt{2}}{3}$, $\sin 2\alpha = \frac{4\sqrt{2}}{9}$ and $\cos 2\alpha = \frac{7}{9}$.

Therefore the sum of the moments of the forces about A is approximately

$$mga \left(\frac{4\sqrt{2}}{9} - \frac{16\sqrt{2}}{9} + \frac{12\sqrt{2}}{9} \right) - mga\varepsilon \left(-8 \cdot \frac{1}{9} + 8 \cdot \frac{8}{9} + \frac{2}{3} - \frac{14}{9} \right) = -\frac{16mga\varepsilon}{3}$$

$$\Rightarrow (2\ddot{\varepsilon}) \approx \frac{-\frac{16mga\varepsilon}{3}}{\frac{4ma^2}{3}} = -\frac{2g(2\varepsilon)}{a}.$$

The motion is therefore approximately SHM of period
$$T = 2\pi \sqrt{\left(\frac{a}{2g}\right)}.$$

Exercise 10.3

1. Four uniform thin rods, each of mass m and length $2a$, are joined rigidly together to form a square frame of side $2a$. Show that the moment of inertia of the frame about a horizontal axis through the mid-point of one of the rods and perpendicular to the plane of the frame is $28ma^2/3$. The frame performs, under gravity, small oscillations in its plane about this axis. Find the period of these oscillations. (L.)

2. Three particles of the same mass m are fixed to a uniform circular hoop of mass M and radius a at the corners of an equilateral triangle. The hoop is free to swing in a vertical plane about any point in its circumference. Prove that the equivalent simple pendulum is equal in length to the diameter of the circle. (O.C.)

3. A thin uniform rod AB, of mass m and length $2a$, is free to rotate in a vertical plane about a fixed smooth horizontal axis through A. A uniform circular disc, of mass $3m$ and radius $a/3$, is clamped to the rod with its centre O on the rod so that it lies in the plane of rotation. If $AO = x$, show that the moment of inertia of the system about the axis is $3m(a^2 + 2x^2)/2$.

The system is slightly displaced from its position of stable equilibrium and oscillates freely under gravity. Find the period of small oscillations and show that this is least when $x = (\sqrt{22} - 2)a/6$. (L.)

4. A uniform rod of mass m and length $2a$ is free to turn in a vertical plane about a pivot through one end. A particle of mass $6m$ is attached to the rod at distance c from the pivot. Prove that, if θ is the inclination of the rod to the downward vertical at time t and the system is released from rest with $\theta = \alpha$, then in the subsequent motion
$$(2a^2 + 9c^2)\dot\theta^2 = (3a + 18c)g(\cos\theta - \cos\alpha).$$

Deduce the period of small oscillations of the system and prove that as c varies the period is a minimum when $c = a/3$. (O.C.)

5. Two uniform rods, AOB, COD, each of length $2a$, are fixed to a uniform circular hoop of radius a so as to form two perpendicular diameters. The mass of each rod is equal to the mass of the hoop. The system is suspended from A and swings freely about A in its own plane. Find the time of a small oscillation. (O.C.)

6. Find the moment of inertia of a uniform circular hoop of mass M and radius a (i) about an axis through its centre and perpendicular to its plane, (ii) about a diameter of the hoop.

A particle of mass m is fixed to a point A on the circumference of this hoop and O is the point at the opposite end of the diameter through A. The system can turn freely about a horizontal axis through O and perpendicular to the plane of the hoop. Prove that the periodic time of small oscillations of the system about its position of stable equilibrium is $2\pi\sqrt{(2a/g)}$. (N.)

7. A uniform rod, of mass m and length $2a$, is rigidly fastened at its middle point A to a uniform circular disc, of mass M and radius a, so that the rod is in the plane of the disc and tangential to it. If the system can swing freely in the vertical plane of the disc about a horizontal axis through the point O, which is on the circumference of the disc at the other end of the diameter passing through A, prove that the periodic time of a small oscillation is
$$2\pi\sqrt{\left(\frac{9M + 26m}{6M + 12m} \cdot \frac{a}{g}\right)}.$$
(N.)

8. A uniform circular disc of radius a has a particle of mass equal to that of the disc fixed to a point of its circumference. The disc can turn freely about a fixed horizontal axis through its centre at right angles to its plane. Assuming that the radius of gyration of the disc about this axis is $a/\sqrt{2}$, show that the length of the equivalent simple pendulum for small oscillations of the system about its position of stable equilibrium is $3a/2$. (O.C.)

9. A thin uniform wire is in the form of a semicircular arc of radius a. Find its radius of gyration about an axis through one end perpendicular to its plane. The wire is free to swing about this axis which is horizontal. Find the period of small oscillations about this position of stable equilibrium.

If the wire is slightly disturbed from its position of unstable equilibrium find the angular speed with which it passes through its position of stable equilibrium.   (L.)

10. A uniform circular disc of radius a swings in its own plane about a point of its circumference. Prove that the length of the equivalent simple pendulum is $3a/2$.

If any point of the disc may be taken as the point of suspension, prove that the time of a small oscillation is least when the distance of the point of suspension from the centre of the disc is $a/\sqrt{2}$.   (O.C.)

11. A uniform rod AB, of length $2l$ and mass M, is welded to a uniform circular disc, of radius $a(<2l)$ and mass M, in the plane of the disc and with the end B coinciding with the centre of the disc. The system can rotate freely in a vertical plane about a horizontal axis through A perpendicular to the plane of the disc. If θ denotes the angle between AB and the downward vertical and the system is released from rest when $\theta = \alpha$, prove that the angular speed ω of the system in the subsequent motion is given by

$$(32l^2 + 3a^2)\omega^2 = 36gl(\cos\theta - \cos\alpha).$$

Find also the period of small oscillations if $4l = 3a$.   (O.C.)

12. A thin uniform wire of length $6a$ is bent into the form of a plane rectangle $ABCD$ whose longer side, AB, is of length $2a$. Find the radius of gyration about (i) AB and (ii) XY (the line joining the mid-points of AB and CD).

Hence, or otherwise, find the length of the equivalent simple pendulum, for small oscillations of the rectangle under gravity in a vertical plane about a horizontal axis through X.   (L.)

13. A compound pendulum consists of a thin uniform rigid rod of length $2a$ and mass m, with a heavy particle of mass $2m$ at its middle point. Show that there are two points of suspension on each side of the mid-point of the rod, about each of which the small oscillations will have the same period as those of a simple pendulum of length a, and show that the distance between these points is $0.745a$ approximately.   (N.)

14. Find the moment of inertia of a uniform thin circular disc, of radius a and mass m, about an axis through its centre perpendicular to its plane.

Two such discs, in the same plane, have their centres connected by a light rigid bar of length $4a$. Find the period of small oscillations in a vertical plane, when the system is freely pivoted at a point of the bar distant x from its middle point.

For what value of x is this period least?   (L.)

15. A thin uniform rod AB, of mass m and length $2a$, can turn freely in a vertical plane about a fixed horizontal axis through the end A. A uniform thin circular disc, of mass $24m$ and radius $\frac{1}{2}a$, has its centre C clamped to the rod so that $AC = x$ and the plane of the disc passes through the axis of rotation. Show that the moment of inertia of the system about this axis is $2m(a^2 + 12x^2)$.

The system oscillates under gravity. Write down the energy equation when AB makes an angle θ with the downward vertical. Find the period T of small oscillations. Show that, if x is varied, the least value of T is $2\pi(a/2g)^{1/2}$.   (N.)

16. A solid is made up of two uniform spheres A and B, each of radius r, and a thin uniform rod of length $2r$. The rod connects the spheres so that the centres of the spheres are at a distance $4r$ apart. Sphere A is of mass $2M$ and sphere B is of mass $4M$. The mass of the rod is M. The solid is pivoted at the mid-point of the rod and is free to perform small oscillations, with the rod in a vertical plane and B below A. Show that the period is

$$2\pi\sqrt{\left(\frac{401r}{60g}\right)}.$$ (L.)

17. A compound pendulum consists of a thin uniform circular disc, of mass $12m$ and radius a, rigidly attached to a uniform rod AB, of mass m and length $6a$, so that a diameter of the disc lies along the rod and the centre is at a distance x from A. Show that the period of small oscillations of the pendulum about a smooth horizontal axis through A perpendicular to the plane of the disc is

$$2\pi\sqrt{\left[\frac{6a^2 + 4x^2}{(a + 4x)g}\right]}.$$ (L.)

10:4 THE FORCE EXERTED ON THE AXIS OF REVOLUTION

A special case of equation (9.45) is
the motion of the centre of mass G of a rigid body is that of a particle of mass equal to the mass of the body under the vector sum of the external forces acting on the body and transferred to act at G.

In the case of a rigid body of mass M rotating in a vertical plane under gravity about a fixed horizontal axis, Fig. 10.10, the external forces acting on the body are its weight and the action of the hinge upon it. If the components of this force are X, Y in directions at right angles to and along GP, then, since the corresponding accelerations of G are $h\ddot{\theta}$ and $h\dot{\theta}^2$,

$$X - Mg \sin \theta = Mh\ddot{\theta}, \tag{10.10}$$
$$Y - Mg \cos \theta = Mh\dot{\theta}^2. \tag{10.11}$$

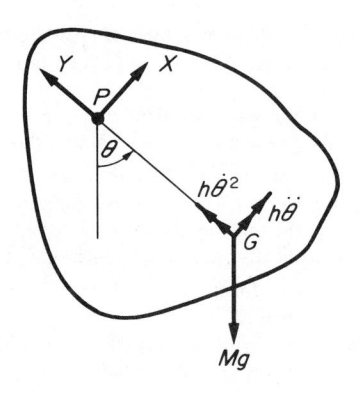

FIG. 10.10.

Together with the energy equation and the initial (boundary) conditions these two equations are sufficient to determine X and Y. In practice the use of the equation of motion discussed in § 10:3 obviates the necessity of differentiating the energy equation and may shorten the calculation.

Example 1. A thin uniform rod swings in a vertical plane about a smooth pivot at one end, through an angle $\pi/2$ on each side of the vertical. Prove that when the rod makes an angle θ with the vertical, the reaction on the pivot makes an angle ϕ with the rod, such that

$$\tan \theta = 10 \tan \phi. \tag{L.}$$

The moment of inertia of the rod about the axis is $4ml^2/3$, where $2l$ is the length of the rod and m is the mass of the rod.

The equations of motion for the centre of mass of the rod are

$$X - mg \sin \theta = ml\ddot{\theta},$$
$$Y - mg \cos \theta = ml\dot{\theta}^2,$$

where θ is the angle made by the rod with the downward vertical.

The energy equation is

$$\frac{2ml^2}{3} \dot{\theta}^2 - mgl \cos \theta = C,$$

where C is a constant such that $\dot{\theta} = 0$ when $\theta = \tfrac{1}{2}\pi$,

$$\Rightarrow C = 0 \quad \text{and} \quad \dot{\theta}^2 = \frac{3g\cos\theta}{2l}.$$

The equation of motion for the rod is

$$\frac{4ml^2}{3}\ddot{\theta} = -mgl\sin\theta$$

$$\Leftrightarrow \ddot{\theta} = -\frac{3g\sin\theta}{4l}$$

$$\Rightarrow X = \frac{mg\sin\theta}{4}, \quad Y = \frac{5mg\cos\theta}{2}$$

$$\Rightarrow \tan\phi = \frac{X}{Y} = \frac{\tan\theta}{10}$$

$$\Leftrightarrow \tan\theta = 10\tan\phi.$$

Example 2. A light rod of length $4a$ has particles of masses m and $2m$ attached to its ends A and B respectively. The rod can rotate freely in a vertical plane about C, where C is the point on the rod such that $AC = a$. The rod is released from rest when AB is horizontal. Show that when the rod has turned through an angle θ, $19a(d\theta/dt)^2 = 10g\sin\theta$. Find the stresses in the parts AC and CB of the rod when it is passing through the vertical position. (N.)

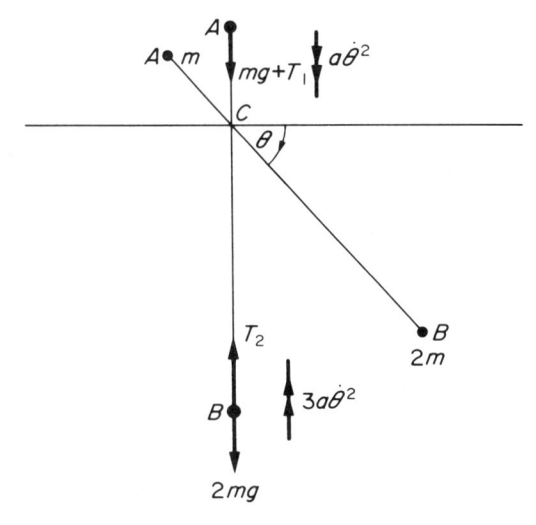

FIG. 10.11.

The M. of I. of the rod and particles about the axis is

$$ma^2 + 2m(3a)^2 = 19ma^2.$$

The energy equation is, therefore,

$$\frac{19ma^2}{2}\dot{\theta}^2 + mga\sin\theta - 2mg\,3a\sin\theta = K,$$

where K is a constant such that $\dot{\theta} = 0$ when $\theta = 0$,

$$\Rightarrow K = 0 \quad \text{and} \quad 19a\dot{\theta}^2 = 10g\sin\theta.$$

Therefore when

$$\theta = \frac{1}{2}\pi, \quad \dot{\theta}^2 = \frac{10g}{19a}.$$

Figure 10.11 shows the forces acting on the particles at A and B respectively, and the accelerations of those particles when the rod is vertical. The equation of motion for the particle at A is

$$mg + T_1 = \frac{ma.10g}{19a} \quad \Leftrightarrow T_1 = -\frac{9mg}{19}.$$

The equation of motion for the particle at B is

$$T_2 - 2mg = 2m.3a.\frac{10g}{19a} \quad \Leftrightarrow T_2 = \frac{98mg}{19}.$$

The rod AC is in compression with a force $9mg/19$ and BC is in tension with a force $98mg/19$.

Exercise 10.4

1. A light rod, which can turn freely in a vertical plane about one end A, carries particles of mass P and Q at distances a and b from A. If the rod is held in a horizontal position and released, prove that its angular speed when vertical is given by

$$\omega^2 = 2g\left(\frac{Pa + Qb}{Pa^2 + Qb^2}\right).$$

If the particles, instead of being attached at different points, are attached at the same point, prove that the pull on the point of support when the rod is vertical is $3(P+Q)g$.

2. Prove that the moment of inertia of a uniform rod, of length $2a$ and mass m, about an axis perpendicular to the rod at a distance x from its mid-point is

$$\tfrac{1}{3}m(a^2 + 3x^2).$$

A uniform rod, of weight W and length $6b$, is smoothly pivoted to a fixed point at a distance $2b$ from one end and is free to swing in a vertical plane. The rod is held horizontal and released. Prove that after it has rotated through an angle θ its angular speed is

$$\sqrt{\left(\frac{g \sin \theta}{2b}\right)},$$

and that when the rod is vertical the reaction on the pivot is $3W/2$. (N.)

3. A uniform circular lamina, centre C, is freely pivoted at a point O of the circumference and is held with its plane vertical and with OC making an angle of $60°$ with the downward vertical through O. If the lamina is released, prove that, when OC is vertical, the reaction on the pivot at O is $5/3$ of the weight of the lamina. (N.)

4. A uniform rectangular lamina $ABCD$ of mass m in which $AB = 2a$, $BC = 4a$ is free to rotate about a fixed horizontal axis which is perpendicular to its plane and passes through O, the mid-point of AB. Show that the moment of inertia of the lamina about this axis is $17ma^2/3$.

When the lamina is at rest with AB horizontal and uppermost, a particle of mass $2m$ is attached to it at B without impulse. Show that in the subsequent motion the lamina will be momentarily at rest when AB is vertical, and find an expression for ω, the angular speed of the lamina when AB is inclined at angle θ to the horizontal. Show also that when AB is vertical, the vertical component of the reaction at O is $57\,mg/23$, and find the horizontal component. (N.)

5. A light rod OA of length $2a$ is rigidly fixed at A to a uniform square plate $ABCD$ of mass m and side $2a$, so that OAB is a straight line. The end O of the rod is pivoted to a fixed horizontal axis which is perpendicular to the plane of the square and the system can swing freely about this axis. If the system is released from rest when OAB is horizontal and CD is below AB, show that when OA has turned through an angle θ, the angular speed of OA is ω, where

$$16a\omega^2 = 3g(3 \sin \theta + \cos \theta - 1).$$

Show also that when the centre of gravity of the square plate is vertically below O, the action at O is vertical and of magnitude of $mg(46 - 3\sqrt{10})/16$. (N.)

6. A uniform rod AB, of mass m and length $2a$, is free to rotate in a vertical plane about a horizontal axis through A. The rod is slightly disturbed from the position in which B is vertically above A. Show that in the subsequent motion

$$2a\dot\theta^2 = 3g(1 - \cos\theta)$$

where θ is the angle made by AB with the *upward* vertical. Calculate the horizontal and vertical components of the force exerted by the rod on the axis of rotation. (L.)

7. Prove that the moment of inertia of a straight uniform rod AB, of mass M and length $2a$, about an axis through A and perpendicular to AB is $\frac{4}{3}Ma^2$.

If the rod, which is free to rotate in a vertical plane about a horizontal axis through A, is released from rest when AB is horizontal, prove that the angular speed ω of the rod about A when AB has moved through an angle θ is given by

$$a\omega^2 = \tfrac{3}{2}g \sin\theta$$

and find the component of the reaction on the rod at A perpendicular to AB. (O.C.)

8. A rod of length $2l$ is held almost vertically with one end resting on a horizontal plane, which is rough enough to prevent slipping, and is then released. Prove that in the subsequent motion the normal reaction of the plane vanishes when the rod makes an angle $\cos^{-1}(1/3)$ with the vertical, and that the angular speed of the rod is then $\sqrt{(g/l)}$. (N.)

9. A uniform rod AB of mass m and length $2a$ has a particle of mass m fixed to the end B. It is held in a horizontal position with its middle point resting on a fixed small rough peg. If the rod is allowed to move, prove that it will turn through an angle $\tan^{-1}(5\mu/14)$, where μ is the coefficient of friction, before slipping takes place. (N.)

10. A uniform circular disc, of mass m and radius a, can turn freely in a vertical plane about a horizontal axis which is perpendicular to the plane of the disc and which passes through a point P on the rim of the disc. The disc is released from rest when the radius through P is horizontal. Show that, when the disc has turned through an angle θ,

$$3a\dot\theta^2 = 4g \sin\theta.$$

Find the magnitude of the force exerted by the disc on the axis when $\theta = \pi/2$. (L.)

10:5 IMPULSE AND ANGULAR MOMENTUM

When a rigid lamina, free to rotate about a fixed axis, is acted upon by an impulse, the theorems (9.49) and (9.50) can be applied. Note that they may be applied in the forms:
(1) *The initial motion of the centre of mass of a rigid body is the same as that of a particle of mass equal to the sum of the masses of the constituent particles acted upon by an impulse which is the vector sum of the external impulses acting on the particles.*
(2) *Moment of impulse about the axis = change in angular momentum about the axis.*
The angular momentum (moment of momentum) is, by definition,

$$\sum r_i \times m_i r_i \dot\theta = \left(\sum m_i r_i^2\right)\dot\theta = I_0 \dot\theta = I_0 \omega.$$

Figure 10.12 represents a rigid body capable of rotation under gravity about a horizontal axis through a fixed point O. The body hanging freely at rest receives a *horizontal* blow B (i.e. it is acted upon by a horizontal impulse B) in the vertical plane containing G, its centre of mass. There will, in general, be an impulsive reaction of the axis on the body, denoted in the diagram by components X and Y in directions at right angles to and along GO. The line of action of B meets OG at N and $ON = x$. Then

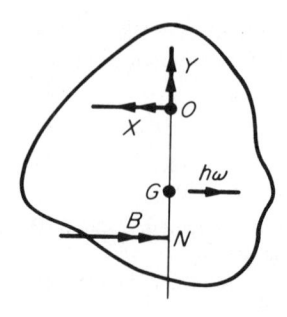

(i) the moment of the impulse of B about the axis is equal to the change in the angular momentum about the axis of the rigid body

$$\Rightarrow Bx = M(k^2 + h^2)\omega,$$

where M is the mass of the body, k its radius of gyration about a parallel axis through G, $GO = h$ and ω is the angular speed of the body immediately after receiving the blow, and

(ii) the initial motion of G is the same as that of a particle of mass M acted upon by a horizontal impulse $B - X$, a vertical impulse Y and starting to move horizontally with speed $h\omega$

$$\Rightarrow B - X = Mh\omega, \quad Y = 0.$$

Therefore eliminating B,

$$X = \frac{M(k^2 + h^2)\omega}{x} - Mh\omega = M\omega\left\{\frac{k^2 + h^2}{x} - h\right\}.$$

It follows that $X = 0$ when $x = (k^2 + h^2)/h$. In this case N is called the *centre of percussion* corresponding to the axis. The distance ON is the length of the E.S.P. for this axis, so that the centre of oscillation discussed in § 10:3 and the centre of percussion are the same point for a particular axis.

Conservation of angular momentum. When two bodies interact on each other with impulsive forces and there are no external impulses which have moments about a particular axis, then the total angular momentum about that axis remains constant. This is a direct consequence of Newton's third law. The impulsive forces between the bodies are equal in magnitude, opposite in direction, and act for the same time. The changes in the angular momenta about the axis of the two bodies are therefore equal and opposite and the total change in angular momentum about the axis is zero.

Example 1. A straight uniform rigid rod of mass m and length l is freely hinged at one end and is held in a vertical position with the free end uppermost. The rod is now slightly displaced and allowed to fall. When the rod has turned through an angle θ prove that the angular speed is

$$\sqrt{\left(\frac{3g(1-\cos\theta)}{l}\right)}.$$

At the instant when $\theta = \tfrac{1}{3}\pi$ the rod is brought suddenly to rest by an impulse applied at the free end at right angles to the rod. Calculate this impulse in terms of m and l. (N.)

The M. of I. of the rod about the axis is $\frac{1}{3}ml^2$. When the rod has turned through an angle θ, the equation of energy is

$$\tfrac{1}{2}mgl(1-\cos\theta) = \tfrac{1}{2}\cdot\tfrac{1}{3}ml^2\dot\theta^2$$
$$\Leftrightarrow \dot\theta = \sqrt{\{3g(1-\cos\theta)/l\}}.$$

Therefore when $\theta = \tfrac{1}{3}\pi$, $\dot\theta = \sqrt{\{3g/2l\}}$.
If B is the impulse which brings the rod to rest,

$$Bl = \frac{ml^2}{3}\sqrt{\left(\frac{3g}{2l}\right)} \Leftrightarrow B = m\sqrt{\left(\frac{gl}{6}\right)}.$$

Example 2. A thin uniform disc, of centre O, radius r and mass M, is free to rotate in a vertical plane about a horizontal axis through a point P of its circumference. When the disc is hanging freely at rest it is struck an impulsive blow B in its plane and along a radius QO where $P\hat{O}Q = 3\pi/4$. Calculate, in terms of B, the impulsive reaction of the axis on the disc, and the angular speed with which the disc begins to move.

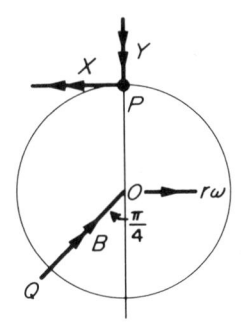

FIG. 10.13.

Figure 10.13 shows the impulses (double arrows), and the initial velocity (single arrow) of the centre of mass O of the disc. The moment of inertia of the disc about the axis is $3Mr^2/2$.
The equations of initial motion for the disc are:
(a) for the angular momentum of the disc,

$$Br\sin\tfrac{1}{4}\pi = 3Mr^2\omega/2,$$

(b) for the initial motion of O,

$$B\sin\tfrac{1}{4}\pi - X = Mr\omega,$$
$$B\cos\tfrac{1}{4}\pi - Y = 0.$$

These equations give

$$X = \frac{B\sqrt{2}}{6},\quad Y = \frac{B\sqrt{2}}{2},\quad \omega = \frac{B\sqrt{2}}{3Mr}.$$

Therefore the impulsive reaction of the axis on the disc is

$$B\sqrt{[2\{(\tfrac{1}{6})^2+(\tfrac{1}{2})^2\}]} = B\sqrt{(\tfrac{5}{9})}$$

at an angle $\tan^{-1}\tfrac{1}{3}$ with the downward vertical on the same side of PO as Q. The initial angular speed of the disc is $\omega = (B\sqrt{2})/(3Mr)$.

Example 3. Two gear wheels are spinning about parallel axles. Their moments of inertia about the axles and their effective radii are I_1, I_2, r_1, r_2, and their angular speeds measured in the same sense are ω_1, ω_2. The axles are then moved so that the gears engage. Find the resultant angular speed of each wheel, and show that the impulse acting upon the first at the instant of contact has a moment

$$-\frac{I_1 I_2 (r_1\omega_1 + r_2\omega_2) r_1}{r_1^2 I_2 + r_2^2 I_1}$$

about the axle of this wheel. (N.)

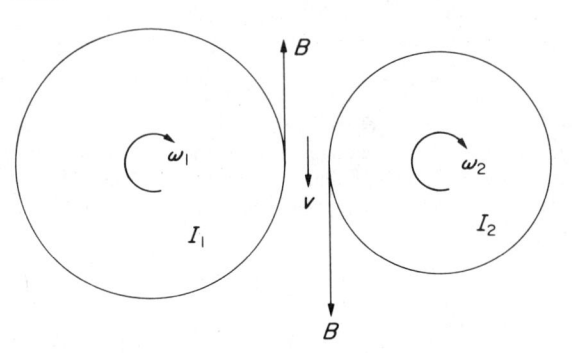

FIG. 10.14.

Let the impulse acting on each wheel at the instant of contact be B and let the final, common, linear velocity of the point of contact of the two wheels be v, Fig. 10.14. (After the wheels enmesh they will continue to move so that their angular speeds are in opposite senses.)

1. The original angular momentum of the first wheel *about its axis* is $I_1\omega_1$.
2. The original angular momentum of the second wheel *about its axis* is $I_2\omega_2$.
3. The final angular momentum of the first wheel *about its axis* is $I_1 v/r_1$.
4. The final angular momentum of the second wheel *about its axis* is $-(I_2 v/r_2)$, angular momentum being taken in this case as positive in the clockwise sense,

$$\Rightarrow Br_1 = I_1\omega_1 - \frac{I_1 v}{r_1},$$

$$Br_2 = I_2\omega_2 + \frac{I_2 v}{r_2}.$$

Hence, equating values of B,

$$\frac{I_1\omega_1}{r_1} - \frac{I_1 v}{r_1^2} = \frac{I_2\omega_2}{r_2} + \frac{I_2 v}{r_2^2}$$

$$\Rightarrow v = \frac{r_1 r_2 (I_1\omega_1 r_2 - I_2\omega_2 r_1)}{I_2 r_1^2 + I_1 r_2^2}$$

$\Rightarrow$ the new angular speeds of the wheels are

$$\frac{r_2 (I_1\omega_1 r_2 - I_2\omega_2 r_1)}{I_1 r_2^2 + I_2 r_1^2} \quad \text{and} \quad -\frac{r_1 (I_1\omega_1 r_2 - I_2\omega_2 r_1)}{I_1 r_2^2 + I_2 r_1^2}.$$

The moment of the impulse about its axle acting on the first wheel at the instant of contact is

$$-Br_1 = -\left\{ I_1\omega_1 - \frac{I_1 r_2 (I_1\omega_1 r_2 - I_2\omega_2 r_1)}{I_1 r_2^2 + I_2 r_1^2} \right\}$$

$$= -\frac{I_1 I_2 (\omega_1 r_1 + \omega_2 r_2) r_1}{r_1^2 I_2 + r_2^2 I_1}$$

the negative sign indicating that the sense of the impulse-moment is opposite to the sense of the original rotation.

Exercise 10.5

1. A uniform circular lamina of mass M and radius a is free to move in its own plane (which is vertical) about the extremity A of a diameter AB. The lamina hangs in equilibrium with B vertically below A, and is struck by a blow applied horizontally at B in the plane of the lamina. If the lamina first comes to rest when AB has turned through an angle of $60°$, calculate, in terms of M and a, the impulse of the blow. (N.)

2. A uniform lamina of mass m in the form of a square $ABCD$ of diagonal $2d$ is suspended freely from a corner A. An impulsive force P is applied to the lamina at C in the direction CB. Prove that the kinetic energy communicated to the lamina is $3P^2/4m$. Show also that the lamina will come to instantaneous rest after turning through an angle α, where

$$\sin \frac{\alpha}{2} = \frac{P}{2m} \sqrt{\left(\frac{3}{2gd}\right)}.$$ (N.)

3. A particle of mass m is fixed to one end of a uniform rod of mass m and length $2a$, the other end of the rod being smoothly pivoted to a point O. The rod is released from rest when it is horizontal; find its angular speed when it is vertical.

If, at the instant when the rod is vertical, its angular speed is suddenly changed by the action of a horizontal impulsive force, prove that there will be no impulsive action at O if the line of action of the impulse is at a depth $16a/9$ below O. (N.)

4. A particle of mass m is placed at the lowest point of a smooth circular tube of mass M and radius a, which is free to turn about a fixed vertical axis coincident with a diameter. The tube is given an initial angular speed and the particle is slightly displaced from its position of rest. In the subsequent motion the radius to the particle when it reaches its highest point makes an angle α with the downward vertical. Find the initial angular speed of the tube. (N.)

5. A uniform rod AB, of mass m and length $2a$, is freely pivoted to a fixed point at A and is initially hanging in equilibrium. A particle of mass m, moving horizontally with speed u, strikes the rod at its middle point and rebounds from it. If there is no loss of energy at the impact, show that immediately afterwards the speed of the particle is $u/7$, and find the angular speed with which the rod begins to rotate. Deduce that, if $u^2 < 49ag/12$, the rod will come to rest at an inclination θ to the downward vertical at A, where

$$\cos \theta = 1 - \frac{24}{49} \frac{u^2}{ag}.$$ (N.)

6. A target consists of a uniform plane square lamina, of side a and mass M, freely hinged along one edge to a fixed horizontal axis. The target hangs at rest with its plane vertical and is struck in its centre by a bullet of mass m moving with speed v perpendicular to the plane of the target. If the bullet becomes embedded in the target, prove that the speed of the bullet is instantaneously reduced to $3mv/(3m+4M)$ and that the fractional loss of kinetic energy in the impact is $4M/(3m+4M)$.

Prove that the target and bullet will make complete revolutions about the hinge in the subsequent motion if

$$3m^2v^2 > 2ag(m+M)(3m+4M).$$ (N.)

7. Two uniform rods AB and BC, each of length $2a$ and of masses m and $3m$ respectively, are smoothly pivoted to a fixed point B. The rods are initially in a horizontal straight line ABC and are released from rest so that they collide when A and C are vertically below B. Find the angular speed of the rods just before the collision. If the rods adhere to each other on meeting, find the angle which the combined rod makes with the vertical when its angular speed is zero. (N.)

8. Show that the moment of inertia of a uniform rod, of mass m and length $2a$, about an axis through its mid-point O and perpendicular to its length, is $\tfrac{1}{3}ma^2$.

The rod is free to rotate in a horizontal plane about O which is fixed. Two smooth small rings, each of mass $\tfrac{1}{2}m$, are free to slide on the rod. The rings are initially on opposite sides of O and each at a distance $a/\sqrt{3}$ from O. The rod is given an initial angular speed $2\sqrt{(g/a)}$, the rings being initially at rest relative to the rod. Show that, when the rings are about to slip off the rod, its angular speed is $\sqrt{(g/a)}$. Show also that at this instant the speed of either ring is $\tfrac{1}{3}\sqrt{(21ga)}$. (N.)

9. Two equal wheels, mounted on the same horizontal axis, have each radius a and moment of inertia ma^2 about the axis, on which they can rotate independently of each other. To the rim of one wheel a particle P of mass m is attached at a point on the same level as the axis. This wheel then starts rotating from rest, the other meanwhile rotating uniformly in the opposite direction with angular speed ω. When P reaches the position vertically below the axis, the two wheels are suddenly locked together. Find the value of ω if the wheels are brought to rest immediately after the locking.

If, however, the value of ω is so great that the locked wheels just carry P back to its starting point, show that

$$\omega = (2 + \sqrt{6})\sqrt{(g/a)}. \tag{N.}$$

10:6 NOTE ON THE RELATIONSHIP BETWEEN THE EQUATIONS OF ANGULAR MOTION OF A RIGID BODY AND THE EQUATIONS OF MOTION OF A PARTICLE MOVING IN A STRAIGHT LINE

The dynamical equations for a particle moving in a straight line are transformed into the corresponding equations for a rigid body rotating about a fixed axis if *mass* is replaced by *moment of inertia about the axis*, *force* by *moment of force about the axis* and the linear quantities *distance*, *speed* and *acceleration* by the quantities *angular distance*, *angular speed*, and *angular acceleration*.

Thus:

	For a particle moving in a straight line	For a rigid body rotating about an axis
Force-Acceleration	$P = m\ddot{x}$	$L = I\ddot{\theta}$
Work-Energy	$\int_{x_1}^{x_2} P\, dx =$ Change in $\tfrac{1}{2}mv^2$	$\int_{\theta_1}^{\theta_2} L\, d\theta =$ Change in $\tfrac{1}{2}I\dot{\theta}^2$
	(Change in kinetic energy)	(Change in kinetic energy)
Impulse-Momentum	$\int_{t_1}^{t_2} P\, dt =$ Change in mv	$\int_{t_1}^{t_2} L\, dt =$ Change in $I\dot{\theta}$
	(Change in linear momentum)	(Change in angular momentum about the axis)

Note on dimensions. It is of interest at this point to consider the dimensions of the quantities with which we are concerned in this chapter.

Angles (measured in circular measure) have zero dimensions.

Angular speed has dimensions T^{-1}.

Angular acceleration has dimensions T^{-2}.

Moment of inertia has dimensions ML^2.

Kinetic energy. The quantity expressed by either $\tfrac{1}{2}mv^2$ or $\tfrac{1}{2}I\dot{\theta}^2$ has dimensions ML^2T^{-2}.

Impulse and linear momentum (mv) have dimensions MLT^{-1}.

Impulsive couple and angular momentum ($I\dot{\theta}$) have dimensions ML^2T^{-1}.

10:7 THE MOTION OF A LAMINA IN ITS OWN PLANE—INSTANTANEOUS CENTRE OF ROTATION

In Fig. 10.15, P_1, P_2, P_3 are three particles of a rigid lamina which is free to move in its own plane and AB is an arbitrary line, *fixed in the frame of reference*, in the plane of the lamina. The angle between AB and P_1P_2 is θ, the angle between AB and P_1P_3 is ϕ, and the angle between P_1P_3 and P_1P_2 is α, each angle being measured in the anticlockwise sense from the first-named direction.

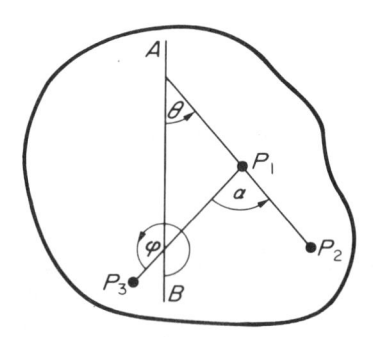

FIG. 10.15.

We have, in this case, $\phi - \theta = 2\pi - \alpha$, and from the definition of a rigid body, α is constant (In general, for all positions of P_1, P_2, P_3 relative to the axis, $\phi - \theta$ is constant.)

$$\Rightarrow \dot{\phi} = \dot{\theta} \quad \text{and} \quad \ddot{\phi} = \ddot{\theta}.$$

Similarly it can be shown that $\dot{\theta}$ and $\ddot{\theta}$ are each independent of the choice of the fixed direction AB. It follows that all lines of the lamina have the same angular speed and the same angular acceleration and these quantities are respectively defined as the *angular speed* and the *angular acceleration* of the lamina.

Translation and Rotation. If the angular speed of a rigid lamina moving in its plane is zero, the motion is described as *translation*. In the case of the motion of a rigid lamina about a fixed axis, each particle of the lamina moves in a circle about the fixed axis and the motion in this case is described as *rotation*. For the purposes of this discussion the moving lamina is assumed to be of infinite extent; the motion we consider resembles the sliding of a sheet of glass on a table top. If we wish to consider the motion of a special body, such as a rod, or a wheel, this body is imagined to be part of an extensive lamina and the rod, or wheel, is imagined to be drawn on the sheet of glass. (The discussion also applies to the motion of a three-dimensional body when every particle of the body moves parallel to one plane; in this case we may imagine the body as rigidly attached to the sheet of glass.)

We show now that for a lamina moving in any manner in its plane there is *at any instant* a point of the lamina about which the motion of the lamina is rotation. To show that this point exists it is necessary to show

(i) that there is a point C of the lamina which, instantaneously, has zero velocity,
(ii) that, at the same instant, every point of the lamina is moving in a circle about C.

If condition (i) is satisfied, condition (ii) is satisfied directly from the definition of a rigid body which states that the distance between any two particles Q_1, Q_2 of the body is constant and so each describes a circle relative to the other, i.e. the velocity of Q_2 relative to Q_1 is perpendicular to $Q_1 Q_2$.

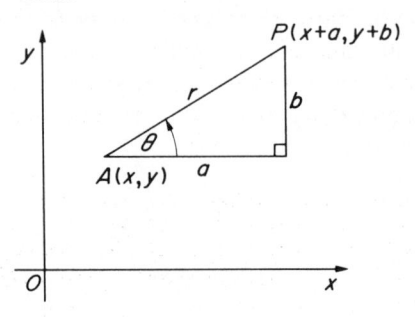

FIG. 10.16.

Figure 10.16 shows points A and P of the lamina. Referred to rectangular axes in the plane of the lamina, and fixed in the frame of reference, the coordinates of A are (x, y) and the coordinates of P are $(x + a, y + b)$. The angle between the positive direction of the x-axis and AP is θ so that $\dot{\theta}$ is the angular speed of the lamina at any instant. If $AP = r$, $a = r \cos \theta$ and $b = r \sin \theta$, and, *from the definition of a rigid body*, r is constant. Therefore, at any instant, the components of the velocity of P parallel to the axes are $\dot{x} - r\dot{\theta} \sin \theta$ and $\dot{y} + r\dot{\theta} \cos \theta$, and thus the velocity of P at this instant is zero if

$$\dot{x} - b\dot{\theta} = 0 \quad \text{and} \quad \dot{y} + a\dot{\theta} = 0.$$

This argument establishes the fact that, unless $\dot{\theta} = 0$ when the motion is one of translation, there is a point C in the plane of the lamina, whose coordinates are $(x - \dot{y}/\dot{\theta}, y + \dot{x}/\dot{\theta})$, which is instantaneously at rest. This point of the body is defined as the *instantaneous centre of rotation of the body* (associated with one particular motion of the body). The instantaneous centre of rotation is unique. This follows at once from the fact that only one point of the lamina can be instantaneously at rest. (The velocity of Q_1 relative to Q_2 is $Q_1 Q_2 \times \dot{\theta}$ perpendicular to $Q_1 Q_2$ and, unless $\dot{\theta} = 0$, this relative velocity cannot vanish.)

Condition for rolling

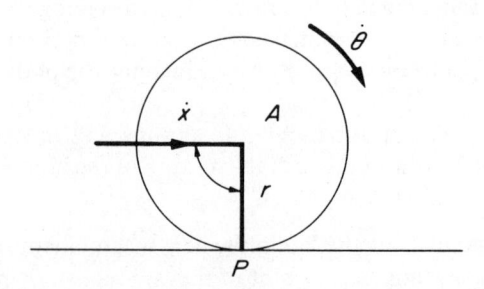

FIG. 10.17.

INSTANTANEOUS CENTRE OF ROTATION

If A is the centre of a circular disc of radius r which is moving along a horizontal plane with angular speed $\dot{\theta}$, so that the disc remains in a fixed vertical plane, Fig. 10.17, the horizontal and vertical components of the velocity of A are $\dot{x}$, 0 respectively. The corresponding components of the velocity of P, the point of contact with the plane, are $(\dot{x} - r\dot{\theta})$, 0. If the plane is sufficiently rough to prevent slipping at P, then P is at rest relative to the plane, so that P is the instantaneous centre of rotation of the disc. The relation $\dot{x} = r\dot{\theta}$ between linear speed and angular speed is the condition for pure rolling of the disc.

Example. A rod passes through a small fixed ring at O and is constrained so that one end A moves on a fixed straight line. The distance of O from the line is c. Find the instantaneous centre of rotation C of the rod, and show that the path of C in the plane containing the straight line and O is a parabola.

If the rod is turning with uniform angular speed ω, show that the speed of C, at the instant when the angle between the rod and the fixed line is $\frac{1}{4}\pi$, is $2c\omega\sqrt{5}$. (N.)

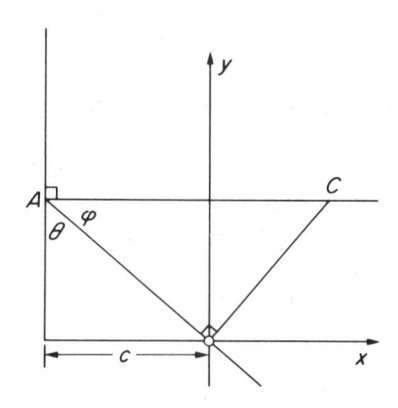

FIG. 10.18.

Consider the rod in a position inclined at an angle θ to the given straight line, Fig.10.18. The instantaneous centre is found by drawing perpendiculars to the direction of motion of the rod at the points A and O of the rod. This centre, C, is thus at the intersection of the perpendicular to the rod at O, where the rod is moving along OA, and the perpendicular to the given straight line at A. If rectangular axes perpendicular and parallel respectively to the given straight line are taken through O and if (x,y) are the coordinates of C,

$$y = c \cot \theta, \quad x = y \cot \theta = c \cot^2 \theta$$
$$\Rightarrow y^2 = cx$$

is the equation of the locus of C. Therefore the path of C in the plane containing the straight line and O is a parabola.

Also we have

$$\dot{y} = -c\dot{\theta} \operatorname{cosec}^2 \theta, \quad \dot{x} = -2c\dot{\theta} \operatorname{cosec}^2 \theta \cot \theta.$$

Hence the speed of C is

$$\sqrt{(\dot{y}^2 + \dot{x}^2)} = \dot{\theta} \sqrt{(c^2 \operatorname{cosec}^4 \theta + 4c^2 \operatorname{cosec}^4 \theta \cot^2 \theta)},$$

and when $\dot{\theta} = \omega$ and $\theta = \frac{1}{4}\pi$ the speed of C is

$$\omega \sqrt{(4c^2 + 16c^2)} = 2c\omega \sqrt{5}.$$

Note. For a lamina moving in its own plane the locus of the instantaneous centre in space is called the *space-locus* or *space-centrode*, the locus relative to the moving lamina is called the *body-locus* or *body-centrode*.

In the above example the space-centrode is the parabola $y^2 = cx$. To find the equation of the body-centrode we must find the locus of C referred to axes *fixed relative to the rod*. We let $r = AC$, $\phi = O\hat{A}C$ be the polar coordinates of C referred to A as pole and AO as initial line (both fixed relative to the rod). Then, since

$$AC = c + x = c(1 + \cot^2\theta) = c\,\mathrm{cosec}^2\,\theta, \quad \phi = \tfrac{1}{2}\pi - \theta,$$

the polar equation of the body-centrode is

$$r = c \sec^2 \phi.$$

Exercise 10.7

In each of exercises 1–4 draw a diagram to show the position of the instantaneous centre of rotation of the lamina. Find the space-centrode and the body-centrode in each case.

1. A uniform rod AB of length $2a$ moving in a plane containing rectangular axes Ox, Oy so that A moves along Ox and B along Oy.

2. A triangular lamina moving so that its vertices are always in contact with the circumference of a fixed circle with centre O.

3. A triangular lamina moving so that two of its sides always touch a fixed circle with centre O.

4. PX is a tangent at P to a fixed circle with centre O. A rod AB moves in the plane of the circle so that it is always a tangent to the circle and so that A moves on PX.

5. A rod AB moves with its extremities upon two fixed perpendicular lines Ox, Oy so that the mid-point C of AB describes a circle about O with uniform angular speed ω. Describe the motions of A and B.

Find the instantaneous centre of rotation of AB, and show that, if D, E are two points on AB such that $AD = EB$, the velocity of D at any instant is perpendicular to OE. (N.)

10:8 THE GENERAL MOTION OF A RIGID LAMINA IN ITS OWN PLANE

The results of § 10:7 and the theorems of § 9:8 when applied to a rigid lamina moving in its own plane, may be summarised thus:

1. *The motion of a rigid lamina in its plane can be analysed into a motion of translation of the centre of mass, G, together with a motion of rotation about G.*

2. *The motion of translation of G is the same as that of a particle of mass equal to the total mass, M, of the lamina acted upon by a force which is the vector sum of the external forces acting on the lamina.*

3. *The motion of rotation about G is the same as the motion of rotation would be if the lamina rotated about a fixed axis through G and the same external forces acted on the lamina.*

4. *The kinetic energy of a lamina moving freely in its own plane is equal to the kinetic energy due to the motion of translation of G together with the kinetic energy due to the motion of rotation relative to G about G*:

$$\text{K.E.} = \tfrac{1}{2}MV^2 + \tfrac{1}{2}I_G\omega^2, \tag{10.12}$$

where I_G is the M. of I. about an axis through the centre of mass perpendicular to the plane of the lamina, ω is the angular speed of the lamina about this axis, and V is the speed of translation of the centre of mass of the lamina.

Note. If C is the instantaneous centre of the lamina and P is a point of the lamina distant r from C, the speed of the particle of mass m at P is $r\omega$ and the kinetic energy of the lamina is

$$T = \Sigma \tfrac{1}{2} m r^2 \omega^2 = \tfrac{1}{2} \omega^2 \Sigma m r^2 = \tfrac{1}{2} I_C \omega^2. \tag{10.13}$$

5. *The angular momentum of a lamina moving in its own plane about a fixed origin in that plane is equal to the sum of the angular momentum relative to G about G and the angular momentum about the origin of the whole mass concentrated at G and moving with the velocity of translation of G:*

$$\text{Angular momentum about the origin} = I_G \omega + MVp, \tag{10.14}$$

where p is the perpendicular distance from the origin to the vector representing the velocity of translation of G.

6. *The change in velocity of G is the same as the change in velocity of a particle of mass M at G under the action of all the external impulses transferred to act at G.*

7. *The change in the moment of momentum (angular momentum) of the lamina about any point A equals the momentum of the external impulses about A.*

Example 1. A uniform rod AB, of length $2a$ and mass M, moves in a vertical plane with the upper end A against a smooth vertical wall and the lower end B on a smooth horizontal floor. Initially the rod was at rest in a vertical position against the wall and the end B was slightly displaced. Calculate the angular speed ω of the rod when it makes an angle $\tfrac{1}{4}\pi$ with the wall.

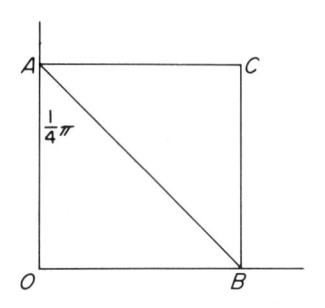

Fig. 10.19.

The instantaneous centre of rotation C is at the intersection of the perpendicular to the wall at A with the perpendicular to the floor at B, Fig. 10.19.

When $O\hat{A}B = \tfrac{1}{4}\pi$, the moment of inertia of the rod about a horizontal axis through C is

$$\frac{Ma^2}{3} + Mu^2 = \frac{4Ma^2}{3}.$$

Therefore the energy equation for the rod in this position is

$$Mga\left(1 - \tfrac{1}{2}\sqrt{2}\right) = \frac{2Ma^2}{3}\omega^2$$

$$\Leftrightarrow \omega = \sqrt{\left[\frac{3g(2-\sqrt{2})}{4a}\right]}.$$

Example 2. A uniform thin disc, of centre O, radius a and mass m, has a particle of mass M attached to a point P of its circumference. The disc stands at rest in a vertical plane and on a rough horizontal plane. P is its highest point. The disc is now slightly displaced from rest so as to continue to move in contact with the plane and in a vertical plane. Find an expression for the velocity of the particle when OP is horizontal. (Assume that the disc rolls on the plane.)

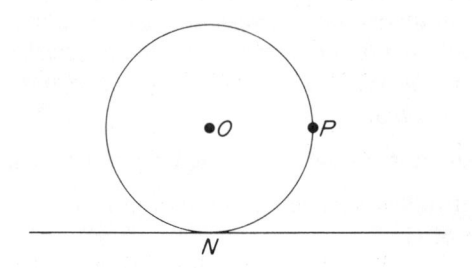

FIG. 10.20.

The instantaneous centre of the motion is at N the point of contact of the disc with the horizontal plane, Fig. 10.20. The moment of inertia of the disc and particle combined about a horizontal axis through N is
$$\frac{3ma^2}{2} + M.PN^2.$$

Therefore, when $P\hat{O}N = \frac{1}{2}\pi$, $\qquad I_N = \frac{3ma^2}{2} + 2Ma^2.$

The energy equation for the system in this position is
$$Mga = \frac{1}{2}\left(\frac{3ma^2}{2} + 2Ma^2\right)\omega^2,$$
where ω is the angular speed of the system,
$$\Leftrightarrow \omega^2 = \frac{4Mg}{(3m+4M)a}.$$

Therefore the velocity of the particle when OP is horizontal is
$$a\omega\sqrt{2} = 2\sqrt{\left(\frac{2Mga}{3m+4M}\right)}$$
in a direction at right angles to NP.

Example 3. A uniform rod AB, of mass m and length $2a$, lies at rest on a smooth horizontal table. A horizontal impulse P, making an angle β with AB, is applied to the end A. Find the kinetic energy generated by the impulse.

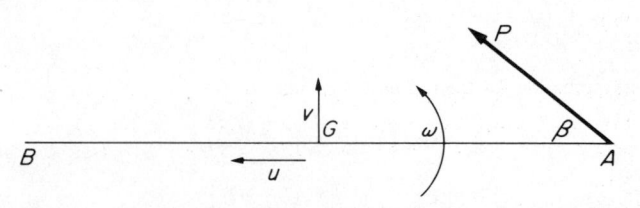

FIG. 10.21.

Let the components of the velocity of the centre of mass G be u, v and the angular speed be ω just after the application of the impulse, Fig.10.21. Then the (impulsive) equations of motion of G along and perpendicular to AB are respectively

$$mu = P \cos \beta, \qquad mv = P \sin \beta. \tag{1}$$

The equation for the (impulsive) change of angular momentum about G is

$$\tfrac{1}{3}ma^2\omega = aP \sin \beta$$

$$\Rightarrow u = \frac{P \cos \beta}{m}, \quad v = \frac{P \sin \beta}{m}, \quad \omega = \frac{3P \sin \beta}{ma}$$

and the kinetic energy generated by the impulse is

$$\tfrac{1}{2}m(u^2 + v^2) + \tfrac{1}{2}\frac{ma^2}{3}\omega^2 = \frac{(1 + 3\sin^2 \beta)P^2}{2m}.$$

Note. Instead of taking moments about G we could take moments about A, writing down the fact that the change in angular momentum about A is zero, thus

$$\tfrac{1}{3}ma^2\omega - amv = 0.$$

This technique can be useful in avoiding the introduction of unknown internal impulses into the equations of a problem. See Examples 4, p. 391, and 6, p. 393, and Example 4 below.

Example 4. A uniform circular disc, of radius a and mass m, is rolling along a horizontal line with speed V and its plane is vertical. The highest point of the disc is then suddenly fixed by passing a smooth spindle parallel to the axle of the disc through a small hole at the circumference. Show that the disc will make complete revolutions about the spindle if $V^2 > 24ga$.

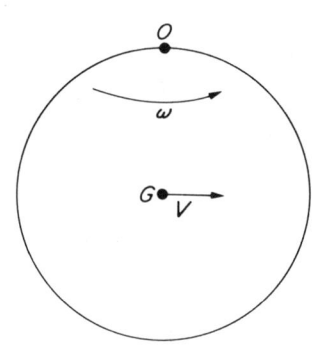

FIG. 10.22.

Let ω, Fig. 10.22, be the angular speed of the disc about the spindle, O. Just before the spindle is inserted the counter-clockwise angular momentum about O is

$$amV - \tfrac{1}{2}ma^2(V/a) = \tfrac{1}{2}amV.$$

(The first term on the l.h. side is the angular momentum due to the motion of G, the second is that due to the motion about G.) When the disc begins to turn about O, its angular momentum about O is $3ma^2\omega/2$ counter-clockwise. Since there is no moment of the impulsive forces about O, angular momentum about O is conserved and

$$\tfrac{1}{2}maV = \tfrac{3}{2}ma^2\omega$$

$$\Leftrightarrow \omega = \tfrac{1}{3}V/a.$$

The disc will execute complete revolutions about O if the kinetic energy just after the spindle is inserted is *more than sufficient* to raise the centre of mass through a height $2a$, i.e. if

$$\frac{1}{2}\frac{3ma^2}{2}\omega^2 > 2mga$$
$$\Leftrightarrow V^2 > 24ga.$$

10:9 APPLICATION TO MISCELLANEOUS PROBLEMS

The principles of motion, kinetic energy and momentum stated in this chapter make possible the solution of problems concerning the motion of a rigid body. In general, the motion is analysed into motion of the centre of mass and motion about an axis through the centre of mass. The principles of particle dynamics are applied to the former and the principles discussed earlier in this chapter are applied to the latter. In some cases, use can be made of the instantaneous centre of rotation.

Example 1. A uniform cylinder, of mass M and radius r, rolls, without slipping, down a fixed plane of inclination α to the horizontal. Use the energy equation to find the time taken by the cylinder to move through a distance x from rest down the plane. Find the minimum value of the coefficient of friction between the cylinder and the plane for which the motion is possible.

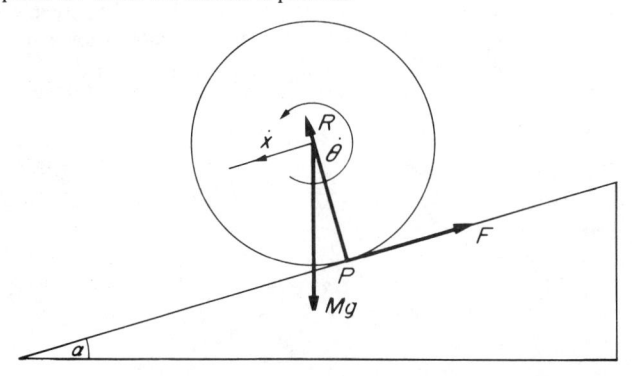

FIG. 10.23.

Figure 10.23 shows the forces acting on the cylinder, the angular speed $\dot{\theta}$ of the cylinder about its axis, and the linear speed $\dot{x}$ of the axis. The direction of $\dot{x}$ is determined by the geometrical constraints of the system. Further, since the generator of the cylinder in contact with the plane is instantaneously at rest, $\dot{x} = r\dot{\theta}$.

The energy equation is

$$\frac{1}{2}M\dot{x}^2 + \frac{1}{2}\frac{Mr^2}{2}\dot{\theta}^2 = Mgx\sin\alpha$$
$$\Rightarrow \tfrac{3}{4}M\dot{x}^2 = Mgx\sin\alpha$$
$$\Leftrightarrow \dot{x}^2 = \frac{4gx\sin\alpha}{3}.$$

Differentiating this equation with respect to t

$$\Rightarrow 2\dot{x}\ddot{x} = \frac{4g\sin\alpha}{3}\dot{x}$$
$$\Leftrightarrow \ddot{x} = \tfrac{2}{3}g\sin\alpha.$$

The time taken for the axis of the cylinder to move through a distance x from rest with this constant acceleration is given by

$$x = \frac{1}{2}\frac{2g \sin \alpha}{3} t^2$$

$$\Leftrightarrow t = \sqrt{\left(\frac{3x}{g \sin \alpha}\right)}.$$

The equations of motion for the cylinder are:
(1) for the motion of *translation* of its axis *parallel* to the plane,

$$Mg \sin \alpha - F = M\ddot{x},$$

where F is the force of friction, not necessarily limiting, acting on the cylinder up the plane,
(2) for the motion of *translation* of its axis *perpendicular* to the plane

$$Mg \cos \alpha - R = 0,$$

where R is the normal action of the plane on the cylinder,
(3) for the motion of rotation about the axis

$$Fr = \frac{Mr^2}{2}\ddot{\theta}.$$

From these equations and from the values of $\ddot{x}$ and $\ddot{\theta}$ obtained above we have,

$$R = Mg \cos \alpha, \quad F = \tfrac{1}{3} Mg \sin \alpha.$$

For the motion to be possible $F/R \leqslant \mu$, i.e. $\mu \geqslant \tfrac{1}{3} \tan \alpha$.

Note. Here we do not assume limiting friction; the frictional force brought into play is just sufficient to prevent slipping. It should only be assumed that friction is limiting if slipping is known to be taking place (and the direction of this limiting frictional force must then oppose the direction of relative slipping.)

Example 2. A circular disc AB rolls without slipping up a line of greatest slope of an inclined plane, being drawn up by a light string which is wound round the disc. The string passes over a smooth peg P and carries a heavy particle of mass m' at its free end C. If AP is parallel to the plane and the motion takes place in a vertical plane through the line of greatest slope, find the acceleration of the disc. The plane is inclined at an angle α to the horizontal, the mass of the disc is m, its radius is a, and the radius of gyration about an axis through its centre and perpendicular to its plane is k.　　　　　　　　　　　　　　　　　　　　　　　　　　　　(N.)

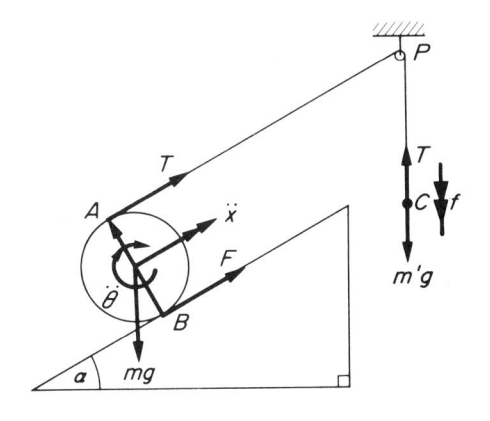

FIG. 10.24.

Figure 10.24 shows the forces acting on the disc, the linear acceleration, $\ddot{x}$, of its centre and the angular acceleration, $\ddot{\theta}$, about its centre. The downward acceleration of C is f and the tension in the string is T. The equations of motion for the disc are

$$Ta - Fa = mk^2\ddot{\theta}, \tag{1}$$
$$T + F - mg\sin\alpha = m\ddot{x}, \tag{2}$$

where F is the force of friction between the plane and the disc. The equation of motion for the particle is

$$m'g - T = m'f. \tag{3}$$

Since B is the instantaneous centre of rotation of the disc,

$$a\ddot{\theta} = \ddot{x}. \tag{4}$$

Since the acceleration of C is equal to the linear acceleration of A,

$$f = \ddot{x} + a\ddot{\theta} = 2a\ddot{\theta}. \tag{5}$$

From these equations,

$$m'g - T = 2m'a\ddot{\theta},$$
$$T - F = \frac{mk^2\ddot{\theta}}{a},$$
$$T + F = ma\ddot{\theta} + mg\sin\alpha$$
$$\Rightarrow 2T = \ddot{\theta}\left(\frac{mk^2}{a} + ma\right) + mg\sin\alpha$$
$$\Rightarrow 2m'g - 4m'a\ddot{\theta} = \ddot{\theta}\left(\frac{mk^2}{a} + ma\right) + mg\sin\alpha$$
$$\Rightarrow \ddot{\theta} = (2m'g - mg\sin\alpha)a/(4m'a^2 + mk^2 + ma^2).$$

Hence the acceleration of the disc is

$$(2m'g - mg\sin\alpha)a^2/(4m'a^2 + mk^2 + ma^2).$$

Example 3. A uniform rod is held inclined to the vertical at an angle β, with one end on a smooth horizontal table, and is released. Find the angular speed of the rod just before it strikes the plane, and show that the angular acceleration at that instant is independent of β.

Prove also that, at the same instant, the reaction of the plane on the rod is one quarter of the weight of the rod.

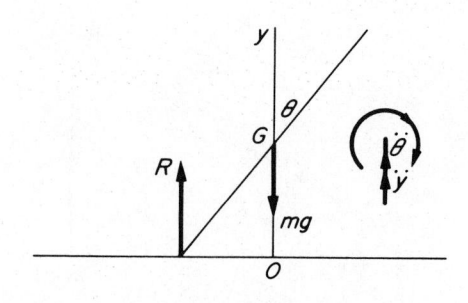

FIG. 10.25.

Because there is no horizontal force acting on the rod, its centre of gravity G does not acquire any horizontal component of velocity and therefore remains in the same vertical line Oy, Fig. 10.25. If m is the mass of the rod, $2a$ the length of the rod, y the height of G above the plane and θ the inclination of the rod to the vertical, the kinetic energy of the rod is

$$\tfrac{1}{2}m(\dot{y}^2 + k^2\dot{\theta}^2),$$

where $k^2 = \tfrac{1}{3}a^2$ and $y = a\cos\theta$. Also
$$\dot{y} = -a\dot{\theta}\sin\theta.$$
Therefore the kinetic energy of the rod is $\tfrac{1}{2}ma^2(\tfrac{1}{3}+\sin^2\theta)\dot{\theta}^2$.
The energy equation is therefore
$$\tfrac{1}{2}ma^2(\tfrac{1}{3}+\sin^2\theta)\dot{\theta}^2 + mga\cos\theta = C,$$
where C is constant. Also $\dot{\theta}=0$ when $\theta=\beta \Rightarrow C = mga\cos\beta$. The energy equation becomes
$$\tfrac{1}{2}ma^2(\tfrac{1}{3}+\sin^2\theta)\dot{\theta}^2 + mga(\cos\theta-\cos\beta) = 0. \qquad (1)$$
When $\theta = 90°$, $\dot{\theta} = \sqrt{\left(\dfrac{3g\cos\beta}{2a}\right)}$

$\Rightarrow$ the angular speed of the rod just before it strikes the plane is
$$\sqrt{\left(\dfrac{3g\cos\beta}{2a}\right)}.$$

The equations of motion for the rod are
$$R - mg = m\ddot{y} = -ma(\dot{\theta}^2\cos\theta + \ddot{\theta}\sin\theta),$$
$$Ra\sin\theta = \tfrac{1}{3}ma^2\ddot{\theta}.$$
When the rod is about to strike the plane $\theta = \tfrac{1}{2}\pi$, and if the appropriate variables are indicated by the suffix $\pi/2$,
$$R_{\pi/2} - mg = -ma\ddot{\theta}_{\pi/2},$$
$$R_{\pi/2} = \tfrac{1}{3}ma\ddot{\theta}_{\pi/2}$$
$$\Rightarrow R_{\pi/2} = \dfrac{1}{4}mg, \quad \ddot{\theta}_{\pi/2} = \dfrac{3g}{4a},$$
both being independent of β. Note that $\ddot{\theta}_{\pi/2}$ can be obtained at once by differentiating (1) with respect to t and putting $\theta = \tfrac{1}{2}\pi$.

Example 4. Two uniform rods AB, BC of equal lengths and masses are smoothly hinged at B and laid in a straight line on a smooth horizontal table. A horizontal impulse perpendicular to its length is applied at the mid-point of AB. Determine the subsequent initial motions, and show that C moves off with one-fifth the initial speed of A and in the opposite direction. (N.)

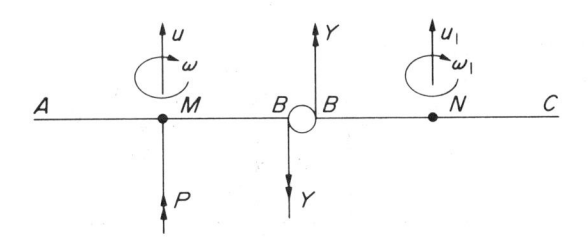

FIG. 10.26.

Let each rod be of mass m and length $2a$ and let the horizontal impulse be P. At the hinge there is an impulsive action and an equal and opposite impulsive reaction between the rods and because initial relative motion in the direction parallel to the rods is not possible these impulses ($= Y$, say) are at right angles to the rods.

The initial velocity of the centre of mass of each rod is at right angles to its length. In Fig. 10.26 initial velocities are marked with a single arrow and impulses with a double arrow.

The equations of motion for the rods are:
for the rod AB, and for the initial motion of translation of its centre of mass,
$$P - Y = mu; \qquad (1)$$
for the rod AB, and for the initial motion of rotation about its centre of mass,
$$Ya = \frac{ma^2}{3}\omega; \qquad (2)$$
for the rod BC and for the initial motion of translation of its centre of mass,
$$Y = mu_1; \qquad (3)$$
for the rod BC and for the initial motion of rotation about its centre of mass,
$$Ya = \frac{ma^2}{3}\omega_1. \qquad (4)$$

In addition, there is an equation which expresses the fact that the initial linear speeds of the ends of the rods at B are equal,
$$u_1 + a\omega_1 = u - a\omega. \qquad (5)$$

From equations (2) and (4), $\quad \omega = \omega_1.$

From equations (3) and (4), $\quad u_1 = \dfrac{a\omega_1}{3} = \dfrac{a\omega}{3}.$

Therefore from equation (5), $\quad u = \dfrac{7a\omega}{3}.$

Therefore from equation (1), $\quad u = \dfrac{7P}{8m}.$

$$\Rightarrow u_1 = \frac{P}{8m}, \quad \omega = \omega_1 = \frac{3P}{8am}.$$

Hence the linear speed of A is
$$u + a\omega = \frac{5P}{4m},$$
and the linear speed of C is
$$u_1 - a\omega = -\frac{P}{4m}.$$

Therefore C starts to move with one-fifth of the initial speed of A in the opposite direction.

Note that this problem can be solved without introducing the internal impulses at B as follows:
Linear momentum of the whole system perpendicular to the line of the rods gives
$$mu + mu_1 = P. \qquad (6)$$
The equations expressing the change of angular momentum about B of AB and BC are respectively
$$amu + \tfrac{1}{3}ma^2\omega = aP, \qquad (7)$$
$$amu_1 - \tfrac{1}{3}ma^2\omega_1 = 0. \qquad (8)$$

Equations (6), (7), (8) together with the kinematical condition (5) are sufficient to determine u, u_1, ω and ω_1.

Example 5. A hollow cylinder of internal radius a is fixed with its axis horizontal. A uniform solid cylinder of radius b, with its axis also horizontal, rolls without slipping on the inner surface of the hollow cylinder. If initially the plane through the two axes is horizontal and the inner cylinder is released, find the angular speed of this plane when it is vertical. Show that the motion of this plane is the same as that of a simple pendulum of length $3(a-b)/2$ swinging through the same angle. (N.)

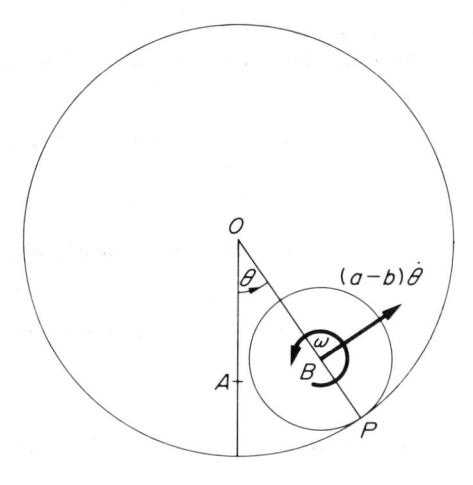

Ch. 10 §10:9   APPLICATION TO MISCELLANEOUS PROBLEMS   393

FIG. 10.27.

Figure 10.27 shows the cylinders when the angle between the plane of the axes and the vertical is θ. The velocity of the axis of the rolling cylinder is $(a-b)\dot\theta$ in the direction at right angles to the plane of the axes, and the angular speed of this cylinder about its axis is ω (say).

Because the generator of contact between the two cylinders is instantaneously at rest

$$b\omega + (a-b)\dot\theta = 0$$

The kinetic energy of the rolling cylinder is

$$\frac{1}{2}m(a-b)^2\dot\theta^2 + \frac{1}{2}\frac{mb^2}{2}\omega^2,$$

where m is its mass, and this is equal to

$$\tfrac{1}{2}m(b^2\omega^2 + \tfrac{1}{2}b^2\omega^2) = \tfrac{3}{4}mb^2\omega^2 \quad \text{or} \quad \tfrac{3}{4}m(a-b)^2\dot\theta^2.$$

The energy equation for the rolling cylinder is therefore

$$\tfrac{3}{4}m(a-b)^2\dot\theta^2 - mg(a-b)\cos\theta = \text{constant}.$$

Therefore, differentiating with respect to t,

$$\tfrac{3}{2}m(a-b)^2\dot\theta\ddot\theta + mg(a-b)\dot\theta\sin\theta = 0$$

$$\Leftrightarrow \ddot\theta = \frac{-2g}{3(a-b)}\sin\theta.$$

The equation of motion for a simple pendulum of length l obtained in § 10:3 is $\ddot\theta = -(g/l)\sin\theta$. The motion is therefore the same as that of a simple pendulum of length $3(a-b)/2$ swinging through the same angle.

Example 6. A uniform hoop, of mass m, radius a and centre A, rolling upright on a horizontal plane with speed v collides with a fixed perfectly rough inelastic rail which is perpendicular to the plane of the hoop and at a height $h(<a)$ above the horizontal plane. Show that, for the hoop to surmount the rail without jumping,

$$\frac{2a\sqrt{\{g(a-h)\}}}{2a-h} \geq v > \frac{2a\sqrt{(gh)}}{2a-h}.$$

After impact the hoop will begin to turn about the rail; let the angular speed just after the impact be ω.

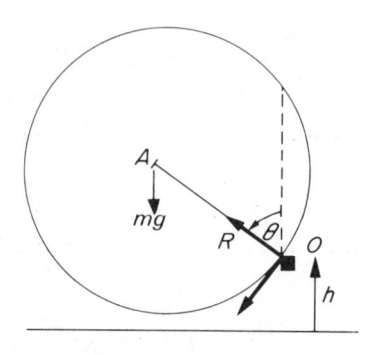

FIG. 10.28.

Then, since the angular momentum of the hoop about the rail is unaltered by the impact,

$$m(a-h)v + ma^2\frac{v}{a} = 2ma^2\omega$$

$$\Leftrightarrow a^2\omega = \tfrac{1}{2}(2a-h)v. \tag{1}$$

Let O be the point of the rail about which the hoop turns and suppose that the hoop remains in contact with the rail and that at time t after the impact OA makes the angle θ with the upward vertical, Fig. 10.28. Then the energy equation for rotation about O is

$$\tfrac{1}{2}2ma^2\dot\theta^2 + mg(h + a\cos\theta) = \tfrac{1}{2}2ma^2\omega^2 + mga. \tag{2}$$

Also, if R is the component along OA of the reaction between the rail and the hoop, the equation of motion of A resolved along AO, is

$$mg\cos\theta - R = ma\dot\theta^2. \tag{3}$$

Equations (2) and (3) give

$$a^2\dot\theta^2 = a^2\omega^2 - g(h + a\cos\theta - a), \tag{4}$$

$$R = m\{g(2a\cos\theta + h - a) - a^2\omega^2\}/a. \tag{5}$$

In order that the hoop should surmount the rail without jumping we must have both $\dot\theta^2 > 0$ and $R \geq 0$ for

$$\frac{a-h}{a} \leq \cos\theta \leq 1.$$

Therefore, from (4) and (5),

$$a^2\omega^2 > gh, \tag{6}$$

$$g(a-h) \geq a^2\omega^2. \tag{7}$$

Here (6) expresses the condition that the hoop has sufficient energy to surmount the rail and (7) expresses the condition that the hoop is not moving so fast *that it leaves the rail immediately after the impact* (when R is least). Substitution for ω^2 from (1) gives the required inequalities.

Example 7. A uniform hoop with centre O, of mass m, radius a, and moment of inertia mk^2 about its axis, is projected with its plane vertical along a rough horizontal table. The coefficient of friction between the hoop and the table is μ. The initial angular speed of the hoop is Ω and its centre has an initial speed V. Discuss the subsequent motion in the three cases (a) $V = a\Omega$, (b) $V > a\Omega$, (c) $V < a\Omega$. [By using mk^2 as the moment of inertia the results can be made to apply to other bodies with circular section by choosing the appropriate value for k, e.g., for a sphere $k = a\sqrt{\tfrac{2}{5}}$.]

Without loss of generality we suppose that $V \geq 0$ throughout this example.

(a) In Fig. 10.29 (i), since $V = a\Omega$ the hoop is rolling at the instant of projection. Two possibilities arise;

APPLICATION TO MISCELLANEOUS PROBLEMS

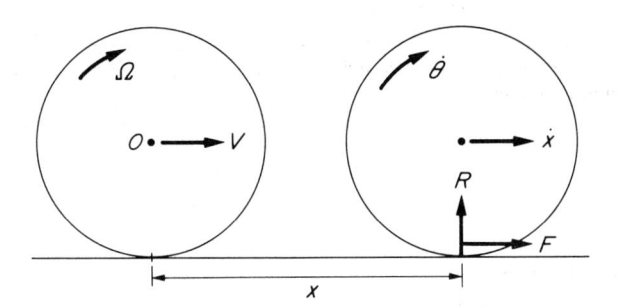

FIG. 10.29 (i)

either the hoop will continue to roll or slipping will commence. Suppose that rolling continues and that at time t the situation is as shown on the right of Fig. 10.29 (i). Since we have assumed rolling, the condition for rolling

$$\Rightarrow \dot{x} = a\dot{\theta}. \tag{1}$$

The equations of motion of the centre O, when resolved horizontally and vertically, and about the centre O

$$\Rightarrow F = m\ddot{x}, \quad R - mg = 0, \quad Fa = -mk^2\ddot{\theta}. \tag{2}$$

Equations (1) and (2)

$$\Rightarrow \ddot{x} = 0, \quad \ddot{\theta} = 0, \quad F = 0, \quad R = mg.$$

Hence the assumption of continued rolling implies that the hoop does so with constant speed $V(\ddot{x} = 0)$ and that $F = 0$. The only alternative assumption, i.e. that slipping commences, implies that F has the limiting value $\mu R = \mu mg$. But the laws of friction imply that the magnitude of the frictional force brought into play is the least possible to prevent relative slipping. Hence we must conclude that rolling continues indefinitely.

If at any stage of the motions to be discussed in cases (b) and (c) rolling takes place instantaneously, the above result allows us to assert that rolling must continue from that point onwards.

(b) In this case, $V > a\Omega$, the point of contact with the plane initially slips to the right. Hence the initial frictional force on the hoop is limiting and has magnitude $\mu R (= \mu mg)$ and acts to the left. After time t the situation is shown in Fig. 10.29(ii). During the period for which the point of contact is slipping to the right, i.e. whilst $\dot{y} - a\dot{\phi} > 0$, the equations of motion of O and of moments about O are

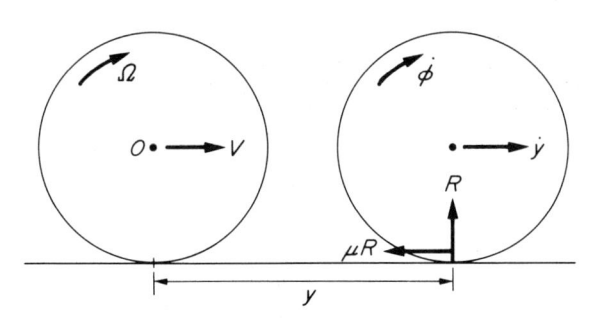

$$m\ddot{y} = -\mu mg, \quad mk^2\ddot{\phi} = \mu mga. \tag{3}$$

FIG. 10.29 (ii)

Using the initial conditions $\dot{y} = V$, $\dot{\phi} = \Omega$, after integration

$$\Rightarrow \dot{y} = V - \mu gt, \quad a\dot{\phi} = a\Omega + \mu g(a^2/k^2)t \tag{4}$$

$$\Rightarrow \dot{y} - a\dot{\phi} = V - a\Omega - \mu g(1 + a^2/k^2)t. \tag{5}$$

Equations (3), (4), (5) hold only so long as $\dot{y} - a\dot{\phi} > 0$ and (5) shows that $\dot{y} - a\dot{\phi}$ decreases as t increases. The equations cease to hold when $\dot{y} - a\dot{\phi} = 0$, i.e. when $t = T_1$, where

$$T_1 = \frac{V - a\Omega}{\mu g(1 + a^2/k^2)}. \tag{6}$$

At this instant

$$\dot{y} = V_1 = \frac{a\Omega + V(a^2/k^2)}{1 + a^2/k^2}, \tag{7}$$

and the hoop rolls instantaneously. The analysis of case (a) then implies that for $t > T_1$ the hoop continues to roll with speed V_1. At the instant $t = T_1$ the frictional force changes instantaneously from limiting friction acting to the left to zero. A further integration of equations (4) shows that when rolling commences

$$y = VT_1 - \frac{1}{2}\mu g T_1^2 = \frac{(V - a\Omega)[V(1 + 2a^2/k^2) + a\Omega]}{2\mu g(1 + a^2/k^2)^2}. \tag{8}$$

If the hoop starts with a back-spin, Ω is negative, and if it has a sufficiently large numerical value so that $V_1 < 0$, i.e. $a\Omega + V(a^2/k^2) < 0$, when rolling commences the hoop is rolling to the left. Further, if $a\Omega + V(1 + 2a^2/k^2) < 0$, the hoop is to the left of its starting point when rolling commences.

(c) In this case, $V < a\Omega$, the point of contact with the plane slips to the left initially, and, whilst $\dot{z} - a\dot{\psi} < 0$, the frictional force on the hoop is μmg to the right and the situation is as shown in Fig. 10.29 (iii). As in the above cases the equations of motion are

$$m\ddot{z} = \mu mg, \qquad mk^2\ddot{\psi} = -\mu mga \tag{9}$$

$$\Rightarrow \dot{z} = V + \mu gt, \qquad a\dot{\psi} = a\Omega - \mu g(a^2/k^2)t \tag{10}$$

$$\Rightarrow a\dot{\psi} - \dot{z} = a\Omega - V - \mu g(1 + a^2/k^2)t. \tag{11}$$

Equations (9), (10), (11) are only valid whilst $a\dot{\psi} - \dot{z} > 0$, i.e. up to the instant $t = T_2$, where

$$T_2 = \frac{a\Omega - V}{\mu g(1 + a^2/k^2)}, \tag{12}$$

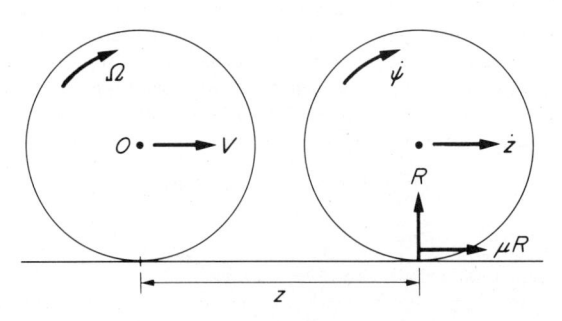

Fig. 10.29 (iii)

when $\dot{z}$ has the value

$$V_2 = \frac{a\Omega + V(a^2/k^2)}{1 + a^2/k^2}. \tag{13}$$

At this point the hoop rolls instantaneously and therefore continues to do so with speed $V_2 > 0$, to the right. As before, the frictional force is discontinuous at this point.

Example 8. One end A of a uniform rod AB, of mass m and length $2a$, is smoothly pivoted to a fixed point O. The rod is released from rest when it is horizontal. Find its angular speed when it is vertical.

At the instant when the rod is vertical the end A is released from the pivot. Discuss the subsequent motion of the rod.

The energy equation gives for the angular speed ω when the rod is vertical

$$\tfrac{1}{2}\cdot\tfrac{4}{3}ma^2\omega^2 = mga$$
$$\Rightarrow \omega = \sqrt{[3g/(2a)]}.$$

At this instant the velocity of the centre, G, of the rod is horizontal and of magnitude V, where

$$V = a\omega = \sqrt{(3ag/2)}.$$

During the subsequent motion G descends with acceleration $\mathbf{g}$ and so describes a parabola under gravity. Further, since the external forces have no moment about G, the angular acceleration of the rod is zero and so it continues to rotate with uniform angular speed ω. Thus the rod describes complete revolutions in periodic time $2\pi/\omega = 2\pi\sqrt{[2a/(3g)]}$.

Exercise 10.9

1. A thin uniform rod AB of mass m and length $2l$ lies on a smooth horizontal table. A particle of mass m moving with speed v in the direction making an angle $\tfrac{1}{4}\pi$ with AB collides with and adheres to the rod at the end B. Find the initial speed of the end A of the rod.

2. Prove that the moment of inertia of a uniform rod of length $2a$ about an axis intersecting the rod at right angles at a distance b from its centre is $M(\tfrac{1}{3}a^2 + b^2)$, where M is the mass of the rod.

A wheel of radius a is formed of a thin uniform rim of mass M and n uniform spokes of length $a - b$, each of mass m, which are fastened to the rim and to an axle of radius b and mass m'. The wheel rolls down an inclined plane of inclination α. Find the acceleration of its centre. (O.C.)

3. A uniform plank AB, of mass m and of length $2a$, rests with A on a smooth horizontal floor and B against a smooth vertical wall. The plank is initially vertical and at rest, and is slightly disturbed so that A slides in a direction at right angles to the wall and B slides vertically downwards. Find the angular speed and angular acceleration of the plank when it has turned through the angle $\tan^{-1}(3/4)$. Find also the reactions at A and B when the plank is in this position. (N.)

4. A uniform rod of mass m and length $2l$ has a small light ring attached to one end and the ring is free to slide on a smooth horizontal wire. When the rod is at rest in the vertical position, it receives at the lower end a horizontal blow of magnitude $2mV$ parallel to the wire. Show that the lower end of the rod starts off with a speed $8V$, and that when it is inclined at an angle θ to the vertical,

$$(1 + 3\sin^2\theta)l^2\dot\theta^2 = 36V^2 - 6gl(1 - \cos\theta).$$ (N.)

5. Prove that the moment of inertia of a uniform cylindrical tube of mass M, about its axis, is equal to

$$\tfrac{1}{2}M(a^2 + b^2),$$

where a, b are the internal and external radii of the tube.

The tube starts from rest and rolls, with its axis horizontal, down an inclined plane (making an angle β with the horizontal). By applying the principle of energy, or otherwise, show that T, the time occupied in travelling a distance l along the plane, is given by

$$l\left(3 + \frac{a^2}{b^2}\right) = gT^2\sin\beta.$$ (O.C.)

6. A uniform rod of length $2a$ is held at an angle α to the vertical with its lower end on a smooth horizontal plane, and is then released. Show that the angular speed of the rod when it becomes horizontal is

$$\sqrt{\{(3g\cos\alpha)/(2a)\}}.$$ (N.)

7. A uniform circular disc of mass m and radius a rolls (without slipping) down a plane inclined at an angle α to the horizontal; the coefficient of friction between the plane and the disc is μ. Prove that the acceleration of the centre of the disc is $\tfrac{2}{3}g\sin\alpha$ and that $\mu \geq \tfrac{1}{3}\tan\alpha$.

If a constant braking couple of moment G is applied to the disc prove that it will continue to roll down the plane if

$$\mu \geq \frac{1}{3}\tan\alpha + \frac{2G\sec\alpha}{3mag}.$$ (N.)

8. A string connects a particle of mass m to the end B of a uniform rod AB of mass M. The system is at rest on a smooth horizontal table with the particle at B. If the particle is projected horizontally with velocity V perpendicular to AB, prove that its speed immediately after the string becomes taut is $4mV/(M+4m)$.

Show also that the loss of kinetic energy of the system due to the tightening of the string is $\tfrac{1}{2}mMV^2/(M+4m)$. (N.)

9. A circular hoop of radius a is rotating and sliding in a vertical plane along a straight line on a rough horizontal table. After a time t the speed of its centre is $V - \mu g t$ and the speed of its highest point is V in the same direction as that of the centre, V being constant. Find the locus of the instantaneous centre of the hoop referred to horizontal and vertical axes through the initial position of its centre. Prove that the hoop will begin to roll after its centre has described a distance $3V^2/8\mu g$. (N.)

10. A uniform solid sphere is projected along a rough horizontal plane with speed V and no spin. Show that it skids for a distance $12V^2/(49\mu g)$, where μ is the coefficient of friction, and that it subsequently rolls with speed $5V/7$. (L.)

11. A uniform circular disc, of mass m and radius a, is at rest with its plane vertical on a rough horizontal table; the coefficient of friction at the point of contact is μ. A constant couple λmga, acting in the plane of the disc, is applied to the disc. Assuming that the disc rolls on the plane show that the acceleration of its centre is $2\lambda g/3$. Show also that in this case $\mu \geqslant 2\lambda/3$. (L.)

12. A uniform rod AB of mass m and length $2a$ slides with its ends on a fixed smooth vertical circular wire whose centre is O. If b denotes the distance of the centre C of the rod from O, and θ the angle which OC makes with the downward vertical, prove that

$$\dot\theta^2 = \frac{6bg}{3b^2 + a^2}(\cos\theta - \cos\phi),$$

where $\phi\,(<\pi/2)$ is the maximum value of θ during the motion. (L.)

13. A particle of mass m rests on a smooth fixed plane inclined at an angle α to the horizontal. A light inextensible string attached to the particle passes up the line of greatest slope of the plane, over a fixed smooth peg at the highest point of the plane, and then vertically downwards to a reel of mass M and radius a round which it is wrapped and to which its end is attached. The reel has symmetry about an axis about which its radius of gyration is k, and the string is free to unwind with this axis remaining horizontal. Find the tension of the string and the acceleration of the particle. If the system is released from rest, show that the subsequent motion of the reel will be upwards if

$$\sin\alpha > \frac{M}{m} + \frac{a^2}{k^2}.$$ (N.)

14. A uniform sphere, of radius a, is projected with speed u and without rotation up a rough plane of inclination α. The coefficient of friction between the sphere and the plane is $\tfrac{1}{7}\tan\alpha$. Show that friction acts down the plane for a time $2u/(3g\sin\alpha)$, and then up the plane, and that the angular speed of the sphere when its centre is reduced to rest is $5u/36a$. (O.C.)

15. A uniform heavy sphere of radius a rolls without slipping inside a fixed rough cylinder of inner radius $5a$ and with its axis horizontal. The sphere rolls from the equilibrium position, in a vertical plane perpendicular to the axis of the cylinder, with initial angular speed $2\sqrt{(ng/a)}$.

Show that the sphere will roll completely round the inner surface of the cylinder if $n > 27/7$. Show also that if $n = 4$ the coefficient of friction between the sphere and the cylinder is $\geqslant 2/\sqrt{35}$. (L.)

16. A uniform sphere of radius a, rotating with angular speed ω about a horizontal diameter, is placed on a rough plane inclined at angle α to the horizontal so that the axis of rotation is parallel to the plane and the direction of rotation is such that the sphere tends to roll up the plane. If μ, the coefficient of friction, is greater than $\tan\alpha$ and if the centre of the sphere is initially at rest, show that the sphere will slip for a time $2a\omega/(7\mu\cos\alpha - 2\sin\alpha)g$ before rolling begins, and find the distance through which the centre of the sphere moves while slipping is taking place. Find also the angular speed at the moment when rolling begins. (N.)

17. A uniform rod AB of mass m and length $2a$ is held in the horizontal position and is then allowed to fall freely. After falling through a height $8a$ the end A engages with a fixed smooth hinge. Prove that the rod begins to turn about A with angular speed $3\sqrt{(g/a)}$.

Find the speed of B when it is vertically below A. (L.)

18. Show, by integration, that the moment of inertia of a uniform thin square plate $ABCD$, of mass m and side $2a$, about the edge AB is $4ma^2/3$. Deduce the moment of inertia of the plate about an axis through its centre and perpendicular to its plane.

The plate is rotating freely with angular speed Ω about this axis, which is vertical, when the corner A strikes a stationary particle of mass $2m/3$ which adheres to the plate. Find
(a) the subsequent angular speed of the plate,
(b) the impulse which acts on the particle,
(c) the loss of kinetic energy due to the impact. (L.)

Miscellaneous Exercise 10

1. A uniform semicircular lamina is bounded by a thin uniform wire of mass equal to that of the lamina. When this rigid body performs small oscillations about its own diameter, fixed horizontally, the period is T_1. When it performs small oscillations about a fixed horizontal axis through the mid-point of the diameter and perpendicular to the plane of the rigid body the period is T_2. Prove that

$$T_1^2 : T_2^2 = 9\pi + 6 : 18\pi + 20.$$ (L.)

2. Show that the moment of inertia of a uniform solid right circular cone, of mass m, height h and base radius r, about a diameter of its base is

$$m(3r^2 + 2h^2)/20.$$

The cone is free to rotate about a fixed smooth horizontal axis coinciding with a diameter of its base. Find the period of small oscillations under gravity. (L.)

3. (i) A rod AB passes through a small fixed ring at O and is constrained to move so that the end A lies on a fixed straight line distant c from O and not passing through O. Find the position of the instantaneous centre of rotation, I, of the rod and show that the locus in space of I is a parabola.

(ii) Two particles of masses $5m$ and m are connected by an inextensible string of negligible mass which passes over a pulley of mass $8m$, radius a and radius of gyration $a/\sqrt{2}$ about its axis. The pulley rotates about a frictionless horizontal axis and the friction between the string and the pulley is large enough to prevent any slipping. Show that in the motion of the system the heavier particle descends with an acceleration $2g/5$, where g is the acceleration due to gravity. (L.)

4. A uniform rod AB, of mass m and length $2a$, is free to rotate in a vertical plane about a horizontal axis through A. The rod is slightly disturbed from the position in which B is vertically above A. Show that in the subsequent motion

$$2a\dot\theta^2 = 3g(1 - \cos\theta),$$

where θ is the angle made by AB with the upward vertical. Calculate the horizontal and vertical components of the force exerted by the rod on the axis of rotation. (L.)

5. Three uniform rods, each of length $2a$ and mass m, are rigidly attached end to end to form an equilateral triangle ABC, which can turn freely in a vertical plane about a fixed smooth horizontal axis through A. Show that the moment of inertia of the system about this axis is $6ma^2$.

If the triangle is released from rest with AB horizontal and C below AB, show that when BC is horizontal the angular speed is $\sqrt{[g/(a\sqrt{3})]}$.

If, at this instant, the triangle is brought to rest by a horizontal impulse applied at C, find the magnitude of the impulse. (L.)

6. A pendulum is formed from a uniform circular disc of radius a, centre O and mass M, attached rigidly at a point P of its circumference to the end of a thin uniform rigid rod of mass m whose other end is at A. The points O, P and A are collinear and such that $AP = l$ and $AO = l + a$. The rod is smoothly pivoted at A and the pendulum is free to swing in the vertical plane which contains the disc. When it is hanging in equilibrium the pendulum is set in motion with initial angular speed ω. By considering the energy equation, or otherwise, show that the angular speed $\dot\theta$ when the rod makes an angle θ with the downward vertical is given by $\dot\theta^2 = \omega^2 - k(1 - \cos\theta)$, where k is a constant which should be determined in terms of m, M, l and a.

(i) Find in terms of k the least value of ω which will cause the rod to reach a horizontal position.
(ii) Show that if θ is so small that θ^3 and higher powers may be neglected then $\dot\theta \approx (\omega^2 - \tfrac{1}{2}k\theta^2)^{1/2}$ and

hence, or otherwise, express θ in terms of k, ω and t, the pendulum being set in motion at time $t = 0$.
(N.)

7. Given that the moment of inertia of a uniform lamina of mass m in the form of an equilateral triangle PQR of side $2a$ about QR is $\tfrac{1}{2}ma^2$, deduce, or show otherwise, that the moment of inertia about a parallel axis through P is $3ma^2/2$.

Show that the moment of inertia of a uniform regular hexagonal lamina of mass M and side $2a$ about one edge is $23Ma^2/6$.

The hexagonal lamina is free to rotate about a fixed horizontal axis coinciding with an edge and is released from rest when its plane is horizontal. Calculate the magnitude and direction of the total force on the axis when the lamina is vertical.
(L.)

8. A thin uniform rod AB, of mass m and length $2a$, is free to rotate in a vertical plane about a smooth horizontal axis through A. A particle of mass $m/3$ is attached to the end B. The rod is released from rest when in a horizontal position and moves under gravity and the influence of a constant frictional couple G. The rod first comes to rest instantaneously after rotating through an angle $5\pi/6$. Show that $G = mga/\pi$. Show also that, if AB has turned through an angle θ, where $0 < \theta < 5\pi/6$, at time t after release,

$$4ma^2 \dot{\theta}^2 = 5mga \sin\theta - 3G\theta.$$

Show also that the angular acceleration of the rod first vanishes when $\cos\theta = 3/(5\pi)$.
(L.)

9. A uniform circular disc of mass m and radius a rests on a horizontal table with its centre at a distance $\tfrac{1}{2}a$ from the edge of the table with part of the disc projecting over a straight edge of the table. The disc is given a blow which causes it to start turning about the edge of the table with angular speed ω. Prove that, if the coefficient of friction between the disc and the edge is μ, the disc will not slip initially if $a\omega^2 < \mu g$.

Show also that, as long as the disc does not slip on the edge, its angular speed when its plane makes an angle θ with the horizontal is given by

$$a\dot{\theta}^2 = a\omega^2 - 2g \sin\theta.$$
(L.)

10. A flywheel has moment of inertia I about its axis of rotation which is vertical. When a constant frictional torque is applied to the flywheel it is brought to rest from angular speed Ω_1 in time t_1. When a particle of mass m is attached to the flywheel at a distance a from the axis, the flywheel comes to rest from angular speed Ω_2 in time t_2 against the same frictional torque. Show that

$$I = \frac{ma^2 \Omega_2 t_1}{\Omega_1 t_2 - \Omega_2 t_1}.$$

If the numbers of revolutions of the flywheel in the respective cases are n_1, n_2, show that

$$I = \frac{ma^2 n_2 t_1^2}{n_1 t_2^2 - n_2 t_1^2}.$$
(L.)

11. A smooth straight tube of length a and mass m is free to rotate on a smooth table about one end. A particle also of mass m is placed in the tube at its middle point and the system is given an initial angular speed Ω. Find the angular speed of the tube when the particle reaches the end of the tube, and show that the speed of the particle relative to the tube at that instant is $\tfrac{1}{8}a\Omega \sqrt{21}$.

12. A uniform circular wire ring of radius a is free to revolve about a vertical diameter. A small bead of the same mass as the ring can slide upon the ring without friction. When the bead is at the uppermost point of the ring the latter is given an angular speed Ω about a vertical diameter, and the bead is slightly displaced from its position. Find the angular speed of rotation of the ring when the bead is at the extremity of a horizontal diameter, and show that the square of its speed relative to the ring is then $\tfrac{1}{3}(a^2\Omega^2 + 6ag)$.
(N.)

13. Two circular cogwheels A and B are of radii R and r and masses M and m respectively. They can rotate freely in the same plane about parallel axles through their respective centres, their moments of inertia about these axles being $\tfrac{1}{2}MR^2$ and $\tfrac{1}{2}mr^2$ respectively. Initially they are both rotating anti-clockwise with angular speeds Ω and ω, respectively. They are brought together so that the teeth mesh, giving rise to a tangential impulse of magnitude J. The wheels then continue to rotate in mesh; the direction of rotation of B is reversed and the angular speeds of A and B are respectively Ω' and ω'. Prove that $\Omega' = (MR\Omega - mr\omega)/\{(M+m)R\}$, and find an expression of similar form for ω'.

Show that $J = mM(R\Omega + r\omega)/\{2(M+m)\}$. Show further that the loss of kinetic energy is $mM(R\Omega + r\omega)^2/\{4(M+m)\}$.
(N.)

14. A uniform rod AB of mass M and length $2l$ is at rest in a vertical position with its lower end B on a smooth horizontal plane, when an impulse of magnitude J is applied to the rod at B in a direction making an acute angle θ with the upward vertical. Find J such that the rod has turned through an angle π when A is at the level of the plane, and by choosing a suitable value of θ show that the least value of J for which this occurs is $M\sqrt{(\pi gl/3)}$.

If $\theta = \pi/2$, prove that B leaves the plane if $J > \frac{1}{3}M\sqrt{(gl)}$. (L.)

15. A uniform square lamina $ABCD$ of mass M and side $2a$ lies at rest on a smooth horizontal table. The lamina is given a horizontal impulse of magnitude MV at the vertex A in the direction AB. Find the distance travelled by the centre of the lamina in the time taken by the lamina to make one complete revolution. Find also the space and body centrodes of the lamina. (L.)

16. A uniform disc, of mass m and radius a, is rolling with speed V, without slipping and with its plane vertical, on a rough horizontal plane. The coefficient of friction between the disc and the plane is μ. At time $t = 0$ a braking couple of magnitude $5\mu amg/3$ is applied to the disc and is subsequently maintained. Show that the disc skids immediately. Show also that rotation will instantaneously cease at time $3V/(4\mu g)$ while the disc is still moving forward. (L.)

17. A uniform hoop has mass m, radius a and moment of inertia ma^2 about its axis. It is projected with its plane vertical along a rough horizontal table. The coefficient of friction between the plane and the hoop is $\frac{1}{2}$. Initially the hoop has no spin and its centre has speed V. Show that the hoop skids for a time V/g and find the angular speed with which it subsequently rolls. (L.)

18. A uniform circular disc, of mass m and radius r, starts from rest and rolls, without slipping and with its plane vertical, down a line of greatest slope of a fixed plane of inclination α to the horizontal. Find the time taken for the disc to move through a distance x down the plane.

Show that the coefficient of friction between the disc and the plane during this motion is at least $\frac{1}{3}\tan\alpha$. (L.)

19. Show that the moment of inertia of a uniform solid sphere of mass M and radius a about a diameter is $2Ma^2/5$.

Such a sphere, resting on a smooth horizontal table, is given a horizontal blow J at a height $3a/4$ above the table in a vertical plane containing the centre. Calculate the linear speed of the centre and the angular speed of the sphere immediately after the blow. Calculate also the kinetic energy generated by the blow. (L.)

20. A rod lying on a smooth table is free to rotate about one end O which is freely pivoted to the table. The rod, at rest, is struck at a point distant d from O by a particle of mass m moving along the table with velocity V at right angles to the rod. The collision brings the particle to rest. Prove that the moment of inertia of the rod about O is equal to md^2/e, where e is the coefficient of restitution. Prove also that when the rod strikes the particle again it gives the particle a speed eV. (O.C.)

21. A rigid hoop of radius a, mass m and moment of inertia (about an axis through its centre and perpendicular to its plane) ma^2 rolls in a vertical plane along a horizontal road until it comes to a step of height ka (where $k < 1$) at right angles to its path. When the hoop comes into contact with the step at A, there is no slipping at A nor any subsequent loss of contact with A. If the hoop was rolling with angular speed ω immediately before striking the step, show that the hoop begins to mount the step with an angular speed of $\omega(1 - \frac{1}{2}k)$.

Find the condition that the hoop surmounts the step and begins to roll across it. (O.C.S.M.P.)

22. Prove that the moment of inertia of a uniform lamina of mass m, in the shape of an equilateral triangle ABC of height h, about an axis through one vertex and perpendicular to the plane ABC, is $\frac{5}{9}mh^2$.

The lamina is free to move in a vertical plane about a smooth horizontal axis through A. While it is in equilibrium, with BC below A, the mid-point of BC is given a velocity u in the direction BC. Prove that the lamina will make complete revolutions if $u^2 > \frac{24}{5}gh$. (O.C.)

23. A uniform rod of length $2a$ and mass m is lying on a smooth horizontal table. One end is given a velocity V upwards. Prove that while the other end is still in contact with the table the energy equation is

$$\tfrac{1}{6}ma^2(1 + 3\cos^2\theta)\dot\theta^2 = \tfrac{1}{6}mV^2 - mag\sin\theta,$$

θ being the angle the rod has turned through.

Prove that, if $6ag < V^2 < 7ag$, the rod turns right over without losing contact with the table. (O.C.)

24. Two small smooth beads of masses m and $2m$ can slide along a thin circular hoop of radius a. The beads are connected to each other by a light rigid straight rod of length a. The hoop is pivoted so that it can rotate freely about a fixed vertical diameter, the moment of inertia of the hoop about the diameter being $\frac{1}{4}ma^2$. The rod is supported vertically with the lighter bead uppermost. The hoop is then given an angular speed Ω and simultaneously the rod is released from its position of rest relative to the hoop. If in the subsequent motion the rod reaches its lower horizontal position, find the angular speed of the hoop and the speeds of the beads relative to the hoop in this configuration.

Hence show that this configuration cannot be reached if

$$15a\Omega^2 > 4g(3\sqrt{3}-1).$$

Prove also that the rod cannot reach its upper horizontal position for any value of Ω.

25. A uniform circular disc has mass $4M$, radius $4a$ and centre O, and AOB is a diameter. A circular hole of radius $2a$ is made in the disc, the centre of the hole being at a point C on AB, where $AC = 3a$. The resulting lamina is free to rotate about a fixed smooth horizontal axis through A perpendicular to the plane of the lamina. Show that the moment of inertia of the lamina about this axis is $85Ma^2$.

Find the period of small oscillations of the lamina about the position of stable equilibrium.

While the lamina is hanging down in equilibrium from A it is given an impulse J along the horizontal line through O in the plane of the lamina. Find the angular speed with which the lamina starts to rotate.

Show that the diameter AB will reach the horizontal position provided that

$$J^2 \geqslant \frac{1105}{8} M^2 ga.$$

Verify the dimensional correctness of this inequality.

26. A uniform rod PQ of length $2a$ and mass m hangs in a horizontal position, being suspended by two light inextensible strings of length $2a$ attached to its ends and to two fixed points A, B. The strings PA, QB are vertical. The rod is rotated about the vertical through its centre, through $180°$, while the strings remain taut, thus bringing P just below B and Q just below A. From this position the rod is released. Prove that the energy equation can be written

$$\tfrac{1}{2}ma^2\dot\theta^2\{\cos^2\tfrac{1}{2}\theta + \tfrac{1}{3}\} = mgz,$$

where θ is the angle turned about the vertical and z is the distance dropped at time t.

Prove that the tension in each string as the rod passes through its lowest position is $\tfrac{7}{8}mg$. (O.C.)

27. A uniform smooth rod of length $2l$ is lying on a smooth horizontal table, passing through a ring O pivoted freely to the table. The ring is a distance a ($a < l$) from the centre of the rod. The centre of the rod is suddenly given a velocity V at right angles to the rod and along the table. Write down equations expressing that in the subsequent motion the kinetic energy and the angular momentum about O are constant.

Prove that when the end of the rod reaches the ring the speed of this end is

$$\{V/(2al)\}\{(3a^2+l^2)(l^2-a^2)\}^{1/2}.$$

28. Two cog wheels A and B are spinning on parallel shafts with angular speeds ω and 2ω respectively in opposite senses. A and B may be regarded as uniform circular discs of radii $r, 2r$ and masses $m, 2m$ respectively. The shafts are moved towards one another so that the wheels engage and thereafter rotate in mesh. Find the angular speeds of the wheels immediately after they engage.

The subsequent motion is such that each wheel is acted upon at any instant by a frictional couple of magnitude λ times the angular speed of the wheel at that instant. Show that A again has angular speed ω after a time interval

$$\tau = \frac{6mr^2}{5\lambda}\ln 3,$$

during which it is rotated through an angle

$$\alpha = \frac{12mr^2\omega}{5\lambda}$$

while in mesh with B.

Find the total energy lost due to the frictional couples when A and B have been brought to rest.

Find also the dimensions of λ and verify the correctness of the dimensions of the above expressions for τ and α.

29. A uniform heavy sphere of radius a is rolling without slipping along a smooth horizontal table with speed V. The highest point P of the sphere is suddenly fixed. Given that $9V^2 > 140\,ag$, show that the sphere will make complete revolutions around P.

30. A uniform rod AB, of length $2a$ and mass $2m$, lies at rest on a smooth horizontal plane. A particle of mass m moving with horizontal velocity V perpendicular to the rod hits the rod at one end and remains attached to the rod. Show that the angular speed of the rod immediately after the impact is $V/(2a)$ and that the loss of kinetic energy due to the impact is $mV^2/6$. (L.)

31. A uniform solid sphere, initially rotating about a horizontal diameter with angular speed Ω, is gently placed on a rough horizontal plane. The radius of the sphere is a, and the coefficient of friction between the sphere and the plane is μ. Prove that the sphere moves a distance $2a^2\Omega^2/(49\mu g)$ before slipping ceases, and find the angular speed at that instant.

Prove that in the subsequent motion the sphere rolls on the plane. (L.)

32. A uniform rod AB, of length $2a$ and mass M, is held inclined to the vertical at an angle β with the end A on a smooth horizontal table. The rod is then released from rest. Assuming that A does not leave the table, find the angular speed and the angular acceleration of the rod just before it strikes the table. Prove also that, at the same instant, the reaction of the table on the rod is of magnitude $Mg/4$. (L.)

33. A circular disc of radius a rolls with constant angular speed ω along a straight line without slipping. State the velocity of the instantaneous centre of rotation of the disc.

Two particles of equal mass fixed to the ends of a light rod AB lie at rest on a smooth horizontal plane. The particle at A is given a horizontal impulse at right angles to AB. Show that the instantaneous centre of rotation of the rod moves with constant velocity, and sketch the paths of the two particles. (L.)

34. A uniform thin hollow cylinder, of mass $2m$ and radius a, is open at both ends. A particle P, of mass m, is attached to the inner surface of the cylinder at a point of the cross-section perpendicular to the axis of the cylinder through the centre of mass of the cylinder. The cylinder is placed on a rough horizontal plane with the particle P level with the axis of the cylinder and is released. Assuming that no slipping occurs, show that, when the cylinder has turned through an angle θ,

$$a(3 - \sin\theta)\dot\theta^2 = g\sin\theta.$$

Find the angular speed and angular acceleration when the cylinder has turned through an angle $\pi/6$. Show that the speed of the particle is then $\sqrt{(ag/5)}$. (L.)

35. Four equal uniform rods, each of mass M and length $2a$, form a rigid square frame $ABCD$. The frame is free to rotate in a horizontal plane about a fixed vertical axis through its centre of mass and perpendicular to the plane of the frame. A mouse, of mass $M/9$, is at E, the mid point of AB, and frame and mouse are initially at rest. If the mouse now runs along the frame from E towards B, show that, when the mouse is at a point P on EB, where $EP = x$, and the frame has rotated through an angle θ, then the components of the velocity of the mouse parallel to AB and CB are $\dot x - a\dot\theta$ and $x\dot\theta$ respectively. Hence, by using the conservation of moment of momentum, show that

$$\dot\theta = a\dot x/(49a^2 + x^2).$$

Deduce that, when the mouse reaches B, the frame will have turned through an angle ϕ where $7\tan 7\phi = 1$. (L.)

36. A uniform solid sphere of mass M and radius a is projected without rotation along a rough horizontal plane with speed U. Show that the sphere slides for a time $t = 2U/(7\mu g)$ during which it covers a distance $12U^2/(49\mu g)$, where μ is the coefficient of friction between the sphere and the plane. Show also that, when the sphere begins to roll, its angular speed is $5U/(7a)$. (L.)

37. Show that the moment of inertia of a uniform solid sphere of radius a and mass M about a diameter is $2Ma^2/5$.

The sphere is rotating freely about a fixed horizontal diameter with angular speed ω when a particle of mass $M/20$ is attached to the lowest point of the sphere. Show that the angular speed of the sphere and particle immediately after attaching the particle is $8\omega/9$.

If the sphere and particle subsequently come to instantaneous rest, find the angle that the radius to the particle makes with the horizontal at this instant. Hence show that the sphere will make complete revolutions about its axis if $16a\omega^2 > 9g$. (L.)

38. A uniform circular disc, of mass m and radius r, is on a rough plane which is inclined to the horizontal at an angle α, the disc being in a vertical plane through a line of greatest slope. A couple of moment $kmgr \sin \alpha$ acts in the plane of the disc, tending to roll the disc up the plane. Given that the coefficient of friction between the disc and the plane is $2 \tan \alpha$, find the acceleration of the centre of the disc in each of the cases $k = 1$, $k = 2$, $k = 3$.

Sketch a graph showing how this acceleration varies with k for $0 \leqslant k \leqslant 3$. (L.)

39. A uniform circular disc, of mass m, radius a and centre O, lies at rest on a smooth horizontal table. The disc is given a horizontal blow at a point A of its circumference, such that the initial velocity of A is of magnitude V at an angle of $45°$ to AO. Show that the magnitude of the blow is $mV(\sqrt{5})/3$, and that the kinetic energy given to the disc is $mV^2/3$.

Find the distance travelled by the centre O while the disc makes one revolution. (L.)

40. Prove that the moment of inertia of a thin uniform circular disc of mass m and radius a about an axis through its centre perpendicular to its plane is $ma^2/2$.

The disc is pivoted about a point O of its circumference and can move freely in its own plane which is vertical. When the disc is hanging in equilibrium, a particle of mass m moving with velocity v horizontally in the plane containing the disc strikes the disc at one end of its horizontal diameter and adheres. Show that the angular speed of the disc after the collision is $2v/7a$.

In the subsequent motion the radius of the disc through O, which is initially vertical, just rises to the horizontal. Find v in terms of a and g. (L.)

11 Further Probability

11:1 THE BINOMIAL AND GEOMETRIC PROBABILITY DISTRIBUTIONS

We briefly recall what was said in Chapter 6 regarding two particular types of discrete probability distribution. Firstly, suppose that the probability of success in a single trial is constant and equal to p, so that the probability of failure is $1 - p = q$. Then, in a series of n separate trials, the number of successes is a discrete variable $r \in \{0, 1, 2, \ldots, n\}$. It was shown in Chapter 6 that the probability $P(r)$ of just r successes in n independent trials is equal to the coefficient of t^r in the binomial expansion of $(pt + q)^n$

$$\Rightarrow P(r) = {}_nC_r \, p^r \, q^{n-r}. \tag{11.1}$$

This is therefore known as a *binomial distribution*. The mean $\mu = np$ and the variance $\sigma^2 = npq$.

Secondly, suppose that a single trial can result in one of just two outcomes A or A', so that if $P(A) = p$, $P(A') = 1 - p$. Then the probability of A' occurring for the first time on the $(n+1)^{th}$ trial is $p^n(1-p)$. This represents a *geometric distribution*.

Example 1. It may be assumed that 5% of a greengrocer's stock of apples is bad. If a customer buys six apples, chosen at random, find the probability that
(a) half of them will be bad,
(b) more than two of them will be bad.

The probability that r apples out of six will be bad is the coefficient of t^r in $(\frac{1}{20}t + \frac{19}{20})^6$;

(a) $P(3 \text{ bad}) = {}_6C_3 (\frac{1}{20})^3 (\frac{19}{20})^3 = (\frac{1}{20})^2 (\frac{19}{20})^3 \approx 0.0021$.

(b) $P(\text{more than 2 bad}) = 1 - P(\text{none bad}) - P(\text{one bad})$

$$= 1 - (\tfrac{19}{20})^6 - 6(\tfrac{19}{20})^5 (\tfrac{1}{20}) = 1 - (\tfrac{19}{20})^5 (\tfrac{5}{4}) \approx 0.033.$$

Example 2. Three marksmen A, B, and C consider that their chances of scoring a bullseye with a single shot are $\frac{2}{5}$, $\frac{1}{3}$ and $\frac{1}{4}$ respectively. They are to shoot in the order A, B, C in a competition, the winner being the one to score the first bullseye. Find the probability of A winning, and show that A's chance of winning is halved if he shoots last instead of first.

The probability of A winning with his first shot is $\frac{2}{5}$. The probability of A winning with his second shot, implying that the previous three shots were misses, is $\frac{3}{5} \times \frac{2}{3} \times \frac{3}{4} \times \frac{2}{5}$.

Hence $P(A \text{ wins}) = \frac{2}{5} + \frac{3}{5} \times \frac{2}{3} \times \frac{3}{4} \times \frac{2}{5} + (\frac{3}{5})^2 \times (\frac{2}{3})^2 \times (\frac{3}{4})^2 \times \frac{2}{5} + \ldots$

$= \frac{2}{5}[1 + \frac{2}{5} \times \frac{3}{4} + (\frac{2}{5} \times \frac{3}{4})^2 + \ldots]$

$= \frac{2}{5} \times \dfrac{1}{1 - (\frac{2}{5} \times \frac{3}{4})} = \frac{4}{7}.$

[Here we have used the formula $a/(1-r)$ for the sum to infinity of a convergent geometric series.]
If A shoots last, the probability of winning with his first shot is $\frac{2}{3} \times \frac{3}{4} \times \frac{2}{5}$ (Clearly it makes no difference in which order B and C shoot.)

$\Rightarrow P(A \text{ wins}) = \frac{2}{3} \times \frac{3}{4} \times \frac{2}{5}[1 + (\frac{2}{3} \times \frac{3}{4} \times \frac{3}{5}) + (\frac{2}{3} \times \frac{3}{4} \times \frac{3}{5})^2 + \ldots]$

$= \frac{1}{5} \times \dfrac{1}{1 - \frac{3}{10}} = \frac{2}{7}.$

Hence A's chance of winning is halved.

Exercise 11.1 contains revision examples on topics covered in Chapter 6.

Exercise 11.1

1. In a mixed school there are 800 boys of mean height 175 cm with standard deviation 10 cm, and 400 girls of mean height 166 cm with standard deviation 7 cm. Find the mean height and standard deviation for all the pupils in the school.

2. Two unbiased coins are tossed; for each, a score of 1 is recorded for a head and 2 for a tail. The two scores are multiplied together to give a product P. Find the mean value of P, and the standard deviation of P.
(O.C.S.M.P.)

3. A population consists of n_1 males and n_2 females. The mean heights of the males and females are μ_1 and μ_2 respectively and the variances of the heights are σ_1^2 and σ_2^2 respectively. Show that the mean height of the whole population is $w_1\mu_1 + w_2\mu_2$ and the variance is $w_1\sigma_1^2 + w_2\sigma_2^2 + w_1w_2(\mu_1 - \mu_2)^2$, where $w_1 = n_1/(n_1 + n_2)$ and $w_2 = n_2/(n_1 + n_2)$.

Hence, or otherwise, show that, if a single observation taking the value x is added to a population of size n with mean μ and variance σ^2, the new variance will be larger than the old if

$$|\mu - x| > \left(\frac{n+1}{n}\right)^{1/2} \sigma. \qquad \text{(O.)}$$

4. A crossword puzzle is published in *The Times* each day of the week, except Sunday. A man is able to complete, on average, eight out of ten of the crossword puzzles.
  (i) Find the expected value and the standard deviation of the number of completed crosswords in a given week.
  (ii) Show that the probability that he will complete at least five in a given week is 0·655 (to three significant figures).
  (iii) Given that he completes the puzzle on Monday, find, to three significant figures, the probability that he will complete at least four in the rest of the week.
  (iv) Find, to three significant figures, the probability that, in a period of 4 weeks, he completes four or less in only one of the 4 weeks. (C.)

5. An ordinary dice is unbiased and has six sides numbered from 1 to 6.
Explain what is meant by the statement that if five ordinary dice are thrown simultaneously the number of sixes appearing has a Binomial distribution, explaining carefully what this means.
Show that the average and the standard deviation of the number of 6's are both equal to 5/6. (If you use a formula for the average or the standard deviation you should prove that it is correct.)
Find also the probability of obtaining
  (i) four or more 6's,
  (ii) exactly three even numbers (not necessarily all different). (O.C)

6. An unbiased die is thrown six times. Calculate the probabilities that the six scores obtained will
   (i) consist of exactly two 6's and four odd numbers,
   (ii) be 1, 2, 3, 4, 5, 6 in some order,
   (iii) have a product which is an even number,
   (iv) be such that a 6 occurs **only** on the last throw and that exactly three of the first five throws result in odd numbers. (L.)

7. Nine discs numbered from 1 to 9 are placed in a bag and three discs are then drawn at random without replacement. The number on the first disc drawn is denoted by n, and the sum of the numbers on the three discs is denoted by S.
   (a) Find the probability that $S = 10$.
   (b) Find the probability that both $n = 2$ and $S = 10$.
   (c) Given that $n = 2$, find the probability that $S = 10$.
   (d) Given that $n \neq 2$, find the probability that $S = 10$. (L.)

8. Given that events A and B are independent, events B and C are mutually exclusive,

$$P(A) = 0\cdot 4, \quad P(A \cup B) = 0\cdot 58, \quad P(C) = 2P(B),$$

find,
   (a) the value of $P(B)$,
   (b) the greatest possible value of $P(A \cap C)$,
   (c) the greatest possible value of $P(A \cup C)$. (L.)

9. A trial consists of selecting a card at random from a pack of 52 playing cards and then replacing it. Obtain an estimate, to two significant figures, of the probability that, in 104 such trials, the ace of spades is selected at least three times. (L.)

10. A hand of five cards is drawn from a pack of playing cards and consists of an Ace, a King, a Queen and a pair of Jacks. The pair of Jacks is retained, the other three cards are discarded, and three more are drawn from the forty-seven cards remaining in the pack.
Calculate the probability that the hand now contains
   (a) a trio of Jacks (but not four Jacks),
   (b) two different pairs (but not a trio),
   (c) a trio and a pair. (L.)

11. A man rolls a die to select one of three boxes. If he rolls a 6, he selects the red box; if he rolls a 5 or a 4, he selects the blue box; and if he rolls a 3 or a 2 or a 1, he selects the yellow box. He opens the box he has chosen and selects a coin at random from it. The red box contains three gold coins, the blue box two gold coins and one silver coin, and the yellow box one gold coin and two bronze coins. Using a tree diagram, or otherwise, find the probability that
   (a) he selects the silver coin,
   (b) he selects a gold coin,
   (c) having selected and retained a gold coin, if he now selects at random a second coin from the same box, it will also be gold. (L.)

12. The probability of success in a certain event is p. If n trials are made, prove that, for the corresponding binomial distribution, the mean number of successes is np.

A sampling bottle contains a very large number of balls, coloured white or red or green but otherwise indistinguishable. Samples of ten balls can be examined without removal from the bottle. The distribution of red balls per sample, in a hundred trials, is as follows:

Number of red balls	0	1	2	3	4	5 or more	Total
Frequency	55	33	9	2	1	0	100

Calculate the mean number of red balls per sample, and estimate the proportion of red balls in the bottle. If $0\cdot 4\%$ of the balls are known to be green, and a hundred trials are made, calculate the expected number of trials in which exactly one green ball appears. (C.)

13. In a firm's car park, at leaving time, there are eight *Ford*, six *Vauxhall* and ten *Leyland* cars. If no more cars enter the car park and cars leave one at a time and at random, find the probabilities that
   (a) the first car to leave is a *Ford*,
   (b) the last car to leave is a *Leyland*,

(c) all *Vauxhall* cars leave one after another,
(d) when there are only three cars left in the car park there is just one of each make. (L.)

14. (a) Two cards are drawn at random and without replacement from a pack of twenty cards which are numbered 1, 2, ..., 20. Find the probabilities
(i) that one is a 7 and the other an odd number,
(ii) that one is not greater than 5 and the other not less than 5.
(b) Two players compete by drawing in turn and without replacement one ball at random from a bag containing four red and four white balls, and the player who first draws a red ball is the winner. What is the probability that the player who makes the first draw wins?
(Answers to this question may be given as fractions.) (O.C.)

15. Suppose that letters sent by first- and second-class post have probabilities of being delivered a given number of days after posting according to the following table (weekends are ignored):

Days to delivery	1	2	3
1st class	0·9	0·1	0
2nd class	0·1	0·6	0·3

The secretary of a committee posts a letter to a committee member who replies immediately using the same class of post. What is the probability that four or more days are taken from the secretary posting the letter to receiving the reply if (a) first-class, (b) second-class post is used?
The secretary sends out four letters and each member replies immediately by the same class of post. Assuming the letters move independently, what is the probability that the secretary receives (a) all the replies within 3 days using first-class post, (b) at least two replies within 3 days using second-class post? (N.)

16. A box contains one red ball and one blue ball. A trial consists of drawing one ball at random from the box, replacing it in the box, and again drawing a ball and replacing it. A trial is successful if the blue ball is drawn twice and is unsuccessful otherwise. After an unsuccessful trial an additional red ball is placed in the box and another trial is carried out; after a successful trial, no further trials are made. Given that the first $(n-1)$ trials are unsuccessful, show that the probability that the nth trial is unsuccessful is $n(n+2)/(n+1)^2$.
Deduce that the probability of having at least n unsuccessful trials is $\frac{1}{2}(n+2)/(n+1)$. Hence show that the probability that the nth trial is successful is $1/[2n(n+1)]$. (L.)

17. (a) How many times should an unbiased die be thrown if the probability that a 6 should appear at least once is to be greater than $\frac{9}{10}$?
(b) The odds *against* a certain student solving a problem are assessed at 3 to 1 (i.e. probability of solving it is $\frac{1}{4}$). The odds *in favour* of a second student solving it are 7 to 5. Find the probability that the problem will be solved if both students separately attempt it.
(c) A batch of fifty articles contains three which are defective. The articles are drawn in succession (without replacement) from the batch and tested. Show that the chance that the first defective met will be the rth article drawn is

$$(50-r)(49-r)/39\,200.$$ (N.)

18. When a biased coin is tossed the probability of obtaining a head is p. Show that the probability that exactly t tosses of the coin are needed to obtain h heads is

$$\binom{t-1}{h-1} p^h (1-p)^{t-h}.$$

For an unbiased coin find the probability that
(a) exactly six tosses are needed to obtain three heads,
(b) more than nine tosses are needed to obtain three heads (L.)

19. (i) Two cards are taken, without replacement, from a well shuffled pack of playing cards. Calculate the probability that
(a) both are Aces,
(b) at least one King or Queen is obtained.
(ii) In a game a turn consists of drawing a card from a pack of playing cards and rolling one of two dice. If the card is a heart, the die rolled is a red one numbered 2, 4, 6, 8, 10, 12; if not, a blue die numbered 1, 2, 3, 4, 5, 6 is rolled. Find the probability of obtaining a *total score* of 6 on the dice in two turns. (L.)

11:2 CONTINUOUS PROBABILITY DISTRIBUTIONS

In the distributions considered so far, r has been a discrete variate; but, if x varies continuously in the interval (a, b), we can no longer consider "the probability of x being x_1". However, we can expect the probability distribution to be a function of x, such that the probability of x lying in the interval $(x - \frac{1}{2}\delta x, x + \frac{1}{2}\delta x)$ is $\phi(x)\delta x$, and such that

$$\int_a^b \phi(x)\,dx = 1. \tag{11.2}$$

Thus for the mean or expected value and variance we have

$$\mu = \int_a^b x\phi(x)\,dx,$$

$$\sigma^2 = \int_a^b x^2\phi(x)\,dx - \mu^2. \tag{11.3}$$

Though we cannot consider the probability of a continuous variate x taking one particular value, it is often convenient to consider the probability of x being not greater than a certain value. We can define a *cumulative probability function* $F(X)$ such that, for X in (a, b),

$$F(X) = P(a \leq x \leq X) = \int_a^X \phi(x)\,dx, \tag{11.4}$$

and clearly $F'(x) = \phi(x)$.

The simplest example of a continuous probability distribution occurs when all values of x in (a, b) are equally likely, so that $\phi(x) = $ constant (k)

$$\Rightarrow \int_a^b k\,dx = 1 \Rightarrow k = \frac{1}{b-a}.$$

This is known as a *rectangular distribution*.

Example 1. The variate x is uniformly distributed in the range $0 \leq x \leq a$. Find the mean and variance of x.
Show that for a fixed y ($0 \leq y \leq a$) the probability that $x \leq a - y$ is $1 - y/a$. If y varies independently of x and is uniformly distributed in the range $0 \leq y \leq b < a$, show that the probability that $x + y \leq a$ is $1 - b/2a$.
Two pieces of wood are cut, each to be of length 30 cm; the first may be any length between 29 and 31 cm, the second any length between 28 and 32 cm, with uniform probability in each case. Find the probability that their combined length should be greater than 61 cm.

$$F(X) = \int_0^X k\,dx. \quad \int_0^a k\,dx = 1 \Rightarrow k = 1/a.$$

$$\mu = \frac{1}{a}\int_0^a x\,dx = a/2. \quad \sigma^2 = \frac{1}{a}\int_0^a x^2\,dx - (a^2/4) = a^2/12.$$

For fixed y,

$$P(x \leqslant a - y) = \frac{1}{a}\int_0^{a-y} dx = 1 - y/a.$$

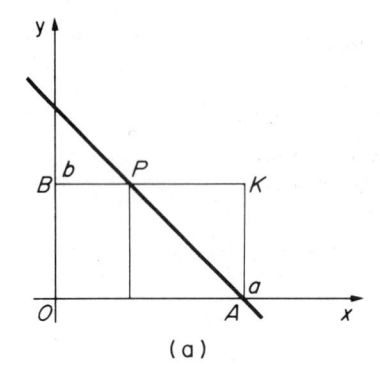

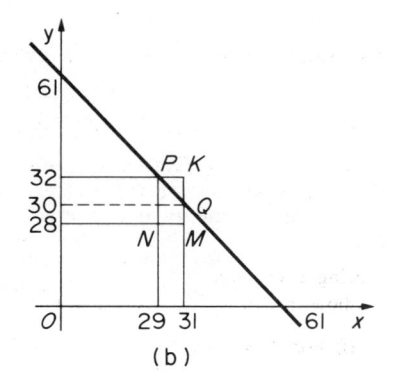

FIG. 11.1.

When $0 \leqslant x \leqslant a$, $0 \leqslant y \leqslant b$, the point (x, y) lies within the rectangle $OAKB$, Fig. 11.1(a).
If $x + y \leqslant a$ the point lies within the trapezium $OAPB$, where $AP \equiv x + y = a$.
Hence

$$P(x + y \leqslant a) = \text{Area } OAPB/\text{area } OAKB$$
$$= \tfrac{1}{2}b(a + a - b)/(ab) = 1 - b/2a.$$

If the lengths of the two pieces of wood are x, y respectively, the point (x, y) must lie within the rectangle $PKMN$, Fig. 11.1(b).
$x + y > 61 \Rightarrow (x, y)$ lies within the triangle PKQ, where $PQ \equiv x + y = 61$. Hence

$$P(x + y > 61) = \text{Area } PKQ/\text{area } PKMN$$
$$= \frac{1}{2} \cdot \frac{2 \times 2}{4 \times 2} = \frac{1}{4}.$$

Example 2. X is a continuous random variable with probability density function given by

$$f(x) = \begin{cases} ke^{-2x} & (x \geqslant 0), \\ 0 & (x < 0). \end{cases}$$

(i) Prove that $k = 2$.
(ii) Calculate the mean and variance of X.
(iii) *The median of a continuous distribution is defined to be the number m such that $P(X \geqslant m) = \tfrac{1}{2}$. Prove (do not merely verify) that in this case $m = \tfrac{1}{2}\ln 2$.*

(i) $\int_0^\infty ke^{-2x}dx = 1 \Rightarrow k\left[-\tfrac{1}{2}e^{-2x}\right]_0^\infty = 1 \Rightarrow \tfrac{1}{2}k = 1 \Rightarrow k = 2.$

(ii) $E(x) = 2\int_0^\infty xe^{-2x}dx = 2\left[-\tfrac{1}{2}xe^{-2x}\right]_0^\infty + 2 \cdot \tfrac{1}{2}\int_0^\infty e^{-2x}dx = 0 + \left[-\tfrac{1}{2}e^{-2x}\right]_0^\infty = -\tfrac{1}{2}(0 - 1) = \tfrac{1}{2}.$

$\text{Var}(X) = 2\int_0^\infty x^2 e^{-2x} - (\tfrac{1}{2})^2 = \left[-x^2 e^{-2x}\right]_0^\infty + 2\int_0^\infty xe^{-2x}dx - \tfrac{1}{4} = 0 + \tfrac{1}{2} - \tfrac{1}{4} = \tfrac{1}{4}.$

(iii) $\int_m^\infty 2e^{-2x}dx = \frac{1}{2} \Rightarrow -[e^{-2x}]_m^\infty = \frac{1}{2}$

$\Rightarrow e^{-2m} = \frac{1}{2} \Rightarrow -2m = -\ln 2 \Rightarrow m = \frac{1}{2}\ln 2.$

Example 3. (i) An integer takes the value r with probability P_r defined by

$P_r = 0 \quad (r < 0);$ $P_r = 0 \quad (r > 5);$
$P_r = ar^2 \quad (0 \leqslant r \leqslant 2);$ $P_r = a(5-r)^2 \quad (3 \leqslant r \leqslant 5),$

where a is constant. Find the mean and variance of the variate x where $x = r^2$.
(ii) The continuous Gamma distribution has probability density $f(x)$, where

$$f(x) = ce^{-x}x^{p-1} \quad (0 \leqslant x < \infty, \ p \in \mathbb{N})$$

c being a constant.
Show that the mean and variance are each equal to p.

(i) For $P_r \neq 0, r \in \{1, 2, 3, 4\} \Rightarrow x \in \{1, 4, 9, 16\}.$

$$\sum_1^4 P_r = 1 \Rightarrow a(1 + 4 + 4 + 1) = 1 \Rightarrow a = \tfrac{1}{10}.$$

Mean of $x = \sum_1^4 xP(x) = \tfrac{1}{10}(1.1 + 4.4 + 9.4 + 16.1) = 6.9$

$\Rightarrow$ Variance of $x = \tfrac{1}{10}(1.1 + 16.4 + 81.4 + 256.1) - 6.9^2 = 16.89.$

(ii) $I_n = \int_0^\infty e^{-x}x^n dx \Rightarrow I_n = \left[-e^{-x}x^n\right]_0^\infty + n\int_0^\infty e^{-x}x^{n-1}dx$

$\Rightarrow I_n = nI_{n-1} \qquad (n \in \mathbb{N})$

$\Rightarrow I_n = n!$

$\int_0^\infty ce^{-x}x^{p-1}dx = 1 \Rightarrow c = 1/(p-1)!$

$\mu = \int_0^\infty ce^{-x}x^p dx = (p!)/(p-1)! = p.$

$\sigma^2 = \int_0^\infty ce^{-x}x^{p+1}dx - \mu^2 = p(p+1) - p^2 = p.$

Exercise 11.2

1. The random variable X takes values between 0 and 4 so that

$$P(X \leqslant x) = \begin{cases} 0 & , x < 0, \\ \dfrac{1}{96}x^2(10-x), & 0 \leqslant x \leqslant 4, \\ 1 & , x > 4. \end{cases}$$

Obtain the probability density function $f(x)$ and the mean value of X. Sketch the graph of $f(x)$. (L.)

2. (a) A positive integer variate takes the value r with probability $P(r) = a$ (a constant) for $0 < r \leq n$ and $P(r) = 0$ for $r > n$. Find the value of a and the mean and variance of r.

(b) The probability distribution of a continuous variate x is given by $P(x) = a$ (a constant) for $0 \leq x \leq n$ and $P(x) = 0$ elsewhere. Find the value of a and the mean and variance of x.

(c) Find the mean and variance of z where $z = 2x$ and x is the variate of (b) above. (O.C.)

3. The probability density function y of a continuous variable x is given by $y = \frac{2}{9}(k^2 - x^2)$ for $-k \leq x \leq k$ and $y = 0$ for all other values of x.

Sketch the graph of y.

Calculate
(a) the value of k,
(b) the probability that $x < \frac{1}{2}$. (C.)

4. Write down the two conditions which must be satisfied by any probability density function.

The function f, where
$$f(x) = \begin{cases} k\{a - (x-b)^2\} & \text{when } -1 \leq x \leq 1, \\ 0 & \text{otherwise,} \end{cases}$$

where k, a and b are constants and where $k > 0$, is a probability density function with zero mean. Show that $b = 0$, express k in terms of a and find the set of possible values of a.

Determine the value of a for which the variance of the distribution has its smallest value. (L.)

5. (i) A continuous probability distribution has density function $f(x) = cx^2(1 - x^2)$ for $0 \leq x \leq 1$ and $f(x) = 0$ elsewhere. Show that the expected value is $5/8$ and the variance $17/448$.

(ii) The negative binomial distribution for the probability of r events (r integer) is given by the rule that $P(r)$ is equal to the coefficient of t^r in the expansion in ascending powers of t of $G(t) = (1-p)^n(1-pt)^{-n}$ for $0 \leq r < \infty$ and $0 \leq p < 1$. Show that the expected value μ of r is $np/(1-p)$ and the variance $np/(1-p)^2$. (O.C.)

6. A random variable x has cumulative probability function $F(x)$, where
$$F(x) = \begin{cases} 0 & (x \leq a) \\ \dfrac{x-a}{b-a} & (a < x < b), \\ 1 & (x \geq b). \end{cases}$$

Find the probability density function $\phi(x)$, and sketch the graph of $\phi(x)$. Obtain the mean and variance of x.

7. A chance of failure distribution is given for time t by the probability density function
$$p(t) = (1/a)e^{-t/a} \quad (0 < t < \infty).$$

Show that a is the mean time to failure and that the variance is a^2.

Two components in a machine have failure time distributions corresponding to means a and $2a$ respectively. The machine will stop if either component fails and the failures of the two components are independent. Show that the chance of the machine continuing to operate for a time a from the start is $e^{-3/2}$. (O.C.)

8. For certain electrical bulbs, the time x (hours) to failure is thought to have a distribution with probability density function $f(x)$ such that
$$f(x) = \begin{cases} \lambda e^{-\lambda x} & x \geq 0, \\ 0, & x < 0. \end{cases}$$

A number of bulbs are put on test and after 200 hours it is found that 10% have failed. Estimate to two significant figures the value of λ and the probability that a new bulb of this type will last for more than 500 hours. (L.)

9. A continuous random variable X has probability density function $f(x)$ satisfying
$$f(x) = \begin{cases} \lambda e^{-\lambda x}, & x \geq 0, \\ 0, & x < 0. \end{cases}$$

Given that $b > a$, calculate $P(X \geqslant a)$ and $P(a \leqslant X \leqslant b)$ and show that $P(X \leqslant b | X \geqslant a)$ depends only on the value of $b - a$, being otherwise independent of a.

Interpret this property in connection with problems such as those involving waiting time or replacement of parts. (L.)

10. A random variable X is uniformly distributed over the interval $(0, 1)$, and n independent values of X are chosen. Write down the probability that all are less than x and hence show that the probability density function for the greatest of a set of n independent values of X is nx^{n-1}.

Find the mean of this distribution and show that the variance is

$$\frac{n}{(n+1)^2 (n+2)},$$

[You may use the fact that the probability density function is $F'(x)$ where $F(x) = P(X < x)$.] (N.)

11. The variable x is uniformly distributed in an interval of length l. Two independent observations x_1 and x_2 are made of x. By representing the joint distribution of x_1, x_2 on a suitable diagram, or otherwise, show that if $z = x_1 - x_2$, the probability density function $f(z)$ of z is

$$f(z) = (l + z)/l^2, \quad -l \leqslant z \leqslant 0,$$
$$f(z) = (l - z)/l^2, \quad 0 \leqslant z \leqslant l,$$

and $f(z) = 0$ elsewhere.

Hence show that $E(|x_1 - x_2|) = \frac{1}{3}l$.

From the distribution of $|x_1 - x_2|$ obtain values z_L and z_U such that

$$P(|x_1 - x_2| \leqslant z_L) = P(|x_1 - x_2| \geqslant z_U) = 0.05.$$ (O.C.)

11:3 THE POISSON DISTRIBUTION

The binomial distribution arises from considering a sample of definite size n for which the number of times an event with probability p did and did not occur can be calculated. The Poisson distribution is the limiting case of the binomial distribution when n becomes very large and p becomes very small, e.g. when we consider isolated events in time. Such events have a small probability of happening in any one trial but a large number of trials takes place. In illustration consider the (radioactive) disintegration of uranium-238. The probability that any one atom disintegrates, and so activates a Geiger counter, within a small interval of time is extremely small. However, in say 10^{-3} g of uranium there are about 2.5×10^{18} atoms and the result is that approximately 12 atoms disintegrate per second.

We consider the binomial distribution with mean m so that $np = m$ and consider the probability of a (usually small) finite number k of successes when $n \to \infty$, $p \to 0$ but m is a non-zero finite constant. Then

$$P(k \text{ successes}) = \lim_{\substack{n \to \infty \\ p \to 0}} \frac{n!}{k!(n-k)!} p^k (1-p)^{n-k}$$

$$= \lim_{n \to \infty} \frac{n(n-1) \ldots [n-(k-1)]}{k!} \left(\frac{m}{n}\right)^k \left(1 - \frac{m}{n}\right)^{n-k}$$

$$= \lim_{n \to \infty} \frac{1[1-(1/n)][1-(2/n)] \ldots [1-(k-1)/n]}{k!} \cdot \frac{m^k [1-(m/n)]^n}{[1-(m/n)]^k}.$$

Since $\dfrac{m}{n} \to 0$, $\qquad [1-(m/n)]^k \to 1$ when k is finite.

But $\qquad\qquad\qquad \lim\limits_{n\to\infty} [1-(m/n)]^n = e^{-m}$.

Hence $\qquad\qquad\qquad P(k \text{ successes}) = \dfrac{m^k e^{-m}}{k!}$. \hfill (11.5)

In fact we have derived the *Poisson distribution* which can be described thus: If the average number of occurrences of a rare event in a large number of trials is m, then

$$P(k \text{ successes}) = \dfrac{m^k e^{-m}}{k!}.$$

The mean of the Poisson distribution is

$$\mu = \sum_{k=1}^{\infty} k \dfrac{m^k e^{-m}}{k!}$$

$$= me^{-m} \sum_{k=1}^{\infty} \dfrac{m^{k-1}}{(k-1)!} = me^{-m} e^m = m. \qquad (11.6)$$

The variance is

$$\sigma^2 = \sum_{k=0}^{\infty} \dfrac{k^2 m^k e^{-m}}{k!} - m^2$$

$$= me^{-m} \sum_{k=1}^{\infty} \dfrac{km^{k-1}}{(k-1)!} - m^2$$

$$= me^{-m} \dfrac{d}{dm}(me^m) - m^2 = m$$

$$\Rightarrow \sigma = \sqrt{m}. \qquad (11.7)$$

Note that, for the Poisson distribution, the variance is equal to the mean. It is the diagnostic which is used to check on the hypothesis that a frequency distribution is Poisson (i.e. sample variance $\approx$ sample mean).

Example 1. A car takes one minute to cross a bridge, and 270 cars cross it each hour. Find, approximately, the probability that there are at least two cars on the bridge at any given time.

The mean number of cars on the bridge at any moment is $270/60 = 9/2$. We *assume* that the distribution is a Poisson one with $m = 9/2$. Then

$$P \{\text{at least two cars}\} = 1 - P\{\text{no car}\} - P\{1 \text{ car}\}$$
$$= 1 - e^{-9/2} - \tfrac{9}{2} e^{-9/2}$$
$$\approx 0\cdot 94.$$

Example 2. The number of emergency calls received at a telephone exchange during the interval from 10 p.m. to midnight was recorded over 100 consecutive days recently, as follows:

Number of calls	0	1	2	3	4	5
Frequency	21	34	26	13	5	1

Show that the mean number of calls was 1·5 and calculate the theoretical Poisson frequency corresponding to each number of calls.

Estimate the probability that during the same interval of 2 hours (i) no calls will be received tonight, (ii) only one call will be received in a period of three such consecutive nights.

The mean number of calls is

$$(0 \times 21 + 1 \times 34 + 2 \times 26 + 3 \times 13 + 4 \times 5 + 5 \times 1)/100$$
$$= 150/100 = 1 \cdot 5.$$

The theoretical Poisson frequency distribution is calculated from the terms

$$100e^{-1 \cdot 5}, \quad 100 \times 1 \cdot 5 e^{-1 \cdot 5}, \quad 100 \times \frac{(1 \cdot 5)^2 e^{-1 \cdot 5}}{2!}, \ldots$$

to be rounded off to the nearest integer.

Number of calls	0	1	2	3	4	5
Frequency	22	33	25	13	5	2

(i) The probability that no calls are received (in one 2-hour period) tonight is $e^{-1 \cdot 5} \approx 0 \cdot 223$.

(ii) The probability that one call will be received in one specified night is $1 \cdot 5 e^{-1 \cdot 5} \approx 0 \cdot 334$. Therefore the probability that only one call is received in three such successive nights is

$$_3C_1 (0 \cdot 223)^2 (0 \cdot 334) \approx 0 \cdot 050.$$

Example 3. During a certain period of the day, an average of four people enter a certain store each minute. Calculate the probability that

(i) two people enter during a particular minute,
(ii) six people enter during a particular minute,
(iii) no people enter during a particular minute,
(iv) no people enter during two successive minutes.

Since the number of trials is large but the mean is small, we *assume* a Poisson distribution. Hence

(i) $P(2) = \dfrac{e^{-4} \times 4^2}{2!} \approx 0 \cdot 147$;

(ii) $P(6) = \dfrac{e^{-4} \times 4^6}{6!} \approx 0 \cdot 104$;

(iii) $P(0) = e^{-4} \approx 0 \cdot 018$;

(iv) $P(0) \times P(0) \approx 0 \cdot 0003$.

Example 4. At a stage in the mass production of lamp holders, random samples, each of 40 articles, are examined and the number of defective articles recorded. The numbers of defective articles in each of 200 samples are shown in the following frequency table:

Defectives	0	1	2	3	4	5	6	7
No. of samples	29	56	42	42	23	7	0	1

Find the mean number of defectives per sample and the variance. Give reasons for thinking that the distribution approximates to a Poisson distribution.

Show that on this assumption there is a probability of about 5 % of a sample containing more than four defectives and a probability of less than 1 % of a sample containing more than six defectives. (O.C.)

Mean $= (0 + 56 + 84 + 126 + 92 + 35 + 0 + 7)/200 = 2$.
Variance $= (0 + 56 + 42 \times 4 + 42 \times 9 + 23 \times 16 + 7 \times 25 + 0 + 49)/200 - 2^2 = 1 \cdot 97$.

Because n is large and μ is small, and $\sigma^2 \approx \mu$, we assume a Poisson distribution.

$$P(5) + P(6) + P(7) = e^{-2}(2^5/5! + 2^6/6! + 2^7/7!) = 2^5 e^{-2}(1 + \tfrac{1}{3} + \tfrac{2}{21})/5!$$
$$\approx 0{\cdot}05 = 5\%.$$
$$P(7) = 2^7 e^{-2}/7! \approx 0{\cdot}003 < 1\%.$$

Example 5. The table shows the number of occasions in a year on which 100 people had to consult their doctor.

Number of times	0	1	2	3	4	5
Number of people	48	32	15	3	1	1

Show that the mean number of times a person consulted his doctor is 0·8. Calculate the Poisson frequencies corresponding to this mean.

Assuming that the Poisson distribution with mean 0·8 is applicable to this situation, find the probability that someone will consult his doctor more than three times in a 4-year period, the number of consultations in any year being taken to be independent of the numbers in other years. (C.)

$$\text{Mean number of consultations} = \frac{1}{100}(0 + 32 + 30 + 9 + 4 + 5) = 0{\cdot}8.$$

Assuming a Poisson distribution, the probability $P(0)$ of a person consulting his doctor no times in a year is $e^{-0{\cdot}8} \approx 0{\cdot}449$.
Similarly

$$P(1) = e^{-0{\cdot}8} \times 0{\cdot}8 \approx 0{\cdot}359, \qquad P(2) = \frac{e^{-0{\cdot}8} \times (0{\cdot}8)^2}{2!} \approx 0{\cdot}144,$$

$$P(3) = \frac{e^{-0{\cdot}8} \times (0{\cdot}8)^3}{3!} \approx 0{\cdot}038, \qquad P(4) = \frac{e^{-0{\cdot}8} \times (0{\cdot}8)^4}{4!} \approx 0{\cdot}008,$$

$$P(5) = \frac{e^{-0{\cdot}8} \times (0{\cdot}8)^5}{5!} \approx 0{\cdot}001.$$

Hence Poisson frequencies are 44·9, 35·9, 14·4, 3·8, 0·8, 0·1, to 1 decimal place.

$$\text{Mean number of consultations in 4 years} = 3{\cdot}2$$

$$\Rightarrow P(>3) = 1 - P(0) - P(1) - P(2) - P(3)$$
$$= 1 - e^{-3{\cdot}2}|1 + 3{\cdot}2 + (3{\cdot}2)^2/2! + (3{\cdot}2)^3/3!|$$
$$= 1 - e^{-3{\cdot}2}(1 + 3{\cdot}2 + 5{\cdot}12 + 5{\cdot}461) \approx 0{\cdot}397.$$

Example 6. The number of alpha particles emitted per unit time by a radioactive substance was counted over a period of time and the number of occasions f on which n particles were emitted is recorded in the following table:

n	0	1	2	3	4	5	6	7	8	9	10
f	80	206	254	212	136	68	30	10	3	1	0

Assuming that the distribution is of Poisson type, calculate correct to two significant figures the probability that the substance will emit (i) exactly five particles, (ii) more than two particles in any unit of time.

The mean number of particles emitted in one period is $2517/1000 = 2{\cdot}517$.
(i) The theoretical Poisson distribution implies that the probability of exactly five particles being emitted is

$$\frac{(2{\cdot}517)^5 e^{-2{\cdot}517}}{5!} \approx 0{\cdot}068.$$

(ii) The probabilities of 0, 1, 2, particles being emitted are

$$e^{-2{\cdot}517}, \qquad (2{\cdot}517)e^{-2{\cdot}517}, \qquad \tfrac{1}{2}(2{\cdot}517)^2 e^{-2{\cdot}517}.$$

These cases are mutually exclusive and so the probability required is

$$P(\text{more than 2 particles emitted})$$
$$= 1 - e^{-2\cdot517} - (2\cdot517)e^{-2\cdot517} - \tfrac{1}{2}(2\cdot517)^2 e^{-2\cdot517} \approx 0\cdot461.$$

Exercise 11.3

1. The number of suspected bone fractures X-rayed at a clinic in a morning is a Poisson variable with mean 4. On what proportion of mornings are no X-rays taken?
   If the number of suspected bone fractures exceeds 5, X-rays have to be taken in the afternoons; on what proportion of days does this occur?
   What is the probability that no afternoons will be needed in a period of 7 days? (L.)

2. Accidents per shift in a factory, recorded over 200 shifts, are shown in the following frequency table:

Accidents per shift	0	1	2	3	4	5
Frequency	140	41	12	6	0	1

   Calculate the mean number of accidents per shift and use the Poisson distribution to estimate the probability of there being three or more accidents in any one shift. (L.)

3. A man cuts a pack of 52 cards ten times, shuffling the pack after each cut. What is the chance of cutting a spade (a) every time, (b) exactly five times, (c) more than five times? Find the mean and standard deviation of the binomial distribution of the number of spades drawn.
   Use the Poisson distribution to calculate the chances of cutting two or more red aces in ten cuts of the pack. (L.)

4. If $P(X = k) = p(k)$ and X has a Poisson distribution with mean μ show that
   $$p(k+1) = \mu p(k)/(k+1).$$
   Given that $\mu = m + h$, where m is a positive integer and $0 < h < 1$, show that the most probable value of X is m. In the case where $h = 0$ show that X has two equally most probable values and find them.
   Records of the telephone calls received at a fire station indicate that the number of false calls received per week has a Poisson distribution with mean 3·0.
   (i) Calculate the probability that no false call will be received during next week.
   (ii) Find the smallest integer n for which there is a probability of at least 0·8 that n or fewer false calls will be received during next week.
   (iii) Find the largest integer s for which the probability of no false calls being received during the next s weeks is greater than 0·0001. (N.)

5. A family owns two cars and the demand for their use each day closely follows a Poisson distribution with mean 1·2. Calculate the expected number of days per year of 365 days on which there will be no demand for a car.
   Calculate also the expected number of days per year when it will not be possible to satisfy the family's demand for cars
   (a) with the two cars the family owns,
   (b) if the family now buys a third car. (L.)

6. The average number of telephone calls received by an office in a working day of 8 hours is 1920. Find the probability that more than four calls will be received during any particular minute, and the probability that no call will be received in an unbroken period of 2 minutes chosen at random. (A Poisson distribution may be assumed.) (L.)

7. The number of errors per page made by a certain trained copy typist has a Poisson distribution with mean 0·2. Calculate, to two significant figures, the probabilities that in a particular assignment by this copy typist
   (i) the first two pages contain no error,
   (ii) the first error occurs on the third page,
   (iii) exactly two of the first five pages contain no error. (N.)

8. In the first year of the life of a certain type of machine, the number of times a maintenance engineer is required has a Poisson distribution with mean four. Find the probability that more than four calls are necessary.

The first call is free of charge and subsequent calls cost £20 each. Find the mean cost of maintenance in the first year. (N.)

9. A man has three cars for hire by the day. On average, he has two requests a day for a car. Calculate, in terms of e, the probability that, on a particular day,
   (a) there are no requests,
   (b) he has to turn away custom. (L.)

10. Explain briefly what is meant by the Poisson distribution of rare events. Sketch the probability polygon of the distribution in the case where the mean m is unity.

The number of road accidents notified to a certain police station in a day is shown in the following frequency table relating to 300 successive days:

Accidents	0	1	2	3	4	5	6	7	Total
Frequency	90	113	64	21	7	3	1	1	300

Calculate the mean number of accidents notified daily.

Use the Poisson distribution with this mean to calculate the chance of four or more accidents being notified on a particular day. (O.C.)

11. The number of defective articles produced by a machine in 1 hour is a Poisson variate with mean 1·2. Show that the probability of there being more than two defectives produced in an hour is 0·12 approximately.

If more than two defectives are produced in an hour, the hour's production is scrapped. Find the chance of this happening more than once in 8 hours. (O.C.)

12. The probability distribution of r, the number of vehicles passing a point on a road in an interval of one minute, can be taken as a Poisson distribution with parameter λ, i.e.

$$P(r) = \frac{e^{-\lambda} \lambda^r}{r!} \quad (r = 0, 1, 2, \ldots).$$

Obtain the mean and variance of this distribution.

If the mean number of vehicles passing per minute is two, calculate the probability that in a given minute
   (i) one vehicle,
   (ii) less than three vehicles,
will pass the point.

Calculate also the least value of v such that the probability of obtaining a value of r less than or equal to v is greater than 0·5. (O.C.)

13. Explain what is meant by the Poisson distribution for events of rare frequency and show that the mean of the distribution is equal to its variance.

Samples of forty articles at a time are taken periodically from the continuous production of a machine and the number of samples containing 0, 1, 2, ... defective articles are recorded in the following table:

No. of defectives per sample	0	1	2	3	4	5	6	Total
No. of samples	30	23	27	14	4	2	0	100

Find the mean number of defectives per sample.

Assuming that this is the mean of the population and that the Poisson distribution applies, find the chance of;
   (a) a sample containing four or more defectives,
   (b) two successive samples containing between them four or more defectives. (O.C.)

14. In the manufacture of a certain article, it is found that on the average, one article in ten has to be rejected. What is the probability that three samples of ten articles contain between them not more than two rejects?

15. In a large lot of electric light bulbs 5 % of the bulbs are defective. Calculate the probability that a random sample of twenty will contain at most one defective bulb.

One-third of the lots presented for inspection have 5 % defective, the rest 10 % defective. If a lot is rejected when a random sample of twenty taken from it contains more than one defective bulb, find the proportion of lots which are rejected.

16. Razor blades of a certain kind are sold in packets of five. The following table shows the frequency distribution of 100 packets according to the number of faulty blades contained in them:

Number of faulty blades	0	1	2	3	4	5
Number of packets	84	10	3	2	1	0

Calculate the mean number of faulty blades per packet and, assuming that the binomial law applies, estimate the probability that a blade taken at random from any packet will be faulty.

17. Prove that the mean and the variance of a Poisson distribution are equal. The frequency of accidents per shift in a factory is shown in the following table:

Accidents per shift	0	1	2	3	4	5
Frequency	300	96	34	9	1	0

Prove that the distribution is approximately Poissonian and find the probability of the occurrence of more than two accidents in a shift.

18. A large batch of manufactured articles is accepted if either
(i) a random sample of six articles contains not more than one defective article, or
(ii) a random sample of six contains two defective articles and a second random sample of six is then drawn and found to contain no defective articles.
If, in fact, 20 % of the articles in the batch are defective, what is the chance of the batch being accepted?

19. Spot blemishes occur randomly along a steel wire. The number counted in consecutive centimetre lengths had the following distribution:

No. of spots	0	1	2	3	4	5	6	7
Frequency	102	150	112	56	21	6	2	1

If the count had been over consecutive 2-cm lengths, estimate the frequency of occurrence of three spots per 2 cm. (You may assume that the frequencies are close to a Poisson distribution.) (O.C.S.M.P.)

20. Describe briefly the Poisson distribution and show that its mean is equal to its variance. Show that if x and y are Poisson variates with means m_1 and m_2 respectively, then $z = x + y$ is also a Poisson variate.

Instruments A and B are set to record radiation from separate and independent sources. The numbers of particles recorded by the instruments in one second have Poisson distributions with means 2 and 4 respectively. Calculate the probabilities that
(a) at least one particle is recorded on one of the two instruments in a second,
(b) a total of four or more particles is recorded on the two instruments in a second. (O.C.)

11:4 THE NORMAL DISTRIBUTION

When a quantity is measured several times, the values obtained are distributed about a mean value. Even when all possible precautions have been taken to eliminate systematic errors, there is still a deviation between this mean value and the result of a single trial measurement. These remaining deviations are usually taken to be due to "random" errors arising from factors either unknown or beyond the control of the experimenter. These random deviations can occur either in integral values, e.g. the deviations about the mean in the number of passengers on a given train every weekday, or they can arise from different values of a continuous variable, e.g. the reading on an instrument dial.

By considering a limiting case of the binomial distribution we can obtain, by a plausible argument, the probability distribution for the errors in the value of a continuous variable.

Consider a variable whose measurement has zero for the mean value. Suppose that there are $2n$ factors each of which can cause a deviation $\pm\eta$ in the reading, and that $n+r$ cause a positive deviation. Then, for this case, the resultant deviation is

$$y = (n+r)\eta - (n-r)\eta = 2r\eta.$$

If we assume, plausibly, that positive or negative deviations are equally likely, the probability of this deviation occurring is

$$P(y) = \frac{(2n)!}{(n+r)!\,(n-r)!} \cdot \frac{1}{2^{n+r} 2^{n-r}} \tag{11.8}$$

and the standard deviation σ is given (from the properties of the binomial distribution) by

$$\sigma^2 = 2n \cdot \tfrac{1}{2} \cdot \tfrac{1}{2} (2\eta)^2 = 2n\eta^2.$$

In order to make the transition to continuous variables we suppose that n and r tend to infinity and η tends to zero, i.e. the number of factors at work is very large and the size of each of the individual deviations is infinitesimally small. For the error to be small we must have r small when compared with n.

Since

$$y = 2r\eta, \quad \sigma^2 = 2n\eta^2$$

we can eliminate η by writing

$$\frac{y^2}{2\sigma^2} = \frac{r^2}{n}.$$

Hence, we consider the limiting case in which r^2/n is finite, i.e. $n = O(r^2)$, and η tends to zero in such a manner that y, σ are comparable to the error found in practice.

To obtain the limiting value of the left-hand side of equation (11.8) we use *Stirling's approximation* for factorials of large numbers

$$N! = (2\pi N)^{1/2} N^N e^{-N} [1 + O(N^{-1/2})].$$

This leads to

$$P(y) = \frac{(2n)!}{(n+r)!\,(n-r)!} \frac{1}{2^{2n}}$$

$$\approx \frac{1}{\sqrt{(\pi n)}} \left(1 - \frac{r^2}{n^2}\right)^{-n} \left(1 + \frac{r}{n}\right)^{-r-1/2} \left(1 - \frac{r}{n}\right)^{r-1/2}$$

Because of the relative orders of magnitude we can replace these terms as follows:

$$\left(1 \mp \frac{r}{n}\right)^{\pm r - 1/2} \approx 1,$$

$$\left(1 - \frac{r^2}{n^2}\right)^{-n} = \left(1 - \frac{r^2}{n^2}\right)^{-(n^2/r^2).(r^2/n)} = e^{-r^2/n}.$$

Then
$$P(y) \approx \frac{1}{\sqrt{(\pi n)}} e^{-r^2/n}. \qquad (11.9)$$

Now the value of y can alter either way only by a step of η. Equation (11.9) gives the probability that y falls inside a range $dy = 2\eta = \sigma\sqrt{(2/n)}$, and we replace r^2/n by $y^2/(2\sigma^2)$. Hence the probability of y falling within the range dy enclosing the value $y = 2r\eta$ is

$$p(y)dy = P(y) = \frac{dy}{\sigma\sqrt{(2\pi)}} e^{-y^2/(2\sigma^2)}. \qquad (11.10)$$

This probability density p(y) is called the *normal distribution*. It is usually taken to give the distribution of the errors around the mean zero with standard deviation σ when these errors arise at random. If the errors are distributed about a mean μ with standard deviation σ, the probability density is

$$p(y) = \frac{1}{\sigma\sqrt{(2\pi)}} \exp\left[-\frac{(y-\mu)^2}{2\sigma^2}\right]. \qquad (11.11)$$

The above discussion, following the lines used by De Moivre, Laplace and Simpson originally, is in no sense a "proof" of the normal distribution. It gives a reason for thinking that errors which arise from unknown causes are distributed according to the normal probability density.

If a population of N measurements is distributed about a mean μ with standard deviation σ, then the number of measurements lying between $x - \frac{1}{2}\delta x$ and $x + \frac{1}{2}\delta x$ is

$$\frac{N}{\sigma\sqrt{(2\pi)}} \exp[-(x-\mu)^2/(2\sigma^2)]\,\delta x. \qquad (11.12)$$

It can be shown by numerical integration that for the above normal distribution the probability of a random measurement lying within a certain distance (*measured in terms of the standard deviation*) from the mean (in either the positive or the negative direction) is given approximately by the following table.

Deviation from the mean (either side) d/σ	Probability of value x in range $\mu - d < x < \mu + d$
0·1	0·08
0·25	0·20
0·5	0·38
0·67	0·50
0·8	0·58
1·0	0·68
1·25	0·79
1·50	0·88
2·0	0·954
3·0	0·997
4·0	0·999

Notes. (1) Almost all the distribution lies within three standard deviations of the mean.

(2) The probability that a measurement differs by more than 1.96σ from the mean is $\frac{1}{20}$ or 5%.

(3) The probability that a measurement differs by more than 2.58σ from the mean is $\frac{1}{100}$ or 1%.

(4) Table 1 on page 423 gives the values of the normal probability integral

$$F(X) = \frac{1}{\sqrt{(2\pi)}} \int_{-\infty}^{X} e^{-t^2/2} dt.$$

However, equation (11.12) implies that the probability of a single measurement lying in the range $x - \frac{1}{2}\delta x$ to $x + \frac{1}{2}\delta x$ is

$$\frac{1}{\sigma\sqrt{(2\pi)}} \exp[-(x-\mu)^2/(2\sigma^2)] \, \delta x,$$

and so the probability that it lies in the range $\mu + a\sigma < x < \mu + b\sigma$ is

$$P(\mu + a\sigma < x < \mu + b\sigma) = \frac{1}{\sigma\sqrt{(2\pi)}} \int_{\mu+a\sigma}^{\mu+b\sigma} \exp[-(x-\mu)^2/(2\sigma^2)] \, dx.$$

The substitution $x = \mu + t\sigma$ gives

$$P(\mu + a\sigma < x < \mu + b\sigma] = \frac{1}{\sqrt{(2\pi)}} \int_{a}^{b} e^{-t^2/2} \, dt = F(b) - F(a). \qquad (11.13)$$

We can therefore say that a random element of a normal population has a probability 0·95 (95 %) of lying within 1.96σ (or approximately 2σ) on either side of the mean, and is 99 % likely to lie within 2.56σ of the mean. These limits, 95 % and 99 %, which of course are quite arbitrary but which are frequently used, are often called "confidence limits". The use of Table 1 is illustrated in Examples 1–4 following.

Example 1. The following are the weights in g of a random sample of fifteen 32-g packets delivered by an automatic packing machine:

32·11	31·97	32·18	32·03	32·25
32·07	32·05	32·14	32·19	31·98
32·07	31·99	32·16	32·03	32·18

Calculate the mean and the standard deviation. Assuming the distribution to be normal, estimate the percentage of underweight packets which the machine is delivering.

The Board of Trade stipulates that such a machine must not deliver more than 5 % underweight packets. If the machine is to comply with this regulation and if its variability cannot be better controlled, calculate the value to which the mean must be raised.

Use of a calculator gives the mean $\mu = 32.09$, the variance $\sigma = 0.084$.

The mean is $\left(\dfrac{9}{8.4}\right)\sigma = 1.07\sigma$ above the critical value of 32·00. Table 1 indicates that 85·77 % of the population lies below $\mu + 1.07\sigma$ and so 14·23 % lies above $\mu + 1.07\sigma$. Since the normal distribution is

TABLE 1

THE NORMAL PROBABILITY INTEGRAL $F(X) = (2\pi)^{-1/2} \int_{-\infty}^{X} \exp(-\tfrac{1}{2}t^2)\,dt$

X	0	1	2	3	4	5	6	7	8	9
0.0	0.5000	0.5040	0.5080	0.5120	0.5160	0.5199	0.5239	0.5279	0.5319	0.5359
0.1	0.5398	0.5438	0.5478	0.5517	0.5557	0.5596	0.5636	0.5675	0.5714	0.5754
0.2	0.5793	0.5832	0.5871	0.5910	0.5948	0.5987	0.6026	0.6064	0.6103	0.6141
0.3	0.6179	0.6217	0.6255	0.6293	0.6331	0.6368	0.6406	0.6443	0.6480	0.6517
0.4	0.6554	0.6591	0.6628	0.6664	0.6700	0.6736	0.6772	0.6808	0.6844	0.6879
0.5	0.6915	0.6950	0.6985	0.7019	0.7054	0.7088	0.7123	0.7157	0.7190	0.7224
0.6	0.7258	0.7291	0.7324	0.7356	0.7389	0.7422	0.7454	0.7486	0.7518	0.7549
0.7	0.7580	0.7612	0.7642	0.7673	0.7704	0.7734	0.7764	0.7794	0.7823	0.7852
0.8	0.7881	0.7910	0.7939	0.7967	0.7996	0.8023	0.8051	0.8078	0.8106	0.8133
0.9	0.8159	0.8186	0.8212	0.8238	0.8264	0.8289	0.8315	0.8340	0.8365	0.8389
1.0	0.8413	0.8438	0.8461	0.8485	0.8508	0.8531	0.8554	0.8577	0.8599	0.8621
1.1	0.8643	0.8665	0.8686	0.8708	0.8729	0.8749	0.8770	0.8790	0.8810	0.8830
1.2	0.8849	0.8869	0.8888	0.8906	0.8925	0.8944	0.8962	0.8980	0.8997	0.9015
1.3	0.9032	0.9049	0.9066	0.9082	0.9099	0.9115	0.9131	0.9147	0.9162	0.9177
1.4	0.9192	0.9207	0.9222	0.9236	0.9251	0.9265	0.9279	0.9292	0.9306	0.9319
1.5	0.9332	0.9345	0.9357	0.9370	0.9382	0.9394	0.9406	0.9418	0.9430	0.9441
1.6	0.9452	0.9463	0.9474	0.9484	0.9495	0.9505	0.9515	0.9525	0.9535	0.9545
1.7	0.9554	0.9564	0.9573	0.9582	0.9591	0.9599	0.9608	0.9616	0.9625	0.9633
1.8	0.9641	0.9649	0.9656	0.9664	0.9671	0.9678	0.9686	0.9693	0.9699	0.9706
1.9	0.9713	0.9719	0.9726	0.9732	0.9738	0.9744	0.9750	0.9756	0.9761	0.9767
2.0	0.9772	0.9778	0.9783	0.9788	0.9793	0.9798	0.9803	0.9808	0.9812	0.9817
2.1	0.9821	0.9826	0.9830	0.9834	0.9838	0.9842	0.9846	0.9850	0.9854	0.9857
2.2	0.9861	0.9864	0.9868	0.9871	0.9875	0.9878	0.9881	0.9884	0.9887	0.9890
2.3	0.9893	0.9896	0.9898	0.9901	0.9904	0.9906	0.9909	0.9911	0.9913	0.9916
2.4	0.9918	0.9920	0.9922	0.9925	0.9927	0.9929	0.9931	0.9932	0.9934	0.9936
2.5	0.9938	0.9940	0.9941	0.9943	0.9945	0.9946	0.9948	0.9949	0.9951	0.9952
2.6	0.9953	0.9955	0.9956	0.9957	0.9959	0.9960	0.9961	0.9962	0.9963	0.9964
2.7	0.9965	0.9966	0.9967	0.9968	0.9969	0.9970	0.9971	0.9972	0.9973	0.9974
2.8	0.9974	0.9975	0.9976	0.9977	0.9977	0.9978	0.9979	0.9979	0.9980	0.9981
2.9	0.9981	0.9982	0.9982	0.9983	0.9984	0.9984	0.9985	0.9985	0.9986	0.9986
3.0	0.9987	0.9987	0.9987	0.9988	0.9988	0.9989	0.9989	0.9989	0.9990	0.9990
3.1	0.9990	0.9991	0.9991	0.9991	0.9992	0.9992	0.9992	0.9992	0.9993	0.9993
3.2	0.9993	0.9993	0.9994	0.9994	0.9994	0.9994	0.9994	0.9995	0.9995	0.9995
3.3	0.9995	0.9995	0.9996	0.9996	0.9996	0.9996	0.9996	0.9996	0.9996	0.9997
3.4	0.9997	0.9997	0.9997	0.9997	0.9997	0.9997	0.9997	0.9997	0.9998	0.9998
3.5	0.9998	0.9998	0.9998	0.9998	0.9998	0.9998	0.9998	0.9998	0.9998	0.9998
3.6	0.9998	0.9998	0.9999	0.9999	0.9999	0.9999	0.9999	0.9999	0.9999	0.9999

symmetrical about the mean it follows that 14·23 % of the population lies below $\mu - 1·07\sigma$, i.e. below 32 g.

In order that only 5 % of the packets shall be underweight the mean must be approximately $1·65\sigma$ above 32 g, i.e. the mean must be increased to 0·139 g above 32 g, i.e. to 32·14 g approximately.

Example 2. Jackets for young men are made in the following sizes, according to chest measurement:

Size	1	2	3	4	5	6
Chest measurement	30–	32–	34–	36–	38–	40–42

The chest measurements of young men in a certain age range are known to be normally distributed with mean 35·63 and standard deviation 2·00. Estimate to the nearest unit the percentages of young men in this age range likely to require each of the six sizes and also the percentages likely to fall above or below the size range.

In this case $\mu = 35·63$, $\sigma = 2·00$ and so we tabulate the range of sizes below, interpreting them in terms of μ and σ and then reading off the percentages from Table 1 as follows:

Jacket size	x	$x-\mu$ in terms of σ	% in range
< 1	< 30	$< -2·82\sigma$	0·24
1	30–32	$-2·82\sigma$ to $-1·82\sigma$	3·20
2	32–34	$-1·82\sigma$ to $-0·82\sigma$	17·17
3	34–36	$-0·82\sigma$ to $0·18\sigma$	36·53
4	36–38	$0·18\sigma$ to $1·18\sigma$	30·96
5	38–40	$1·18\sigma$ to $2·18\sigma$	10·44
6	40–42	$2·18\sigma$ to $3·18\sigma$	1·37
> 6	> 42	$> 3·18\sigma$	0·07
			99·99

Note the effect of rounding off errors on the cumulative percentage.

Example 3. Given that the probability that a child is left-handed is $\frac{1}{5}$, calculate the probabilities that in a random sample of five children,

  (i) just two are left handed,
  (ii) at least two are left handed.

Use the Normal approximation to estimate the probability that of 1600 children in a school, the school contains between 330 and 350 left-handed children.

Using the binomial distribution

  (i) the probability that just two of five children are left-handed is

$$\binom{5}{2}\left(\frac{1}{5}\right)^2\left(\frac{4}{5}\right)^3 \approx 0·205,$$

  (ii) the probability that at least two of five children are left-handed is

$$1 - P(0 \text{ or } 1 \text{ are left-handed})$$

$$= 1 - \left(\frac{4}{5}\right)^5 - \binom{5}{1}\left(\frac{1}{5}\right)\left(\frac{4}{5}\right)^4 \approx 0·263.$$

Using results (6.15) and (6.16) with $p = \frac{1}{5}$, $q = \frac{4}{5}$ and $n = 1600$, the mean μ and variance σ^2 of the left-handedness of the population of 1600 children are

$$\mu = 320, \qquad \sigma^2 = 256 \quad \Rightarrow \sigma = 16.$$

The probability that between 330 and 350 children in the school are left-handed is therefore

$$P\left(\mu + \frac{5}{8}\sigma < x < \mu + \frac{15}{8}\sigma\right) = \frac{1}{\sqrt{(2\pi)}} \int_{5/8}^{15/8} e^{-t^2/2}\, dt$$

$$= F\left(\frac{15}{8}\right) - F\left(\frac{5}{8}\right) = 0.9696 - 0.7340 = 0.2356 \approx 0.24.$$

Example 4. A random variable X is normally distributed with mean 7 and variance 16. Find the mean and variance of $Y = 2X - 5$. Find also

(a) $P(X < 6)$,
(b) $P(6 < Y < 10)$,
(c) $P(Y - X > 10)$. (L.)

$$E(Y) = E(2X - 5) = 2E(X) - 5 = 9.$$
$$V(Y) = V(2X - 5) = 2^2 V(X) = 64.$$

[Note that it is easy to prove from the definitions of mean and variance that, if X and Y are random variables,

$$E(X + Y) = E(X) + E(Y),$$
$$E(aX) = aE(X),$$
$$V(aX + b) = a^2 V(X),$$

where a and b are constants. This is left as an exercise for the reader.]

(a) $\dfrac{|x - \mu|}{\sigma} = \dfrac{|6 - 7|}{4} = 0.25$

$\Rightarrow P(X < 6) = 1 - F(0.25) \approx 0.4013.$

(b) $\dfrac{|y_1 - \mu|}{\sigma} = \dfrac{|6 - 9|}{8} = 0.375;\quad \dfrac{|y_2 - \mu|}{\sigma} = \dfrac{|10 - 9|}{8} = 0.125$

$\Rightarrow P(6 < Y < 10) = F(0.375) + F(0.125) - 1 \approx 0.1959.$

(c) $E(Y - X) = 9 - 7 = 2;\quad \dfrac{|0 - 2|}{4} = 0.5$

$\Rightarrow P(Y - X > 0) = F(0.5) \approx 0.6915.$

Exercise 11.4

1. In each trial of a random experiment it is known that the event A has probability 0.58 of occurring. In 100 independent trials of the experiment, let X denote the number of times that A occurs. Write down an exact expression for the probability $P(X \geq 50)$, but do not evaluate it. Use the Normal approximation to find an approximate value for this probability. (N.)

2. A machine is turning out components, whose lengths are normally distributed, 90% of them being greater than 1.494 cm, 5% being greater than 1.508 cm. Find the mean length and the standard deviation to the nearest 0.001 cm.
Gauges reject the components if they are less than 1.490 cm or greater than 1.510 cm long. What proportion will be rejected? (L.)

3. The weights of fish, of which a large number are swimming in a dam, are normally distributed with mean 1.25 kg and standard deviation 0.40 kg. Fish caught with weights less than 0.5 kg have to be thrown back. If an angler catches a randomly selected fish,

(a) show that the probability that the fish has to be thrown back is approximately 0.03;
(b) find the probability that its weight fails to break the previous record of 2.65 kg.

On a certain holiday some anglers hold a competition, and between them catch 400 of the fish. Estimate

(c) the number of fish thrown back;

(d) the probability that the record is broken on that day. (L.)

4. Steel bolts of circular cross-section are being manufactured to a specification which requires that their lengths be between 8·45 and 8·65 cm and their diameters between 1·55 and 1·60 cm. A machine produces these bolts so that their lengths are normally distributed about a mean of 8·54 cm with standard deviation 0·05 cm and their diameters are independently normally distributed about a mean of 1·57 cm with standard deviation 0·01 cm.

Find (i) the percentage of bolts produced that will not be within the specified limits for lengths,

(ii) the percentage of bolts produced that will not be within the specified limits for diameters,

(iii) the percentage of bolts that will not meet the specifications,

(iv) the chance that in a sample of five bolts taken at random four should meet the specifications and one should fail.

5. An automatic machine produces bolts whose diameters are required to lie within the tolerance limits 0·496 cm to 0·504 cm. A random sample of bolts produced by the machine is found to have a mean diameter of 0·498 cm and a standard deviation of 0·002 cm. Assuming that the diameters are normally distributed, estimate the probability that any bolt produced by the machine will have a diameter outside the tolerance limits.

If the machine is adjusted to produce bolts of mean diameter 0·500 cm, the standard deviation being unaltered, estimate the percentage of bolts likely to be rejected on full inspection.

6. A machine is programmed to fill "250 g net" packets with 255 g of tea. It is known from past experience that the distribution of weights is approximately normal with a standard deviation of 2 g. How many packets in an order of 10 000 packets may be expected to be under weight? (L.)

7. A lake contains a very large number of fish. The length of a fish caught may be taken as a random variable, normally distributed with mean 25 cm and standard deviation 7 cm. A fisherman keeps any fish he catches which is more than 35 cm long, returning others to the lake. Estimate by calculation, using tables as necessary, the probability that he keeps two or more fish out of the first ten fish he catches. (L.)

8. Steel billets are made for an order which requires the manganese content to be between 0·50 % and 0·60 %. Those that meet this specification are sold for £40 per tonne. Billets which fail to meet the specification are sold for £25 per tonne if their manganese content lies between 0·45 % and 0·75 %. Otherwise they are sold as scrap for £18 per tonne. If the actual manganese content varies in a normal distribution with a mean of 0·54 % and a standard deviation of 0·04 %, find:

(i) the proportion sold for £40 per tonne;

(ii) the proportion sold for £25 per tonne;

(iii) the average price per tonne obtained for the whole output. (O.C.)

9. In the manufacture of machined components the proportion of the product which does not pass through a gauge of 4·025 cm is 6 % and the proportion passing a gauge of 3·965 cm is 2 %. What proportion of the product lies in the range 4·000 ± 0·0275 cm if the dimension of the component is Normally distributed?
(N.)

10. It is known that 30 % of an apple crop has been attacked by insects. A random sample of 150 is selected from the crop. Assuming that the distribution of the number of damaged apples in random samples of 150 may be approximated to by a normal distribution, estimate the probability that

(a) more than half the sample is damaged,

(b) less than 10 % is damaged,

(c) the number of damaged apples in the sample lies between 35 and 50, inclusive. (L.)

11. First-class mail not delivered by the first post on the weekday following posting is considered "late". Assuming that 5 % of all letters sent by first-class mail are late, calculate the probability that, of ten letters sent by first-class post, not more than one will be delivered late.

State the mean and standard deviation of the probability distribution of the number of late deliveries out of 200 letters sent by first-class post, and use the normal distribution to estimate the probability that not more than five of the letters will be delivered late. (L.)

12. (i) A and B throw a six-sided die alternately in a game in which the player who first throws a six wins.

Before he makes a throw a player must place 5p in the pool and the winner takes all the money in the pool. A has first throw; find
  (a) his chance of winning,
  (b) the average amount that A will gain if he wins,
   (ii) A symmetrical coin is to be tossed 100 times. Write down the mean and variance of the distribution of the number of 'heads' expected. Use the normal distribution as an approximation to the binomial distribution to estimate the chance of 55 or more 'heads' being tossed. (O.C.)

13. The intelligence of an individual is frequently described by a positive integer known as an IQ (intelligence quotient). The distribution of IQs amongst children of a certain age-group can be approximated by a Normal probability model with mean 100 and standard deviation 15. Write a sentence stating what you understand about the age-group from the fact that $F(2·5) = 0·994$.
A class of thirty children is selected at random from the age-group. Calculate (to 2 significant figures) the probability that at least one member of the class has an IQ of 138 or more.

14. A filling machine pours material into cartons labelled "50 g net". The total material delivered in 1000 fillings is 54·41 kg. If the distribution is Normal with standard deviation 1·83 g, what percentage of cartons is expected to be under weight? (O.C.)

15. In an examination there are 4000 candidates. In Paper I the marks may be taken as a continuous and normal distribution with mean 47·5 and standard deviation 12·5.
  (a) Estimate the numbers of candidates who will
    (i) obtain less than 40 marks,
    (ii) obtain marks between 36 and 44.
  (b) Estimate the mark which will be exceeded by 10% of the candidates.
  (c) If a particular examiner marks 100 scripts taken at random and awards marks whose mean is 49·5, is there evidence that his marking is more lenient than that of the other examiners?
  (d) In Paper II the mean mark is 50·5 and the standard deviation 15 and the total mark for the examination is the average for the two papers. How many candidates will have a total mark less than 40? (O.C.)

16. In a mechanism a plunger moves inside a cylinder and the difference between the diameters of the plunger and the cylinder must be at least one-hundredth of a cm. Cylinders and plungers are manufactured separately. The diameters of the cylinders are normally distributed about a mean 2·025 cm with standard deviation 0·002 cm; the diameters of the plungers are normally distributed about a mean 2·010 cm with standard deviation 0·003 cm.
  (i) Find the proportion of cylinders that will have inadequate clearance for plungers of diameter 2·01 cm.
  (ii) Find the proportion of plungers that will have inadequate clearance in cylinders of diameter 2·025 cm.

17. The systolic blood pressures of 150 individuals between the ages of 50 and 60 are given in the following table:

Central value	95	105	115	125	135	145	155	165	175	Total
Frequency	3	8	25	40	38	15	10	7	4	150

Calculate the mean and the standard deviation.
How do the statistics you have found lend credence to the assumption that the values in the population from which the sample is taken are normally distributed? On this assumption estimate the proportion of the population with blood pressures between 120 and 150.

18. Two types of electrical resistor are being manufactured and in each type the resistance is normally distributed. Type A has a mean resistance of 50 ohms with standard deviation 1·25 ohms; type B has a mean resistance of 40 ohms with standard deviation 1·05 ohms.
  (a) Find upper and lower tolerance limits for resistors of type A so that not more than 1% of the manufactured resistors will be excluded by them.
  (b) Find the percentage of resistors of type B with resistance between 39 and 42 ohms.
  (c) If a resistor of type A is chosen at random and assembled in series with a random resistor of type B, what percentage of the combined resistors will have a total resistance greater than 92 ohms? (O.C.)

19. Two trains (A and B) are due to arrive at a junction at 10·00 and 10·05 respectively. If their actual times

of arrival vary in normal probability distributions with these times as means and with standard deviations of 3 and 4 minutes respectively, find
(i) the probability that if A arrives on time B will not yet have arrived,
(ii) the probability that if B arrives on time A will not yet have arrived.

20. When the flat surface of a certain substance is examined under a microscope, large numbers of small circular holes of various sizes are seen. The diameters of the holes in a square centimetre are measured in micrometres (μm) and the results recorded in the following grouped frequency table:

Centre of interval	20	40	60	80	100	120	140	160
No.	5	9	21	28	26	17	8	2

(i) Find the total number of holes, the mean diameter and the standard deviation of the diameters.
(ii) Assuming this is a sample from a normal distribution with the same mean and variance, estimate the proportion of holes with diameters less than 10 μm.
(iii) What proportion of the surface area consists of holes?
[1 cm = 10 000 μm.] (O.C.)

11:5 TRANSITION MATRICES FOR MARKOV CHAINS

A square matrix **P** of order n with elements p_{ij} such that

$$0 \leqslant p_{ij} \leqslant 1 \quad \forall\, i,j \leqslant n \quad \text{and} \quad \sum_{i=1}^{n} p_{ij} = 1 \quad \forall\, j \leqslant n$$

is called a *stochastic* or *probability* or *transition matrix*, since each element p_{ij} can represent the probability of an event passing from one "state" i to another state j in a *Markov chain*, where the set of states is exclusive and exhaustive. For example, a sales representative, travelling among three countries C_1, C_2 and C_3, may consider his next journey from one country, or state (no pun intended), to another country to be subject to the probabilities shown in the transition matrix **P**, where

$$\mathbf{P} = \text{From} \begin{array}{c} \\ C_1 \\ C_2 \\ C_3 \end{array} \overset{\begin{array}{ccc} \text{To} \\ C_1 & C_2 & C_3 \end{array}}{\begin{pmatrix} 0{\cdot}6 & 0{\cdot}3 & 0{\cdot}1 \\ 0 & 0{\cdot}2 & 0{\cdot}8 \\ 0{\cdot}4 & 0{\cdot}2 & 0{\cdot}4 \end{pmatrix}}.$$

The sum of the elements in each row (or of course in each column, depending how the matrix is constructed) is 1, showing that whichever country the representative is in, he is certain to travel to one of the three; $p_{21} = 0$ means that it is impossible for him to travel directly from C_2 to C_1.

Suppose that the salesman is now due to make a second journey; if he originally arrived in C_1, say, and is to be in C_1 after his second journey, there are three possibilities — C_1 to C_1 and C_1 to C_1, C_1 to C_2 and C_2 to C_1, C_1 to C_3 and C_3 to C_1, with probabilities $0{\cdot}6 \times 0{\cdot}6$, $0{\cdot}3 \times 0$ and $0{\cdot}1 \times 0{\cdot}4$ respectively. Hence the probability of a two-stage journey from C_1 to C_1 is $0{\cdot}36 + 0 + 0{\cdot}04$; this is clearly the element in the first row and column of $\mathbf{P}^2$, and similarly each two-stage journey has a probability which will be the corresponding element of $\mathbf{P}^2$. It follows by induction that the probabilities for an n-stage journey are given by the elements of $\mathbf{P}^n$.

Suppose now that the sales director wishes to contact the representative, knowing that he has made one journey since his arrival in one of the three countries. If the director has no idea from which country the salesman started, he will have to regard each as equally likely. However, suppose he knows that the salesman could not go to C_1 (no direct flight?), but was equally likely to have gone to C_2 or C_3; then to calculate the probability of the first journey being to C_1, C_2 or C_3 he uses a *probability vector* (0 0·5 0·5) thus

$$(0\ 0.5\ 0.5) \begin{pmatrix} 0.6 & 0.3 & 0.1 \\ 0 & 0.2 & 0.8 \\ 0.4 & 0.2 & 0.4 \end{pmatrix} = (0.2\ 0.2\ 0.6).$$

Hence the salesman is most likely to be in C_3 after his first journey.

11:6 STEADY-STATE VECTOR AND MATRIX

If we write a three-state transition matrix in the form

$$\mathbf{P} = \begin{pmatrix} 1 - p_{12} - p_{13} & p_{12} & p_{13} \\ p_{21} & 1 - p_{21} - p_{23} & p_{23} \\ p_{31} & p_{32} & 1 - p_{31} - p_{32} \end{pmatrix},$$

the characteristic equation of the matrix $\mathbf{P}$ is

$$\begin{vmatrix} 1 - p_{12} - p_{13} - \lambda & p_{12} & p_{13} \\ p_{21} & 1 - p_{21} - p_{23} - \lambda & p_{23} \\ p_{31} & p_{32} & 1 - p_{31} - p_{32} - \lambda \end{vmatrix} = 0,$$

of which the roots are eigenvalues of the matrix $\mathbf{P}$. It is clear that for this matrix, and in fact for any n-state transition matrix, one eigenvalue is given by $\lambda = 1$, since then

$$\text{col } 1 + \text{col } 2 + \text{col } 3 = \begin{pmatrix} 0 \\ 0 \\ 0 \end{pmatrix}.$$

Hence there is an eigenvector $(x\ y\ z)$ such that

$$(x\ y\ z)\mathbf{P} = (x\ y\ z).$$

To find $(x\ y\ z)$ we have

$$(1 - p_{12} - p_{13})x + p_{21}y + p_{31}z = x$$

and two similar equations, giving

$$(x\ y\ z) = k(x_1\ y_1\ z_1) \quad \forall k.$$

But, if $(x\ y\ z)$ is a probability vector, then $x + y + z = 1$. Hence we obtain a unique probability vector

$$s = \left(\frac{x_1}{x_1+y_1+z_1} \quad \frac{y_1}{x_1+y_1+z_1} \quad \frac{z_1}{x_1+y_1+z_1} \right).$$

If this is the initial probability vector for the three states, these probabilities are not changed whatever the elements in the transition matrix may be. Thus **s** represents a *steady state* or a *state of equilibrium*.

In our example we require

$$0.6x \qquad\qquad +0.4z = x,$$
$$0.3x + 0.2y + 0.2z = y,$$
$$0.1x + 0.8y + 0.4z = z$$
$$\Rightarrow \quad x \qquad\qquad -z = 0,$$
$$0.3x - 0.8y + 0.2z = 0,$$
$$0.1x + 0.8y - 0.6z = 0.$$

Clearly these three equations are dependent, and give

$$x:y:z = 0.8:0.5:0.8.$$

For $x+y+z = 1$, $x = 8/21$, $y = 5/21$, $z = 8/21$.

It is obvious that

$$k(x_1 \; y_1 \; z_1) \begin{pmatrix} kx_1 & ky_1 & kz_1 \\ kx_1 & ky_1 & kz_1 \\ kx_1 & ky_1 & kz_1 \end{pmatrix} = \begin{pmatrix} kx_1 & ky_1 & kz_1 \\ kx_1 & ky_1 & kz_1 \\ kx_1 & ky_1 & kz_1 \end{pmatrix}$$

since $k(x_1 + y_1 + z_1) = 1$. Hence the steady-state vector **s** gives rise to the steady-state matrix $\mathbf{P}_s$, in which each row is equal to **s**. Since the probabilities after n trials are given by $\mathbf{P}^n$, we might expect that $\mathbf{P}^n = \mathbf{P}_s$ for some n, or at least that $\lim_{n \to \infty} \mathbf{P}^n = \mathbf{P}_s$. This in fact will generally be so, exceptions occurring only if **P** is not a **regular** matrix, i.e. a transition matrix such that there is some value of n for which $\mathbf{P}^n$ contains no zero element. Thus, if there is a steady-state matrix, each of its rows consists of the steady-state vector, which can be found as the eigenvector corresponding to the eigenvalue 1, with the sum of its elements equal to 1.

We will now, for greater simplicity, consider a two-state transition matrix; this will provide further opportunity for revising some of the work on matrices in Volume 2, Part 1, Chapter 15.

Example 1. Given the two-state transition matrix

$$\begin{pmatrix} \frac{1}{2} & \frac{1}{2} \\ \frac{1}{4} & \frac{3}{4} \end{pmatrix},$$

find its steady state, and compare with

$$\lim_{n \to \infty} \begin{pmatrix} \frac{1}{2} & \frac{1}{2} \\ \frac{1}{4} & \frac{3}{4} \end{pmatrix}^n.$$

A steady-state vector will be given by

$$(x \ y) \begin{pmatrix} \frac{1}{2} & \frac{1}{2} \\ \frac{1}{4} & \frac{3}{4} \end{pmatrix} = (x \ y)$$

$$\Leftrightarrow \tfrac{1}{2}x + \tfrac{1}{4}y = x,$$
$$\tfrac{1}{2}x + \tfrac{3}{4}y = y,$$
$$x + y = 1$$
$$\Leftrightarrow y = 2x, \quad x + y = 1$$
$$\Leftrightarrow x = \tfrac{1}{3}, \qquad y = \tfrac{2}{3}.$$

We know that one eigenvalue is 1; the other must be $\tfrac{1}{2} + \tfrac{3}{4} - 1$, since the trace of a matrix is equal to the sum of the roots of its characteristic equation. Eigenvectors corresponding to 1, $\tfrac{1}{4}$ are

$$\begin{pmatrix} 1 \\ 1 \end{pmatrix}, \begin{pmatrix} 2 \\ -1 \end{pmatrix} \text{ respectively;}$$

we can diagonalise the transition matrix by using

$$\begin{pmatrix} 1 & 2 \\ 1 & -1 \end{pmatrix} \text{ and } \begin{pmatrix} 1 & 2 \\ 1 & -1 \end{pmatrix}^{-1},$$

thus

$$\begin{pmatrix} 1 & 2 \\ 1 & -1 \end{pmatrix} \begin{pmatrix} \tfrac{1}{2} & \tfrac{1}{2} \\ \tfrac{1}{4} & \tfrac{3}{4} \end{pmatrix} \begin{pmatrix} \tfrac{1}{3} & \tfrac{2}{3} \\ \tfrac{1}{3} & -\tfrac{1}{3} \end{pmatrix}$$

$$\Rightarrow \begin{pmatrix} \tfrac{1}{2} & \tfrac{1}{2} \\ \tfrac{1}{4} & \tfrac{3}{4} \end{pmatrix}^n = \begin{pmatrix} 1 & 2 \\ 1 & -1 \end{pmatrix} \begin{pmatrix} 1 & 0 \\ 0 & [\tfrac{1}{4}]^n \end{pmatrix} \begin{pmatrix} \tfrac{1}{3} & \tfrac{2}{3} \\ \tfrac{1}{3} & -\tfrac{1}{3} \end{pmatrix} = \begin{pmatrix} 1 & 2(\tfrac{1}{4})^n \\ 1 & -(\tfrac{1}{4})^n \end{pmatrix} \begin{pmatrix} \tfrac{1}{3} & \tfrac{2}{3} \\ \tfrac{1}{3} & -\tfrac{1}{3} \end{pmatrix}$$

$$= \begin{pmatrix} \tfrac{1}{3} + \tfrac{2}{3}(\tfrac{1}{4})^n & \tfrac{2}{3} - \tfrac{2}{3}(\tfrac{1}{4})^n \\ \tfrac{1}{3} - \tfrac{1}{3}(\tfrac{1}{4})^n & \tfrac{2}{3} + \tfrac{1}{3}(\tfrac{1}{4})^n \end{pmatrix}$$

$$\Rightarrow \lim_{n \to \infty} \begin{pmatrix} \tfrac{1}{2} & \tfrac{1}{2} \\ \tfrac{1}{4} & \tfrac{3}{4} \end{pmatrix}^n = \begin{pmatrix} \tfrac{1}{3} & \tfrac{2}{3} \\ \tfrac{1}{3} & \tfrac{2}{3} \end{pmatrix}.$$

This is the steady-state matrix, confirming that each row consists of the steady state vector.

Example 2. A Markov chain with three states A, B, C has the transition diagram shown in Fig. 11.2. Given that the initial state is A, find the probability that the system remains in this state for exactly r transitions ($r \geq 0$).

Prove also that the probability that the system first arrives in state C on the rth transition is equal to $2(2^{-r} - 3^{-r})$. (O.C.S.M.P.)

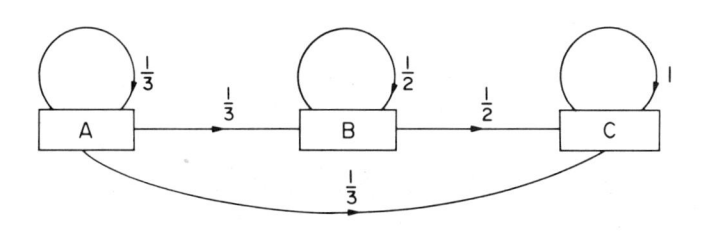

FIG. 11.2.

The transition matrix for the system is

$$\begin{array}{c} \\ A \\ B \\ C \end{array} \begin{array}{ccc} A & B & C \end{array} \\ \begin{pmatrix} \frac{1}{3} & \frac{1}{3} & \frac{1}{3} \\ 0 & \frac{1}{2} & \frac{1}{2} \\ 0 & 0 & 1 \end{pmatrix} = \mathbf{P}$$

and the initial probability vector is (1 0 0), giving the probability vector ($\frac{1}{3}$ $\frac{1}{3}$ $\frac{1}{3}$) for the first transition. Clearly, if the probability vector for the $(k-1)$th transition is (x - -), then the vector for the kth transition is ($\frac{1}{3}x$ - -). Hence the probability that the system remains in state A for the first r transitions is $(\frac{1}{3})^r$, but the probability of a change to B or C on the $(r+1)$th transition is $\frac{2}{3}$. Thus the probability of the system being in state A for *exactly* r transitions is $\frac{2}{3}(\frac{1}{3})^r$.

Since a transition from B to A is impossible, the only possible sets of transitions such that C is first reached at the rth transition are

$$A^{r-1}C, \; A^{r-2}BC, \; A^{r-3}B^2C, \ldots, \; B^{r-1}C.$$

Hence the probability of C being first reached on the rth transition is

$$(\tfrac{1}{3})^{r-1} \cdot \tfrac{1}{3} + (\tfrac{1}{3})^{r-2} \cdot \tfrac{1}{3} \cdot \tfrac{1}{2} + (\tfrac{1}{3})^{r-3} \cdot \tfrac{1}{3} \cdot (\tfrac{1}{2})^2 + \ldots + \tfrac{1}{3} \cdot (\tfrac{1}{2})^{r-1}.$$

This is a G.P. with common ratio $\frac{3}{2}$.

$$\Rightarrow P(C) = \frac{(\tfrac{1}{3})^r [(\tfrac{3}{2})^r - 1]}{\tfrac{3}{2} - 1} = 2(2^{-r} - 3^{-r}).$$

Example 3. Find a steady-state matrix for the system in Example 2, and compare with $\lim_{n \to \infty} \mathbf{P}^n$.

It is clear from Fig. 11.2 that the steady-state vector is (0 0 1), and that the steady-state matrix is

$$\begin{pmatrix} 0 & 0 & 1 \\ 0 & 0 & 1 \\ 0 & 0 & 1 \end{pmatrix}.$$

By inspection, the eigenvalues of $\mathbf{P}$ are $\frac{1}{3}, \frac{1}{2}$ and 1; corresponding eigenvectors are

$$\begin{pmatrix} 1 \\ 0 \\ 0 \end{pmatrix}, \; \begin{pmatrix} 2 \\ 1 \\ 0 \end{pmatrix}, \; \begin{pmatrix} 1 \\ 1 \\ 1 \end{pmatrix}.$$

Hence

$$\mathbf{P}^n = \begin{pmatrix} 1 & 2 & 1 \\ 0 & 1 & 1 \\ 0 & 0 & 1 \end{pmatrix} \begin{pmatrix} (\tfrac{1}{3})^n & 0 & 0 \\ 0 & (\tfrac{1}{2})^n & 0 \\ 0 & 0 & 1 \end{pmatrix} \begin{pmatrix} 1 & -2 & 1 \\ 0 & 1 & 1 \\ 0 & 0 & 1 \end{pmatrix}$$

$$\lim_{n \to \infty} \mathbf{P}^n = \begin{pmatrix} 0 & 0 & 1 \\ 0 & 0 & 1 \\ 0 & 0 & 1 \end{pmatrix} \begin{pmatrix} 1 & -2 & 1 \\ 0 & 1 & 1 \\ 0 & 0 & 1 \end{pmatrix} = \begin{pmatrix} 0 & 0 & 1 \\ 0 & 0 & 1 \\ 0 & 0 & 1 \end{pmatrix}$$

We could not be certain of obtaining this result since $\mathbf{P}$ is not a regular matrix.

11:7 AN EQUIVALENCE RELATION ON TRANSITION STATES

Consider the relation $\mathscr{R}$ defined on a set of transition states such that $A_i \mathscr{R} A_j$ if either $i = j$ or the transitions $A_i \to A_j$ and $A_j \to A_i$ are both possible, though not necessarily as one-stage transitions. The first part of the definition makes this relation reflexive, and the second part ensures that it is symmetric; also if $A_i \leftrightarrow A_j$ and $A_j \leftrightarrow A_k$, then $A_i \leftrightarrow A_k$ through A_j, and the relation is transitive. Thus $\mathscr{R}$ is an equivalence relation, which will partition the set of states into equivalence classes. Since there will be at most a one-way transition between states in different equivalence classes, these classes can be placed in one of two categories. Either the system will eventually leave the class and will then not be able to return to it, in which case the class is called **transient**, or if a single state, a transient state; or the system may eventually enter the class and will then never leave it—the class, or the single state, is then called **absorbing**. If a Markov chain contains at least one absorbing state, and if it is possible to move from every state, directly or indirectly, to an absorbing state, it can be shown that the chain will eventually reach an absorbing state, so that the chain is then absorbed. In Example 2, p. 431 $\{A\}$ and $\{B\}$ are transient states, and $\{C\}$ is an absorbing state, in which the chain is eventually absorbed.

Example. Draw a diagram to illustrate the given transition matrix, and partition the set of states $\{A_1, \ldots, A_6\}$ into transient and absorbing classes.

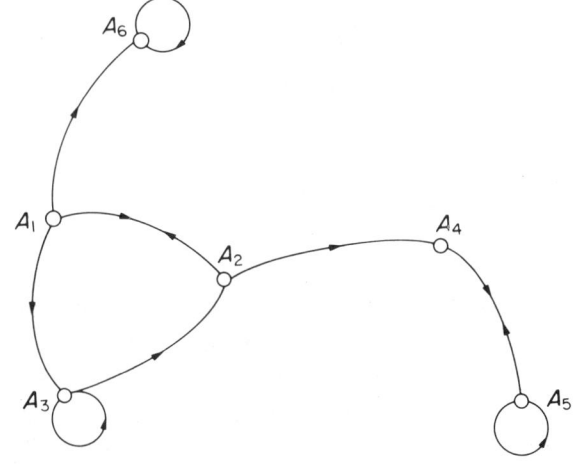

	A_1	A_2	A_3	A_4	A_5	A_6
A_1	0	0.4	0.2	0	0	0.4
A_2	0.3	0	0	0.7	0	0
A_3	0	0.5	0.5	0	0	0
A_4	0	0	0	0	1	0
A_5	0	0	0	0.8	0.2	0
A_6	0	0	0	0	0	1

FIG. 11.3.

$\{A_1, A_2, A_3\}$ is a transient class; $\{A_4, A_5\}$ and $\{A_6\}$ are absorbing classes.

Exercise 11.7

1. The weather on a certain day is classified as either "Fine" or "Wet", and the probabilities for the weather tomorrow, depending on the weather today, are shown by the transition matrix

$$\begin{array}{cc} & \text{Tomorrow} \\ & \begin{array}{cc} F & W \end{array} \\ \text{Today} \begin{array}{c} F \\ W \end{array} & \begin{pmatrix} \frac{2}{3} & \frac{1}{3} \\ \frac{1}{2} & \frac{1}{2} \end{pmatrix}. \end{array}$$

Calculate the probability of the day after tomorrow being wet. If today is Thursday, calculate the probability of at least one fine day over the following weekend (Friday, Saturday, Sunday).

If it is fine on September 25th 1979, what is the probability of a fine day on September 25th 1980?

2. One bag contains two red balls, and another bag contains two blue balls; a transition consists of drawing one ball at random from each bag and interchanging them. Form a transition matrix **P** for the three states A_0, A_1, A_2 corresponding to the number of red balls in one bag. Calculate the probability of this bag containing two blue balls after three transitions.

3. Find the steady-state vector for the following matrices.

(i)
$$\begin{array}{c} \\ A_1 \\ A_2 \\ A_3 \end{array} \begin{array}{c} A_1 \quad A_2 \quad A_3 \\ \begin{pmatrix} \frac{1}{4} & \frac{3}{4} & 0 \\ \frac{2}{3} & 0 & \frac{1}{3} \\ 0 & 1 & 0 \end{pmatrix} \end{array}$$

(ii)
$$\begin{array}{c} \\ B_1 \\ B_2 \\ B_3 \\ B_4 \end{array} \begin{array}{c} B_1 \quad B_2 \quad B_3 \quad B_4 \\ \begin{pmatrix} 0 & 1 & 0 & 0 \\ 0 & 0 & \frac{1}{2} & \frac{1}{2} \\ 1 & 0 & 0 & 0 \\ 0 & 0 & \frac{4}{5} & \frac{1}{5} \end{pmatrix} \end{array}$$

(iii)
$$\begin{array}{c} \\ C_1 \\ C_2 \\ C_3 \\ C_4 \end{array} \begin{array}{c} C_1 \quad C_2 \quad C_3 \quad C_4 \\ \begin{pmatrix} \frac{3}{8} & \frac{1}{4} & \frac{1}{4} & \frac{1}{8} \\ \frac{1}{4} & \frac{1}{4} & 0 & \frac{1}{2} \\ \frac{1}{4} & 0 & \frac{1}{2} & \frac{1}{4} \\ 0 & 0 & 0 & 1 \end{pmatrix} \end{array}$$

4. Three brands of toothpaste labelled A, B and C are on sale in a shop and a woman buys one and only one of the brands on each visit. Advance publicity ensures that on her first visit she has a 50% chance of choosing brand A and equal chances of choosing B or C. When she buys brand A the next purchase is always brand B. However, when she buys brand B the probability of purchasing brand A on the next visit is $\frac{2}{3}$ and that of purchasing brand C is $\frac{1}{3}$. Similarly, after buying brand C the probability of purchasing brand A on the next visit is $\frac{2}{3}$ and of purchasing brand B is $\frac{1}{3}$. Write down the transition matrix that applies for the second and all subsequent visits and find

(a) the probability that she buys brand A on the third visit,
(b) the probability that she buys brand C on the fifth visit.

Assuming that her purchasing habits are unchanged over a long time, find the ultimate proportions in which she buys the brands. (L.)

5. At a certain weather station a simplified record of the weather is kept in which, at the end of each day, the day's weather is recorded as fine (F), dull (D), or wet (W). The record for 44 consecutive days was

FFFF   FFFD   DDFD   WDDW   WDFF   FDDD
DFFD   DFDD   DWDF   DDWW   WWDW

By using a model in which each day's weather depends on, and only on, that of the day before, and in which the transition probabilities are taken directly from the observed proportions, show that the probabilities of a previously fixed day displaying each type of weather when the previous day's weather is not known, are given by F:5/17, D:8/17 and W:4/17. Discuss, without formal tests, the degree of agreement between the frequencies expected from these probabilities and the observed frequencies of each type of weather.

Calculate also the expected number of nights to elapse between the recording of a fine day and the recording of the next wet day, describing carefully the model of the observations you use. (O.C.S.M.P.)

6. The transition matrix of a three-state Markov process is

$$\begin{pmatrix} \frac{1}{3} & \frac{1}{3} & \frac{1}{3} \\ \frac{1}{4} & \frac{1}{2} & \frac{1}{4} \\ \frac{5}{12} & \frac{5}{12} & \frac{1}{6} \end{pmatrix}.$$

Find the limit to which the probability vector will tend.

(ii) An integer a is chosen at random from the set $\{1, 2, 3\}$. Then integers b and c such that $a \leq b \leq c \leq 3$ are chosen in turn at random. Form the transition matrix for the process $a \to b \to c$. Use this matrix to find the probability that $c = 3$.

7. A man on holiday spends his nights at camps A, B, C or D. Each day he walks out and either returns or goes to another camp for the next night. The probabilities of his movements are given by the transition matrix:

$$\text{To} \begin{cases} & \overbrace{\begin{matrix} A & B & C & D \end{matrix}}^{\text{From}} \\ A & \frac{2}{3} & \frac{1}{2} & \frac{1}{4} & \frac{1}{3} \\ B & \frac{1}{3} & 0 & 0 & 0 \\ C & 0 & \frac{1}{2} & \frac{3}{4} & 0 \\ D & 0 & 0 & 0 & \frac{2}{3} \end{cases}$$

Name (i) any closed classes, and (ii) any transient classes, which exist in this system.
Find the expected number of consecutive nights he will spend at A if he has just arrived there.
(O.C.S.M.P.)

8. A small disc is placed on one vertex of a triangle ABC and is moved according to the throw of a die. When the disc is on A, it is moved to B if 1, 2 or 3 is thrown; otherwise it is moved to C. When the disc is on B, it is moved to A if 5 or 6 is thrown; otherwise it is moved to C. When the disc is on C, it is moved to B if 6 is thrown; otherwise it is moved to A.

Form the transition matrix, and show that the probability that the disc will be in its original position after three throws of the die is 11/36.

Show also that the probability that after a large number of throws the disc will be at A is approximately 32/83. (L.)

9. A certain species of wheat occurs in three genetic strains A, B and C. When these strains are grown under conditions which ensure self-pollination the A and C strains always breed true-to-type (that is, an A plant will produce seed only of type A and a C plant will produce seed only of type C), whereas the B strain produces seed of types A, B, C in the proportions 1:2:1. At any given time the proportions of the three strains of wheat in the population are written as the three components of the vector $\mathbf{x}$. Write down the matrix $\mathbf{P}$ which, when applied to $\mathbf{x}$, gives the proportions of each strain in the next generation of wheat plants (assuming that every plant, whatever its strain, produces the same number of seeds).

Show that

$$\begin{pmatrix} 1 \\ 0 \\ 0 \end{pmatrix} \quad \text{and} \quad \begin{pmatrix} 0 \\ 0 \\ 1 \end{pmatrix}$$

are both eigenvectors of $\mathbf{P}$ which correspond to the eigenvalue 1, and give an interpretation of this result in the context of the population of wheat plants. By finding another eigenvalue and a corresponding eigenvector, or otherwise, prove that

$$\mathbf{P}^n = \begin{pmatrix} 1 & \frac{1}{2} - (\frac{1}{2})^{n+1} & 0 \\ 0 & (\frac{1}{2})^n & 0 \\ 0 & \frac{1}{2} - (\frac{1}{2})^{n+1} & 1 \end{pmatrix}.$$

What is the least number of generations that would be required before a population in which the strains A, B, C occurred in equal proportions became one with less than 1% of strain B? (O.C.S.M.P.)

10. My pattern of TV watching is as follows. If I have ITV on one evening, then the probabilities that I will have ITV, BBC or nothing on the next evening are $\frac{1}{6}, \frac{1}{2}, \frac{1}{3}$ respectively; if I have BBC on one evening, then the corresponding probabilities for the next evening are $\frac{1}{4}, \frac{1}{4}, \frac{1}{2}$. If I miss television entirely one evening, then the next evening I shall be certain to watch BBC. I never change channels during an evening. Assuming that in the long run my pattern of watching settles down to a steady state, find in what proportions I shall then watch ITV, BBC or nothing.

If the probabilities of watching ITV, BBC or nothing on the nth night of the year are p_n, q_n, r_n prove that, except perhaps for the first night ($n = 1$) of this pattern of watching, $r_n = 2p_n$. Deduce that, if x_n denotes $10p_n - 3q_n$, then, for $n \geqslant 2$, $x_{n+1} = kx_n$, where k is constant. Hence show that, whatever happens on the first night, my pattern of watching is bound to settle down in time to a steady state. (O.C.S.M.P.)

11:8 FURTHER PROBABILITY PROBLEMS

We conclude this chapter with some more difficult examples on probability.

Example 1. A match between two players A and B is won by whoever first wins 2 games. A's chances of winning, drawing or losing any particular game are p, q and r respectively. Find the probability that A wins in exactly n games. Hence prove that his chance of winning the match is $p^2(p+3r)/(p+r)^3$.

Since each game must be won, drawn or lost by A it follows that

$$p + q + r = 1. \tag{1}$$

In order to win the match in exactly n ($\geqslant 2$) games, A must win the last (nth) game *and* one and only one of the preceding $n - 1$ games whilst B wins only 1 or none of these $n - 1$ games. Using the binomial distribution we find that the relevant probabilities are as follows:

A wins one and the remaining $n - 2$ are drawn,

$$P_1 = (n-1)pq^{n-2}.$$

A wins one, B wins one and the remaining $n - 3$ are drawn,

$$P_2 = (n-1)(n-2)prq^{n-3}.$$

Hence the probability that A wins in exactly n games is

$$P(n) = P_1 + P_2 = p\{(n-1)pq^{n-2} + (n-1)(n-2)prq^{n-3}\}$$
$$= (n-1)p^2q^{n-2} + (n-1)(n-2)p^2rq^{n-3}.$$

The chance that A wins the match is

$$\sum_{r=2}^{\infty} P(r) = p^2 \sum_{2}^{\infty}(n-1)q^{n-2} + p^2 r \sum_{2}^{\infty}(n-1)(n-2)q^{n-3}$$

$$= p^2 \frac{d}{dq}\left(\sum_{2}^{\infty} q^{n-1}\right) + p^2 r \frac{d^2}{dq^2}\left(\sum_{2}^{\infty} q^{n-1}\right)$$

$$= p^2 \frac{d}{dq}\left(\frac{q}{1-q}\right) + p^2 r \frac{d^2}{dq^2}\left(\frac{q}{1-q}\right)$$

$$= p^2 \left\{\frac{1}{1-q} + \frac{q}{(1-q)^2}\right\} + p^2 r \left\{\frac{2}{(1-q)^2} + \frac{2q}{(1-q)^3}\right\}$$

$$= \frac{p^2}{(1-q)^2} + \frac{2p^2 r}{(1-q)^3}.$$

Using equation (1) this gives the required result.

Example 2. Two numbers x and y are chosen at random between 0 and 2. Find the chance that $x^m y^n \leq 1$ in the three cases

(i) $m = n = 1$, (ii) $m = 2$, $n = 1$, (iii) $m = 2$, $n = -1$.

Since each of x, y is chosen at random in the range $(0, 2)$ the point (x, y) in the plane of the cartesian axes Oxy is equally likely to lie anywhere within the square with the sides $x = 0$, $x = 2$, $y = 0$, $y = 2$.

It follows that the probabilities that $x^m y^n \leq 1$ for the stated cases are obtained by dividing the shaded areas of Fig. 11.4 by the area of the square.

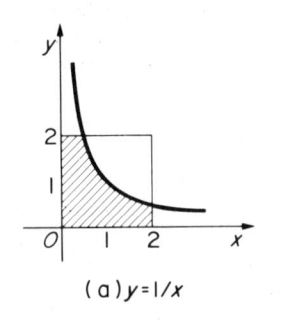

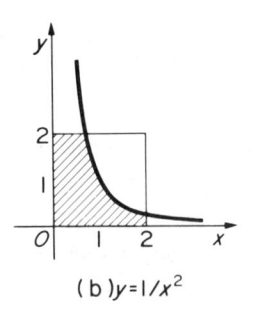

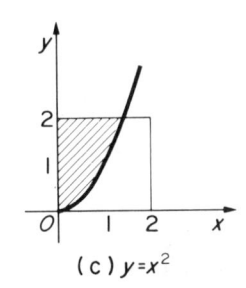

(a) $y = 1/x$

(b) $y = 1/x^2$

(c) $y = x^2$

Fig. 11.4.

The results are

(i) $\dfrac{1}{4}\left[\dfrac{1}{2} \times 2 + \displaystyle\int_{1/2}^{2} \dfrac{1}{x}\,dx\right] = \dfrac{1}{4}(1 + \ln 4)$,

(ii) $\dfrac{1}{4}\left[2 \times \dfrac{1}{\sqrt{2}} + \displaystyle\int_{1/\sqrt{2}}^{2} \dfrac{1}{x^2}\,dx\right] = \dfrac{1}{4}\left(2\sqrt{2} - \dfrac{1}{2}\right)$,

(iii) $\dfrac{1}{4}\displaystyle\int_{0}^{2} y^{1/2}\,dy = \dfrac{\sqrt{2}}{3}$.

Example 3. In a certain population it is estimated that 30 % are fair-haired, also that 70 % of those with fair hair have blue eyes, and 40 % of those with dark hair have blue eyes. Calculate the probability of a person, chosen at random from those having blue eyes, also having fair hair.

If $F =$ "having fair hair", $B =$ "having blue eyes", we have, using Bayes' theorem (Ch.6, §6:3, Theorem 7, p. 198),

$$P(F|B) = \frac{P(F).P(B|F)}{P(F).P(B|F) + P(F').P(B|F')}$$

(assuming, of course, that "having fair hair" and "having dark hair" are exclusive and exhaustive)

$$\Rightarrow P(F|B) = \frac{0.3 \times 0.7}{0.3 \times 0.7 + 0.7 \times 0.4}$$

$$= \frac{0.21}{0.49}$$

$$= \frac{3}{7}.$$

The probability of a blue-eyed person having fair hair is $\frac{3}{7}$.

Example 4. A circular disc of radius r is thrown at random onto a large board divided into squares of side a (where $a \geqslant 2r$). Show that the probability that the disc comes to rest entirely within one square is $[1 - (2r/a)]^2$.

If the disc comes to rest entirely within a square C, the centre O of the disc must lie within a square C_1, concentric with C and of side $a - 2r$. Since O is equally likely to come to rest anywhere on the board, the required probability is

$$\frac{\text{area of } C_1}{\text{area of } C} = \frac{(a-2r)^2}{a^2} = \left(1 - \frac{2r}{a}\right)^2.$$

Exercise 11.8

1. Given the continuous frequency function $f(x) = 2/x^2$, where $1 \leqslant x \leqslant 2$, determine the mean and variance of x and find the probability that x exceeds 1.5. Calculate also the median and quartile values for x and state the inter-quartile range.

2. A point P is taken at random inside an ellipse of eccentricity e. Calculate (in terms of e) the probability that the sum of the focal distances of P should be not greater than the distance from a focus to the opposite end of the major axis.

3. Prove that the total number of ways in which three non-zero positive integers can be chosen to have as their sum a given integer $6N$, where $N = 1, 2, 3, \ldots$, is N^2.

$\Bigg[$ Hint. Consider the coefficient of x^{6N} in the expansion of

$$(x + x^2 + x^3 + \ldots)(x + x^2 + x^3 + \ldots)(x + x^2 + x^3 + \ldots) = \frac{x^3}{(1-x)^3}.\Bigg]$$

4. If A, B, C and D are independent random events and

$$P(A) = 0.1, \; P(B) = 0.2, \; P(C) = 0.3, \; P(D) = 0.4,$$

calculate

$$P(A \cup B \cup C) \text{ and } P(A \cup B \cup C \cup D),$$

giving sufficient explanation to show how your results are obtained. (L.)

5. I have gone to a casino and am faced with two gaming machines, A and B, on which the probabilities of my winning are $2p$ and p respectively (where $0 < p < \frac{1}{2}$). I should, of course, like to play solely on machine A but there is unfortunately no way of distinguishing between the machines. I therefore decide to adopt a policy of playing on one of the machines until I lose, then changing over to the other machine until I again lose, and

so on. Show that with this strategy I will, in the long run, be using machine A for a fraction $(1-p)(2-3p)^{-1}$ of the time.
(O.C.S.M.P.)

6. Two persons X and Y play a game in which X deals from a standard pack of fifty-two cards. Each player receives three cards and the game is won by X if he holds more red cards than there are black cards held by Y.

By considering the number of black cards in play, or otherwise, calculate the probability that the dealer wins a game. What odds should be fixed by the dealer in order that he may expect just to show a profit?
(N.)

7. (i) A box X contains two white balls, one blue ball and three red balls; a box Y contains four white balls, two blue balls and no red balls. An ordinary die is thrown once and, if the score is a 1 or a 6, then a ball is drawn at random from box X, otherwise a ball is drawn at random from box Y. Find the probability of each of the five possible outcomes of the experiment. Find also

(a) the probability that a white ball is drawn,
(b) the probability that box X is used, given that a white ball is drawn.

(ii) It is known that, in a batch of six rods, just two are longer than a specified length. Two rods are chosen at random, without replacement, one after the other. Find

(a) the probability that the two rods chosen both exceed the specified length,
(b) the probability that the length of the second rod chosen exceeds the length of the first.

8. A, B and C are three events such that

$$P(A) = P(B) = P(C) = p_1,$$
$$P(A \cap B) = P(A \cap C) = P(B \cap C) = p_2,$$
$$P(A \cap B \cap C) = p_3.$$

Show that $p_1 \geqslant p_2 \geqslant p_3$.
Express the following in terms of p_1, p_2 and p_3:
  (i) $P(A \cup B \cup C)$;
  (ii) the probability that at least one of A, B, C occur;
  (iii) the probability that exactly two of A, B, C occur.
(M.E.I.)

9. Two rooms L and R each contain p calculating machines; one of the machines in L is damaged; all the others are undamaged. During one morning q ($< p$) of the machines in L are chosen at random and moved to R. During the afternoon of the same day q of the machines in R are chosen at random and moved to L. Find, in the simplest form, the probability that the damaged machine is in L
  (i) at the end of the morning,
  (ii) at the end of the afternoon.
(N.)

10. A bag A contains five similar balls of which three are red and two black and a bag B contains four red and five black balls. A ball is drawn at random from A and placed in B; subsequently a ball is drawn at random from B and placed in A. What is the chance that if a ball is now drawn from A it will be red?
If this ball is in fact red, what is the probability that the first two balls drawn were also red?
(N.)

11. Each of a set of n identical fair dice has faces numbered 1 to 6. Obtain the probability that, when every die is rolled once, all the scores are less than or equal to 5.
Obtain the probability P_r that the highest score in a single throw of all n dice has value r for $r = 1, 2, \ldots, 6$. Determine the mean of r and show that for $n = 3$ its value is $4\frac{23}{24}$.
(C.)

12. A system is made up of N independent components, each of which is used once in a complete cycle of operations. Initially each component has probability p of working successfully at each cycle, but when a component has once failed it cannot work again; when some components have failed, those remaining continue to operate.
Determine the probability that a specified component
  (i) fails at the nth cycle,
  (ii) fails after the nth cycle.
Determine the distribution of the number of components which are still working after the nth cycle is completed. Give the mean of this number.
(C.)

13. (a) Two unbiased six-sided dice carrying numbers 1 to 6 are thrown together; find an expression for the probability $P(r)$ of a score of r in a single throw. If a third die is now thrown, find the probability of a total score of more than 12.

(b) Write down the probability $P(r)$ of r successes in n independent trials when the probability of success in a single trial is p. Show that $P(r) \geqslant P(r+1)$ if $r+1 \geqslant (n+1)p$. Find the most probable number of successes in seven trials when $p = \frac{1}{3}$ and the probability of this number of successes. (O.C.)

14. A game is played with two dice, thrown simultaneously. A symbol is moved, through a distance equal to the sum of the scores shows on the dice, on a track as shown below.

A	B	C	D	E	F	G	H

A player's symbol is situated at A. Landing on E at the first throw of the dice is denoted by E_1; landing on H at the second throw by H_2, and similarly for other results. He continues to throw until he reaches or passes the twelfth square. Calculate

(a) $P(B_1)$, $P(G_1)$ and $P(B_1 \cup G_1)$,

(b) $P(G_2)$ and $P(G)$, where G denotes landing on G on any throw. (L.)

15. (i) Two cards are withdrawn from a pack of playing cards, without replacement. Find the probability that the first is a 2 and just one of them is a heart.

(ii) A coin whose diameter is 2·5 cm is rolled in a random way across a large chequered board whose squares are alternately black and white. When the coin stops rolling and falls on the board, the probability of its lying wholly within a white square is $\frac{1}{8}$. Calculate the length of the sides of the squares. (L.)

16. Two trains, A and B, are scheduled to arrive at a certain station at noon. The probability that train A is late on any one day is 0·1. The probability that train B is late on any one day is 0·27. Assuming that the arrival times of the two trains are independent of each other, calculate the probability that, on a particular day,

(a) both trains will be late,

(b) at least one of the trains will be late.

In practice, the time-keepings are not independent. It is known that the conditional probability of B being late given that A is late is 0·9. Calculate the probability that both trains will be on time on a particular day, assuming once again that the individual probabilities of A and B being late are 0·1 and 0·27 respectively.
(L.)

17. (i) An unbiased coin is tossed repeatedly until just three heads have been thrown. Show that the probability that just n tosses are necessary is $(n-1)(n-2)2^{-n-1}$.

(ii) If the probability of a family having just n children is $P(n)$ and $P(n) = \lambda p^n$ when $n \geqslant 1$ and λ and p are constants, what is the probability $P(0)$? If each child is equally likely to be a boy or a girl, show that the probability of a family having just n boys $(n \geqslant 1)$ is $2\lambda p^n/(2-p)^{n+1}$.

$$\left[(1-x)^{-n} = 1 + nx + \frac{n(n+1)}{2!}x^2 + \ldots \right]$$ (O.C.)

18. Four signalling lamps L_1, L_2, L_3 and L_4 are lit. Each lamp shows at random a red, green or white light, with probabilities in proportions 3:2:1. Calculate the probability that

(a) L_1 and L_2 show green,

(b) at least two lamps show green,

(c) two lamps show one colour and two show another colour. (L.)

19. (i) Drivers are divided by an insurance company into three groups A, B, C, which contain 10%, 60%, 30% respectively of those insured. The probability that a driver in group A will meet with an accident in any one year is 0·008; in group B it is 0·02; in group C it is 0·1. A particular driver insured with this company meets with no accident in two successive years. Find the probability that this driver is in group C.

(ii) A random variable can take the values 1, 2, ..., n each with probability $1/n$. If $G(t)$ denotes the probability generating function of this distribution, show that

$$n(1-t)G(t) = t - t^{n+1}.$$

Find the value of $G''(1)$, and hence show that the variance of this distribution is $(n^2 - 1)/12$. (L.)

Miscellaneous Exercise 11

1. (i) From a bag containing four red, five white, six blue and seven black balls, four are drawn at random without replacement.
Find the chance of their being
(a) all of the same colour,
(b) all of different colours.

(ii) Two six-faced dice with faces numbered 1 to 6 respectively are thrown together. What is the most likely value of the total score and what is the chance of the score being greater than this value?

2. The random events A, B and C are defined in a finite sample space S. The events A and B are mutually exclusive and the events A and C are independent. $P(A) = \frac{1}{5}$, $P(B) = \frac{1}{10}$, $P(A \cup C) = \frac{7}{15}$ and $P(B \cup C) = \frac{23}{60}$.
Evaluate $P(A \cup B)$, $P(A \cap B)$, $P(A \cap C)$ and $P(B \cap C)$ and state whether B and C are independent. (L.)

3. A bag initially contains a red and b green balls. A trial consists of drawing a ball at random from the bag, noting its colour and replacing it together with an additional ball of the same colour.

(i) Find, in its simplest form, the probability of drawing a red ball on the third trial. (You may need to factorise $a^2 + 2ab + b^2 + 3(a+b) + 2$.)

(ii) If by the fifth trial, three red and two green balls have been drawn, what is the probability of drawing a red ball on the eighth trial?

(iii) What is the probability of drawing three red and two green balls in the first five trials?

(O.C.S.M.P.)

4. (i) The event that Mary goes to a dance is M and the event that Linda goes to the same dance is L. The events L and M are independent and $P(L' \cap M') = \frac{1}{4}$, $P(L) + P(M) = \frac{23}{24}$.
Find the probability that both Linda and Mary go to the dance.

(ii) A gold urn contains three red balls and four white balls and a silver urn contains five red balls and two white balls. A die is rolled and, if a six shows, one ball is selected at random from the gold urn. Otherwise a ball is selected at random from the silver urn. Find the probability of selecting a red ball.

The ball selected is not replaced and a second ball is selected at random from the same urn. Find the probability that both balls are white. (L.)

5. A gold urn contains two red balls and three black balls, and a silver urn contains one red ball and five black balls. An urn is selected at random and a ball withdrawn and replaced. Find the probability that this ball is red.

A ball is withdrawn from the gold urn and placed in the silver urn. A ball is then drawn from the silver urn and placed in the gold urn. If all selections are made at random find, using a tree diagram or otherwise, the probability that if two balls are now taken without replacement from the gold urn they will both be red. (L.)

6. A game of chance consists of throwing a disc of radius 6 cm on to a horizontal circular board of radius 66 cm marked with concentric rings of negligible thickness as shown in the diagram. The rings have radii 18 cm, 36 cm and 54 cm respectively and there is a rim round the edge of the board to prevent the disc falling off, Fig. Misc. Ex. 11–6. A player "wins" if, when the disc settles it lies *entirely* within one of the regions

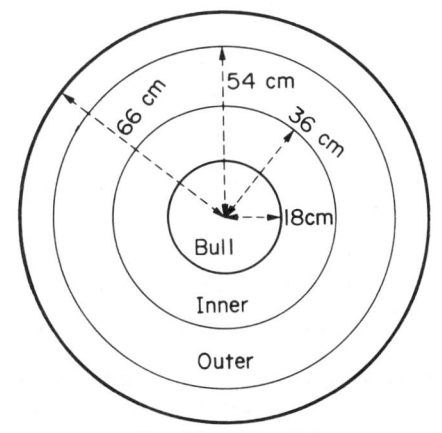

FIG. MISC. EX. 11.6.

marked Bull, Inner, Outer (i.e. it does not lie across any of the bounding rings). Calculate the probabilities that a disc thrown at random will fall entirely within the regions marked (i) Bull, (ii) Inner, (iii) Outer, and deduce that the probability of a "miss" (i.e. not a "win") is 18/25. [Assume that the final position of the centre of the disc is uniformly distributed over the accessible region.]

If prizes of 50p, 25p, 10p are awarded when the disc falls entirely within the regions Bull, Inner and Outer respectively, show that a charge of 6p per throw should enable the owner to make a profit in the long run.

Calculate also the probabilities that a player throwing four times at random will
  (i) score exactly 3 outers,
  (ii) win prizes to the value of exactly 30p,
  (iii) win prizes to the value of exactly 35p.

7. Describe briefly the binomial distribution and derive its mean and variance.

In a white population 15% are of Rhesus negative blood group. Find the chance of there being more than two of this blood group in a random sample of twenty from the population.

If 100 samples each of twenty are taken, find the chance that 70 or more of the samples contain more than two Rhesus negatives.

8. A gun is engaging a target and it is desired to make at least two direct hits. It is estimated that the chance of a direct hit with a single round is 1/4 and that this chance remains constant throughout the firing. A burst of five rounds is fired and if at least two direct hits are scored firing ceases. Otherwise a second burst of five rounds is fired. Find the chance that at least two direct hits will be scored
  (a) only five rounds being fired,
  (b) ten rounds having to be fired.

9. Show that the probability generating function $G(t)$ of the binomial distribution is $(q+pt)^n$, where $p+q=1$. Show also that, when $t=1$,

$$\frac{d}{dt}\{tG'(t)\} = npq + n^2p^2.$$

(ii) In a multiple-choice question five answers are suggested. The correct answer is known by three-quarters of the candidates, while the others make a random guess. A candidate is chosen at random from those who gave the right answer. Find the probability that he did not make a guess. (L.)

10. (i) An integer takes the value r with probability cr, c being a constant, for $0 < r \leqslant 3n$ and the probability is zero elsewhere. Find the value of c and show that the mean is $(6n+1)/3$ and the variance $(3n+2)(3n-1)/18$. Find the probability of r being greater or equal to $2n$.

$$\left[\text{The formula } \sum_{r=1}^{m} r^3 = m^2(m+1)^2/4 \text{ may be used.}\right]$$

(ii) A distribution of positive integers has frequency function $f(r) = \binom{5}{r}$ for $r = 1, 2, 3, 4, 5$, and $f(r) = 0$ for $r > 5$. Prove that the mean is 80/31 and the standard deviation $[4\sqrt{(65)}]/31$. (O.C.)

11. The random variable X comes from a negative exponential distribution with probability density function

$$f(x) = \begin{cases} \lambda e^{-\lambda x}, & x \geqslant 0, \ \lambda > 0, \\ 0, & x < 0. \end{cases}$$

Show that $P(X \geqslant 3.7\lambda^{-1}) = 0.025$ and find L so that $P(X \leqslant L\lambda^{-1}) = 0.025$. Hence find a 95% confidence interval for λ based on a single observation from the distribution. (M.E.I.)

12. The probability density function $p(t)$ of the length of life, t hours, of a certain component is given by

$$p(t)\,dt = ke^{-kt}dt \quad (0 \leqslant t < \infty),$$

where k is a positive constant. Show that the mean and standard deviation of this distribution are each equal to $1/k$.

Find the probability that the life of a component will be at least t_0 hours. Given that a particular component is already t_1 hours old and has not failed, show that the probability that it will last at least a further t_0 hours is e^{-kt_0}.

An apparatus contains three components of this type and the failure of one may be assumed independent of the failure of the others. Find the probability that (i) none will have failed at t_0 hours, (ii) exactly one will fail in the first t_0 hours, another in the next t_0 hours and the third after more than $2t_0$ hours.

13. A and B play a match of five games, each of which must be won or lost. In each of the first three games the probability that A will win is $\frac{2}{3}$ and in the remaining two games the probability is $\frac{3}{4}$. Show that the probability that in the match A will win exactly n games is the coefficient of t^n in the expression

$$(\tfrac{2}{3}t + \tfrac{1}{3})^3 (\tfrac{3}{4}t + \tfrac{1}{4})^2.$$

Find   (i) the chance that A will win the match,
       (ii) the average number of games that A will win.   (O.C.)

14. Show that the probability that exactly k tosses of a coin are needed to obtain r heads, where $k \geq r \geq 1$, is

$$\binom{k-1}{r-1} \frac{1}{2^k}.$$

Find an expression for the probability that l rolls of a die are needed to obtain s sixes where $l \geq s \geq 1$.

The coin is tossed until two heads are obtained and then the die is rolled until two 6's are obtained. Counting each toss as a trial and each roll as a trial, show that the probability that exactly ten trials are needed to obtain two heads and two 6's is

$$\sum_{n=1}^{7} \binom{n}{1} \frac{1}{2^{n+1}} \binom{8-n}{1} \frac{5^{7-n}}{6^{9-n}}. \quad\text{(L.)}$$

15. (i) Explain briefly what is meant by the "Poisson distribution of rare events", and state the mean and the standard deviation.

(ii) Show how the Poisson distribution may sometimes be used as an approximation to the binomial distribution. [You may assume that when n is large $(1 - a/n)^n$ is approximately e^{-a}.]

(iii) In a large consignment of manufactured articles 4% are defective. Use the Poisson distribution to calculate the probability that fifty articles chosen at random will contain five or more defective articles.   (O.C.)

16. Two garages hold between them four cars which are available for hire for a whole day from either garage. The first garage averages one demand for hiring per day, the second two demands per day. Assuming that the demands follow a Poisson distribution, find the probabilities that in any one day
(a) at least one car is hired,
(b) exactly three cars are hired,
(c) some demand for hiring has to be refused.   (O.C.)

17. In an examination 60% of the candidates pass but only 4% obtain distinction. Use the bionomial distribution to calculate the chance that a random group of ten candidates should contain at most two failures.

Use the Poisson distribution to calculate the chance that a random group of fifty candidates should contain more than one distinction.   (O.C.)

18. A counter lies on the x-axis and is initially at $x = 0$. A fair coin is tossed: the counter is moved along the x-axis one unit in the positive direction or one unit in the negative direction according as the result is heads or tails. This process is repeated at each new position and continued indefinitely. Find the probability that after two tosses the counter is again at $x = 0$. State the positions that the counter can be in after three tosses, and the associated probabilities.

Calculate the mean squared distance from $x = 0$ after (i) one toss, (ii) two tosses, (iii) three tosses, (iv) four tosses. Suggest a general result.

If the coin has probabilities p of heads and $(1 - p)$ of tails, show that the mean squared distance from $x = 0$ after three tosses is $9 - 24p + 24p^2$ and determine the value of p for which this is least.   (O.C.S.M.P.)

19. In a town with a population of 30 000 the average number of births per annum is 18·25 per 1000 of the population. Find the average number of births per day (you may take 1 year = 365 days). If the actual number of births per day follows a Poisson distribution find the probability of there being six or more babies born on one day.

How many times would this event be expected to occur in a year?

20. A mathematical model for the fraction x of the sky covered with cloud ($0 < x < 1$) assigns to this a probability density function

$$\phi(x) = k/\sqrt{[x(1-x)]}.$$

Calculate:
(i) the value of k,
(ii) the expected fraction covered by cloud,
(iii) the probability that not more than one-quarter of the sky is covered.

[*Hint.* Your integrations may be made easier by using the substitution $x = \sin^2 \theta$. You may assume that this substitution is valid, even though the functions to be integrated may be discontinuous at the ends of the interval of integration.] (O.C.S.M.P.)

21. The random variable X has the probability density function

$$f(x) = \begin{cases} \dfrac{\lambda}{x^4}, & x > 1, \\ 0, & x \leqslant 1. \end{cases}$$

Find the constant λ, and the mean and variance of the distribution.

If A_1 is the event $\{x : 1 < x < 4\}$ and A_2 the event $\{x : 3 < x < 6\}$, find $P(A_1 \cup A_2)$ and $P(A_1 \cap A_2)$. Also evaluate these two probabilities using a Normal distribution with the same mean and variance, and comment on the result.

22. Sketch the graph of the distribution given by

$$p(x) = \{1/\sqrt{(2\pi)}\} \exp(-\tfrac{1}{2}x^2),$$

and explain the significance of the fact that the integral of $p(x)$ from $-\infty$ to $+\infty$ equals unity.

Calculate approximately the probability that 1600 tosses of an unbiased penny will result in (a) 800 heads and 800 tails, (b) more than 850 heads. (L.)

23. Two types of seed X and Y, difficult to distinguish, are such that in the long run under given conditions 80 % of type X but only 60 % of type Y will germinate. Samples of 100 seeds of type X are selected at random. Find the mean and standard deviation of the number germinating.

A package arrives at a nursery without its label. It contains just one type of seed done up in packets each containing 100 seeds. The nurseryman knows that the seed is either of type X or of type Y, and thinks that it is more likely to be the former. He decides to plant a packet and to accept the seed as type X provided that at least 68 germinate, but otherwise to regard it as type Y. Using the Normal approximation, find the probability that the nurseryman will wrongly label type X seed as type Y.

What is the probability that he will wrongly label type Y seed as type X? (O.C.S.M.P.)

24. In a certain large population of men, heights are distributed Normally about a mean of 180 cm with standard deviation 5 cm. Random samples are taken with three men in each sample and their heights are arranged in increasing order. In 1000 such samples, approximately how many will have
(i) the middle height under 175 cm;
(ii) the least height less than 175 cm;
(iii) the least height between 175 and 180 cm? (O.C.S.M.P.)

25. It is estimated that 1400 commuters regularly aim to catch the 5.30 p.m. train at a certain London terminus, that 50 will have arrived before the platform gate is opened at 5.20 p.m., and that when the train leaves on time 70 arrive too late. Assuming the distribution of arrival times to be Normal, use tables to obtain the mean and standard deviation. Hence estimate
(i) at what time the platform gate should be opened if not more than 20 passengers are to be kept waiting at the gate;
(ii) how many of the commuters will miss the train on a day when (unexpectedly) it leaves 2 minutes late. (O.C.S.M.P.)

26. When a stake of k pence is put into a certain gaming machine the machine pays out $2k$ pence with probability $\tfrac{1}{4}$ (and otherwise pays out nothing). I have twopence in my pocket and decide to gamble with it, in units of 1p, either until I have lost all my money or until I have fourpence in my pocket. Denoting by $S_n (0 \leqslant x \leqslant 4)$ the state of having n pence in my pocket, write down the transition matrix **P** between $S_0, S_1, \ldots, S_4$.

If x_n ($0 \leq n \leq 4$) denotes the probability of my *eventually* reaching state S_4, given that I started in state S_n, explain why $x_0 = 0$ and $x_4 = 1$. Prove (by examining components, or otherwise) that

$$(x_0 \ x_1 \ x_2 \ x_3 \ x_4)\mathbf{P} = (x_0 \ x_1 \ x_2 \ x_3 \ x_4).$$

Use these equations to calculate x_2. Hence show that, if I start in state S_2, I would do better to stake all of my money in a single bet, rather than gamble in units of 1p. (O.C.S.M.P.)

Answers to the Exercises

Exercise 7.1 (p. 234)

1. (a) $-\mathbf{i}+4\mathbf{j}-2\mathbf{k}$, $4\mathbf{i}-5\mathbf{j}+9\mathbf{k}+t(-\mathbf{i}+4\mathbf{j}-2\mathbf{k})$; (b) $\sqrt{798}$.   2. $\sqrt{38}$.
3. $-6\mathbf{i}-\mathbf{j}-\mathbf{k}$; $\sqrt{14}$ N; $\mathbf{r}=-(6\mathbf{i}+\mathbf{j}+\mathbf{k})+t(2\mathbf{i}+3\mathbf{j}-\mathbf{k})$; $2(2\mathbf{i}+3\mathbf{j}-\mathbf{k})$; $2\mathbf{i}+11\mathbf{j}-5\mathbf{k}$.
4. $\mathbf{r}=\mathbf{i}-2\mathbf{j}+3\mathbf{k}+t(11\mathbf{i}+6\mathbf{j}-21\mathbf{k})$.   5. $-7\mathbf{i}+\mathbf{j}+2\mathbf{k}$.   6. $\sqrt{149}$ N.
7. $18\sqrt{26}$ Nm; $9(\mathbf{i}-2\mathbf{j}+2\mathbf{k})$; $(-4\mathbf{i}+\mathbf{j}+3\mathbf{k})/\sqrt{26}$.
8. (a) $7\sqrt{3}$ N; (b) $2\mathbf{i}-\mathbf{j}+3\mathbf{k}$; (c) $\mathbf{r}=2\mathbf{i}-\mathbf{j}+3\mathbf{k}+t(\mathbf{i}+\mathbf{j}+\mathbf{k})$.
9. $(\tfrac{4}{3}a, \tfrac{4}{3}a, a)$; $-6\sqrt{3}aF$.   10. $\mathbf{i}+3\mathbf{j}$; $5\sqrt{21}$, $3\mathbf{j}$; $5\sqrt{5}$.

Exercise 7.2 (p. 238)

1. $Mg(2-\sqrt{3})$.   5. $\mu Mg(\sqrt{2}-1)$.

Exercise 7.3 (p. 247)

9. (i) $R=3W$, $F=5W/(2\sqrt{3})$; (ii) $X=5W/(2\sqrt{3})$, $Y=2W$; (iii) $\mu \geqslant 5/(6\sqrt{3})$.

Exercise 7.4 (p. 251)

1. A thrust $11W/24$.

Exercise 7.6 (p. 260)

4. Stable with BC above the pegs.   5. Stable.
7. If θ is the angle between AB and the upward vertical, $\theta = 0, \pi$ unstable, $\theta = \pm\tfrac{1}{3}\pi$ stable.
8. If θ is the angle between the rod and the vertical, $\theta = 0, \pi$ unstable, $\theta = \tfrac{1}{3}\pi$ stable.
9. Stable.

Exercise 7.7 (p. 267)

2. $24g$ N.   4. $a/\sqrt{3}$.   7. $3l$.   8. $c=4$; $5\tfrac{2}{3}$ m, $4\ln 10$ m.

Exercise 7.8 (p. 269)

1. (a) $We^{4\mu\pi}(e^{\mu\pi}-1)/(e^{4\mu\pi}-1)$; (b) $W(1-e^{-\mu\pi})/(e^{4\mu\pi}-1)$.
2. $(M^2-m^2)/m$.   3. $(M-me^{\mu\pi})g/(M+me^{\mu\pi})$.

Miscellaneous Exercise 7 (p. 270)

1. $-\mathbf{i}+33\mathbf{j}+21\mathbf{k}$.   2. $(\lambda+\mu)(\mathbf{b}+\mathbf{c}+\mathbf{d})-3\lambda\mathbf{a}$.
3. $\mathbf{k}, \mathbf{i}, -\mathbf{j}$; (i) couple $2Fa\sqrt{3}$; (ii) force $2F\sqrt{3}$; (iii) couple $2Fa$.
4. $3\sqrt{14}$; $\mathbf{r}=5\mathbf{i}+2\mathbf{j}+5\mathbf{k}+t(2\mathbf{i}+3\mathbf{j}+\mathbf{k})$; $\dfrac{x-5}{2}=\dfrac{y-2}{3}=\dfrac{z-5}{1}$.
5. $(n+1)a/n$; $\tfrac{1}{3}(n+1)W$.   13. $\sqrt{5M/(2a)}$.
22. Each $13l/18$; $(4l/3)\ln(3/2)$.
23. $\mathbf{F}=(2+s)\mathbf{i}+(2+t)\mathbf{j}+3\mathbf{k}$; $\mathbf{G}=(-8-t)\mathbf{i}+(s-3)\mathbf{j}-3\mathbf{k}$; $s=7/2$, $t=-5/2$.
24. $\tfrac{1}{3}(\mathbf{a}+\mathbf{b}+\mathbf{c})$.   27. $c=3$.

ANSWERS TO THE EXERCISES 447

28. $r = 3i - 3j + 2k + t(6i + 8k)$. **29.** Unstable.
31. $\theta = \frac{1}{3}\pi$ unstable, $\theta = \pi$ stable. **33.** $0 < k < 3\sqrt{3}$.

Exercise 8.1 (p. 279)

1. $\sqrt{2}e^{-\pi/4}$; damped oscillatory.
2. $x = 4e^{-1} \approx 1.5$, $\dot{x} = -2e^{-1} \approx -0.74$; asymptotic to origin.
3. $x = -2e^{-\pi/2}$, $\dot{x} = 0$; damped oscillatory.
4. $x = 2e^{-t} - e^{-2t}$. 5. $x = (u/2k)e^{-kt}\sin 2kt$; $2kue^{-\pi/2}$.
7. $\ddot{x} + 2\lambda n\dot{x} + n^2 x = 0$; $x = \left(ue^{-\lambda nt}\sin[nt\sqrt{(1-\lambda^2)}]\right)\Big/\left[n\sqrt{(1-\lambda^2)}\right]$.
9. $s = \frac{1}{6}\lambda(2\sin t + \sin 2t)$.

Exercise 8.2(a) (p. 282)

1. -4; 6.   2. $s = e^{r^2/(2k)}$.   3. $-\frac{3}{2}$; $\frac{2}{3}(1 - e^{-3}) \cdot [s = \frac{2}{3}(1 - e^{-3t})]$.
5. 5.   6. $e^{1/10}$; $100\ln 100 \approx 460$. $[v = e^{t/100}]$.
7. $10(e^{1/5} - 1)$, $[v = 10(e^{t/5} - 1)]$; $10(5e^{1/5} - 6)$, $[s = 10(5e^{1/5} - t - 5)]$.
8. $v = (V/s)\sqrt{(s^2 - a^2)}$; $f = a^2 V^2/s^3$.
9. $A = 1/80$ m^{-1}s, $B = 1/80^2$ m^{-1}; $x = 6400\ln[(80+t)/80]$; $v = 80e^{-x/6400}$.

Exercise 8.2(b) (p. 290)

1. 2.5m.   7. $4V/\sqrt{7}$.   10. $\frac{1}{2}m\left[V^2 - (g/k)\ln\left(1 + \frac{kV^2}{g}\right)\right]$.

Miscellaneous Exercise 8 (p. 298)

1. $x = V(8 - 9e^{-nt} + e^{-3nt})/(6n)$; $3mn^2 x$.   3. $(V/g)\tan^{-1}(U/V)$.
7. (i) $x = ae^{-5t}(5t+1)$; (ii) $x = \frac{1}{3}ae^{-4t}(3\cos 3t + 4\sin 3t)$.   8. P/R.
11. $H = \left[ku - g\ln\left(1 + \frac{ku}{g}\right)\right]\Big/k^2$; $T = (1/k)\ln\left(1 + \frac{ku}{g}\right)$.
15. $x = (1/k)e^{-kt}\sin 2kt$.   18. $x = (g/10n^2) + \frac{1}{2}ae^{-nt}(2\cos 2nt + \sin 2nt)$.
19. $5v(dv/dx) + v^2 = 50$.   20. $(c/g)\tan^{-1}(2e^2)$.
23. $x = \frac{1}{10}a(5e^{-nt} - 2e^{-2nt} + \sin nt - 3\cos nt)$.   24. $A = 2a$, $B = -a$.
25. $x = e^{-2t}\sin 3t + \cos 3t + 3\sin 3t$.   27. $\frac{1}{8}Mn^2(1-k)[\ln(1-k)]^2$; $1 - e^{-2}$.

Exercise 9.1 (p. 306)

1. $\mathbf{v} = \mathbf{i}(\sin t \pm 3) + \mathbf{j}(1 - \cos t) + \mathbf{k}(t \pm 3)$.
2. $-\hat{\mathbf{a}}\left(\dfrac{d\theta}{dt}\right)^2 + \hat{\mathbf{b}}\dfrac{d^2\theta}{dt^2}$.

Exercise 9.2 (p. 308)

1. $y = (ux/v) - gx^2/(2v^2)$.   2. $(x - 2y)^2 - 2u^2(x - y) = 0$.
3. $x^2 + y^2 = a^2$ (described in the clockwise direction).
4. $x^2/a^2 + y^2/b^2 = 1$ (described in the anticlockwise direction).
5. $y^2 = 4k^2 x$.   6. $xy = 1$.
7. $\int_0^1 \sqrt{[2(t^2 + 2t + 2)]}\,dt = [2\sqrt{5} - \sqrt{2} + \ln(\sqrt{10} + 2\sqrt{2} - 2 - \sqrt{5})]/\sqrt{2}$.
8. $\omega ab/OP$.
9. $\mathbf{r} = (4\sin\omega t)\mathbf{i} + (3\cos\omega t)\mathbf{j}$, $x^2/16 + y^2/9 = 1$; $\omega = \frac{1}{2}\tan^{-1}(3/4)$, $t = 2$.
10. $\mathbf{r} = (\cos t + \sin t)\mathbf{i} + (\cos 2t - \sin 2t)\mathbf{j}$.
11. (a) $2\pi/k$; (b) $4mk^2$; $2(1 - k\pi\sqrt{3})\mathbf{i} + \frac{3}{2}(\sqrt{3} + k\pi)\mathbf{j}$.
12. $2m(\mathbf{i} + 2\mathbf{j} + \mathbf{k})$; $4m(-3\mathbf{i} + \mathbf{j} + \mathbf{k})$.   13. $\frac{4}{3}a\sin nt - \frac{1}{6}a\sin 2nt$.

Exercise 9.3 (p. 312)

1. $2mg$.   2. $\sqrt{(2g)}$; $mg/\sqrt{2}$.   3. (i) $mg\cos\alpha[1+2\ln(2\cos\alpha)]$.
5. $\frac{1}{2}mg\cos\psi(3\cos\psi-2)$.   6. $mg(2+3\cos\theta)$ on the bead towards O.
8. $mg(1+4\sqrt{2})/(3\sqrt{2})$.

Exercise 9.4 (p. 317)

1. $mg\sin\omega t$.   4. $\dot{\mathbf{r}} = -(\omega\sin\omega t)\mathbf{a}+(\omega\cos\omega t)\mathbf{b}$; $\ddot{\mathbf{r}} = -\omega^2\mathbf{r}$; $-m\omega^2\mathbf{r}$.

Exercise 9.5(a) (p. 321)

1. 170 m.   2. 29 m s^{-1} at 9° (above) to the horizontal.   3. 6·0 s.
4. (i) 30 m, (ii) 90 m.   5. $(1/2g\cos\theta)u^2\sin^2(\alpha-\theta)$.
8. (i) $V^2/[g(1+\sin\alpha)]$; (ii) $V^2/[g(1-\sin\alpha)]$.

Exercise 9.5(b) (p. 327)

1. $\beta = (\pi-2\alpha)/4$; $V^2/[g(1+\sin\alpha)]$.   2. $\tan^{-1}\left(\sqrt{[(r+h)/(r-h)]}\right)$.
7. $g\sec^2\theta x^2 - 2V^2\tan\theta x + 2V^2 h = 0$; $\tan^{-1}[V/\sqrt{(V^2-2gh)}]$.

Exercise 9.6 (p. 332)

1. $u_1 = \sqrt{10}$ m s^{-1}, $\alpha = \tan^{-1}(-2)$, $v_1 = \frac{1}{2}\sqrt{10}$ m s^{-1}, $\beta_1 = \tan^{-1}2$.
2. $u_1 = (u\sqrt{34})/8$, $\alpha_1 = \tan^{-1}4$, $v_1 = (3u\sqrt{2})/8$, $\beta_1 = 0$.
3. $u_1 = 5V/4$, $\alpha_1 = \tan^{-1}(-4/3)$, $v_1 = V/4$, $\beta_1 = 180°$.
4. $u_1 = (V\sqrt{111})/9$, $\alpha_1 = \tan^{-1}(-3\sqrt{3}/11)$, $v_1 = (V\sqrt{21})/9$, $\beta_1 = \tan^{-1}(3\sqrt{3})$.
5. $e = 1/n$, $u_1 = (V\sqrt{3})/2$, $v_1 = V/(2n)$, $\beta_1 = 0$.   6. $\alpha_1 = \tan^{-1}8$.
7. $\tan^{-1}[(1+e)\tan\alpha/(2\tan^2\alpha+1-e)]$.   8. $\frac{1}{3}$.   9. 45°.
11. $(1-e)u\mathbf{i}+u\mathbf{j}$; $(1+e)u\mathbf{i}$; $\mathbf{i}+2e\mathbf{j}$; $e = \frac{1}{4}$.   19. $mu(1-e)\sin\alpha$.
20. $4(8\mathbf{i}-15\mathbf{j})$; $8\mathbf{i}-(11/6)\mathbf{j}$, $2\mathbf{i}-(13/3)\mathbf{j}$.
21. $10\mathbf{n}+7\mathbf{t} = (134\mathbf{i}+85\mathbf{j})/13$, $17\mathbf{n}+24\mathbf{t} = (373\mathbf{i}+84\mathbf{j})/13$.
22. $-\mathbf{i}+\mathbf{j}+\mathbf{k}$; $4(\mathbf{i}+2\mathbf{j}-\mathbf{k})$; $\mathbf{i}+2\mathbf{j}-\mathbf{k}$.

Exercise 9.9 (p. 346)

1. $I\sqrt{[5a/(4\lambda m)]}$.   2. $2\ddot{x} = \omega^2(y-x) = -2\ddot{y}$.   3. $\pi/\sqrt{(2\mu)}$; $a\sqrt{(\mu/2)}$.
4. $V_A = 2\lambda V/(m_1 n^2 a)$, $V_B = V - 2\lambda V/(m_2 n^2 a)$; $\lambda V\pi/(m_1 n^3 a)$.
5. $(3mg\sqrt{3})/16$.   6. $(u/a)\sqrt{[2/(1+\cos^2\theta)]}$.   7. $u/\sqrt{2}$.
8. $\frac{1}{2}J\cos\theta$; $(J^2/4m)(1+\sin^2\theta)$.   10. $\frac{1}{2}(3u\mathbf{i}+v\mathbf{j})$, $\frac{1}{2}(-u\mathbf{i}+5v\mathbf{j})$.
11. 1500 J; 50 m s^{-1}, 150 m s^{-1}; $50(4-\sqrt{3})$ m s^{-1}.   12. $\frac{1}{2}u(1-e)$, $\frac{1}{2}u(1+e)$.

Miscellaneous Exercise 9 (p. 352)

1. $6(t^2+4t-5)$; $12m(-15\mathbf{i}-15\mathbf{j}+4\mathbf{k})$.   2. $-2e^{-2t}$.   5. $k = g/a$.
6. $\lambda\mathbf{k}\times\dot{\mathbf{r}}-\mu\mathbf{r} = m\ddot{\mathbf{r}}$.   7. $\mathbf{F} = -mn^2(a\cos nt\,\mathbf{i}+b\sin nt\,\mathbf{j})$; $\mathbf{F}\cdot\dot{\mathbf{r}} = \frac{1}{2}(a^2-b^2)mn^3\sin 2nt$.
8. (i) $\mathbf{F}\cdot\mathbf{v}$; (ii) $\int_0^T \mathbf{F}\cdot\mathbf{v}\,dt$; $\mathbf{v} = -4a\sin 2t\,\mathbf{i}+2a\cos 2t\,\mathbf{j}$;.
$\mathbf{F} = -4am(2\cos 2t\,\mathbf{i}+\sin 2t\,\mathbf{j})$; $12ma^2\sin 4t$; $3ma^2$ when $T = \pi/8$.
9. $\frac{1}{2}\pi$.   10. $x = u\cos\alpha(1-e^{-kt})/k$, $y = [(ku\sin\alpha+g)(1-e^{-kt})-gkt]/k^2$.
11. $\frac{1}{4}\omega$.   13. $2t^2$.   14. $4\mathbf{j}$, $2\mathbf{i}+\mathbf{j}$; 5: $\sqrt{13}$; $\sqrt{5}$ units of distance.
15. For S; $\frac{1}{2}V\cos\beta(1-e)$, $V\sin\beta$; for T; $\frac{1}{2}V\cos\beta(1+e)$, 0.
19. (i) $m\sqrt{(2ga)}$; (ii) $\frac{2}{3}m\sqrt{(2ga)}$.   20. $40\sqrt{5}$ m.
21. $2\gamma a(m_1+m_2)/(2\gamma m_1+2\gamma m_2-av^2)$;
$m_2\{(2\gamma m_1^2+2\gamma m_1 m_2+m_2 av^2)/[a(m_1+m_2)^3]\}^{1/2}$.   23. $\sqrt{(5ag/6)}$.
25. $9V^2/(2a)$.

ANSWERS TO THE EXERCISES   449

26. (i) $10(\mathbf{i}-\mathbf{j}+2\mathbf{k})$, $2(6\mathbf{i}-3\mathbf{j}+8\mathbf{k})$;   (ii) $(3-2e^{-t})\mathbf{i}+(3+5\sin t)\mathbf{j}+(4-3\cos t)\mathbf{k}$,
$(2e^{-t}+3t-1)\mathbf{i}+(3t-5\cos t+6)\mathbf{j}+(4t-3\sin t+1)\mathbf{k}$.
28. $\frac{1}{2}u$.   30. (i) $\omega\sqrt{2}e^{\omega t}$; $2\omega^2 e^{\omega t}$; (ii) ak^2.
31. $52mV/5$; $\tan^{-1}(5/12)$; $112mV^2/25$.
32. (i) $\mathbf{r}=e^{4t}(\mathbf{i}-\mathbf{j}+\mathbf{k})$;   (ii) $\mathbf{r}=e^{-t}(\cos t-\sin t)(\mathbf{i}+\mathbf{k})$;
(iii) $\mathbf{r}=\sin t\,\mathbf{i}+\cos 2t\,\mathbf{j}$.
33. mg; $3a/2$.
34. (i) $\dot{r}^2=2\mathbf{E}\cdot\mathbf{r}+\lambda^2\mathbf{H}^2$,   (ii) $\dot{\mathbf{r}}=\mathbf{E}t+\mathbf{r}\times\mathbf{H}+\lambda\mathbf{H}$.

Exercise 10.2 (p. 364)

1. $\sqrt{(2gl)}$. 2. $\frac{1}{2}Ma^2$; $Ma\pi n^2/(7200F)$. 3. $\frac{3}{2}Ma^2$; $\sqrt{[g/(3a)]}$.
4. $2\sqrt{g}\approx 3.1$ rad s^{-1}. 8. $\frac{1}{2}Ma^2$. 9. $9mga/(2\pi)$.
10. $\sqrt{[21g/(20a)]}$. 11. $Mmg/(M+2m)$. 14. $kmg/(1+2k)$. 15. $3M:(2M+m)$.

Exercise 10.3 (p. 370)

1. $2\pi\sqrt{[7a/(3g)]}$. 3. $2\pi\sqrt{[3(a^2+2x^2)/4(a+3x)g]}$.
4. $2\pi\sqrt{[2(2a^2+9c^2)/3(a+6c)g]}$. 5. $\frac{2}{3}\pi\sqrt{(14a/g)}$.
6. (i) Ma^2; (ii) $\frac{1}{2}Ma^2$.
9. $a\sqrt{2}$; $2\pi\sqrt{\bigl((2a\pi)/[g\sqrt{(\pi^2+4)}]\bigr)}$; $\sqrt{\bigl([2g\sqrt{(\pi^2+4)}]/(\pi a)\bigr)}$.
11. $\frac{2}{3}\pi\sqrt{(14a/g)}$. 12. (i) $2a/3$; (ii) $(a\sqrt{5})/3$; $2\pi\sqrt{(2a/g)}$.
14. $2\pi\sqrt{[(9a^2+2x^2)/(gx)]}$; $3a/\sqrt{2}$.
15. $m(a^2+12x^2)\dot{\theta}^2-mg(a+24x)\cos\theta=$ constant;
$2\pi\sqrt{\bigl([2(a^2+12x^2)]/[(a+24x)g]\bigr)}$.

Exercise 10.4 (p. 374)

4. $\sqrt{[12g(\sin\theta+\cos\theta-1)/(23a)]}$; $12mg/23$.
6. $\frac{3}{4}mg\sin\theta(3\cos\theta-2)$, $\frac{1}{4}mg(1-3\cos\theta)^2$. 7. $\frac{1}{4}mg\cos\theta$.

Exercise 10.5 (p. 379)

1. $\frac{1}{2}M\sqrt{(3ga/2)}$. 2. $3\sqrt{[g/(8a)]}$.
4. $\sqrt{\bigl(2g(M+2m\sin^2\alpha)]/[Ma(1+\cos\alpha)]\bigr)}$. 5. $6u/(7a)$.
7. $\sqrt{[3g/(2a)]}$; $\cos^{-1}(3/4)$. 9. $2\sqrt{(g/a)}$.

Exercise 10.7 (p. 384)

1. Space-centrode (part of) circle centre O radius $2a$, $(x^2+y^2=4a^2)$; body-centrode (part of) circle of radius a and centre the mid-point of AB.
2. I is fixed coinciding with O; the space and body-centrodes are points also coinciding with O.
3. As for 2 above.
4. Space-centrode is parabola focus O directrix Px; body-centrode is parabola focus A, directrix the line through O parallel to the rod.
5. If $(x, 0)$, $(0, y)$ are the coordinates of A, B respectively, then $\dot{x}=\omega y$, $\dot{y}=-\omega x$ so that both A and B move with simple harmonic motion along the axes; I lies on OC produced so that $OI=2OC$.

Exercise 10.9 (p. 397)

1. $(v\sqrt{82})/20$ at $\tan^{-1}(4/5)$ with BA and towards the side of AB from which the particle came.
2. $[(M+m'+nm)ga^2\sin\alpha]/[2Ma^2+m'(\frac{1}{2}b^2+a^2)+(nm/3)(4a^2+ab+b^2)]$.
3. $\sqrt{[3g/(10a)]}$; $9g/(20a)$; $R_B=9mg/50$, $R_A=49\,mg/100$.
9. $2\mu gx(a-y)^2=aV^2(a-2y)$.
13. Tension $=[g(1+\sin\alpha)]/[(1/m)+(1/M)+(a^2/Mk^2)]$;
acceleration $=g\bigl([(1/M)+(a^2/Mk^2)]\sin\alpha-(1/m)]/[(1/m)+(1/M)+(a^2/Mk^2)]\bigr)$.
16. $2a^2\omega^2(\mu\cos\alpha-\sin\alpha)/[(7\mu\cos\alpha-2\sin\alpha)^2g]$;
$2\omega(\mu\cos\alpha-\sin\alpha)/(7\mu\cos\alpha-2\sin\alpha)$.
17. $\sqrt{(42ag)}$. 18. $\frac{2}{3}ma^2$; (a) $\frac{1}{3}\Omega$, (b) $\frac{2}{9}\sqrt{2ma\Omega}$, (c) $\frac{2}{9}ma^2\Omega^2$.

Miscellaneous Exercise 10 (p. 399)

2. $2\pi\sqrt{[(3r^2+2h^2)/(5h)]}$.
4. $\frac{3}{4}mg \sin\theta(3\cos\theta - 2\sin\theta)$, $\frac{1}{4}mg + \frac{3}{4}mg\cos\theta(3\cos\theta - 2)$.
5. $2m\sqrt{(\sqrt{3}ag)}$. 6. (i) $\sqrt{k}$; (ii) $\sqrt{(2/k)}\omega \sin[\sqrt{(k/2)}t]$.
7. $Mg(23+12\sqrt{3})/23$, vertical. 11. $7\Omega/16$. 12. $\frac{1}{3}\Omega$.
13. $(MR\Omega - mr\omega)/(M+m)r$. 14. $M\sqrt{[\pi gl/(3\sin 2\theta)]}$. 15. $\frac{2}{3}\pi a$.
17. $V/(2a)$. 18. $\sqrt{[3x/(g\sin\alpha)]}$. 19. J/M; $5J/(8Ma)$; $37J^2/(64M)$.
21. $\omega^2 > gk/[a(1-\frac{1}{2}k)^2]$.
24. $5\Omega/2$; $\sqrt{[4ga(3\sqrt{3}-1)-15a^2\Omega^2]}/(2\sqrt{3})$.
25. $2\pi\sqrt{(85a/13g)}$, $4J/(85Ma)$. 28. 3ω, $\frac{3}{2}\omega$; $27mr^2\omega^2/4$; ML^2T^{-1}.
31. $2\Omega/7$. 32. $\sqrt{[3g\cos\beta/(2a)]}$; $3g/(4a)$.
34. $\sqrt{[g/(5a)]}$; $3\sqrt{3g/(25a)}$. 37. $\sin^{-1}[(32a\omega^2-9g)/(9g)]$.
39. $\pi\sqrt{2r}$. 40. $\sqrt{(147ag/11)}$.

Exercise 11.1 (p. 406)

1. 172 cm, 10 cm. 2. 2·25, 1·09.
4. (i) 4·8, 0·98; (iii) 0·737; (iv) 0·388. 5. (i) 13/3888; (ii) 5/16.
6. (i) 5/192; (ii) 5/324; (iii) 63/64; (iv) 5/96.
7. (a) 1/21; (b) 1/126; (c) 1/14; (d) 5/126.
8. (a) 0·3; (b) 0·28; (c) 0·88. 9. 0·32.
10. (a) 0·065; (b) 0·168; (c) 0·002.
11. (a) $\frac{1}{9}$; (b) $\frac{5}{9}$; (c) $\frac{1}{3}$. 12. 0·61, 6·1%; 3·86.
13. (a) $\frac{1}{3}$; (b) $\frac{5}{12}$; (c) 0·00014; (d) 0·0395.
14. (a) (i) 9/190; (ii) 32/95; (b) 23/35.
15. (i) (a) 0·01, (b) 0·87; (ii) (a) 0·656, (b) 0·915.
17. (a) 13; (b) 11/16. 18. (a) 5/32; (b) 23/256.
19. (i) (a) 1/221; (b) 220/663; (iii) 53/576.

Exercise 11.2 (p. 411)

1. $x(20-3x)/96$; 22/9.
2. (a) $1/n$, $\frac{1}{2}(n+1)$, $\frac{1}{12}(n^2-1)$; (b) $1/n$, $\frac{1}{2}n$, $n^2/12$; (c) n, $\frac{1}{3}n^2$.
3. (a) 3/2; (b) 20/27. 4. $1\cdot5/(3a-1)$; $a > \frac{1}{3}$; 6/5.
6. $1/(b-a)$ for $a \leq x \leq b$, zero otherwise; $\frac{1}{2}(a+b)$, $\frac{1}{12}(b-a)^2$.
8. 0·0056; 0·062. 10. x^n; $n/(n+1)$.
11. $z_L = 0\cdot25l$, $z_U = 0\cdot776l$.

Exercise 11.3 (p. 417)

1. 0·0183; 0·372; 0·039. 2. 0·44; 0·01.
3. (a) $(\frac{1}{4})^{10}$; (b) $_{10}C_5(\frac{1}{4})^5(\frac{3}{4})^5$; (c) $_{10}C_4(\frac{1}{4})^6(\frac{3}{4})^4 + \ldots + (\frac{1}{4})^{10}$;
$m = 2\cdot5$, $\sigma = 1\cdot37$; 0·057.
4. m, $m-1$; (i) 0·05; (ii) 4; (iii) 3. 5. 110; (a) 44, (b) 12.
6. 0·371; 0·0003. 7. (i) 0·67; (ii) 0·11; (iii) 0·040.
8. 0·37; £80. 9. (a) e^{-2}, (b) $\frac{1}{3}(3-19e^{-2})$.
10. 1·2; 0·034. 11. 0·25.
12. $m = \lambda$, $\sigma = \sqrt{\lambda}$; (i) 0·27; (ii) 0·68; 2.
13. 1·45; (a) 0·055, (b) 0·33. 14. 0·412.
15. 0·735; 49%. 16. 0·26; 0·39.
17. 0·010. 18. 0·78. 19. 50.
20. (a) 0·997(5), (b) 0·847(5).

Exercise 11.4 (p. 425)

1. $\binom{100}{51}(0\cdot58)^{51}(0\cdot42)^{49} + \ldots + (0\cdot58)^{100}$; 0·947.

ANSWERS TO THE EXERCISES   451

2. 1·500, 0·005; 0·046.
4. (i) 5%; (ii) 2·4%; (iii) 7·27%; (iv) 0·269.
6. 31.
8. (i) 77·5%; (ii) 21·3%; (iii) £36·54.
10. (a) 0; (b) 0; (c) 0·776.
12. (i) (a) 6/11; (b) 6·2p; (ii) 50, 25; 0·16.
13. 0·17.
15. (a) (i) 1097, (ii) 844; (b) 63; (d) 713.
16. (i) 0·0062; (ii) 0·048.
18. (a) 53·2, 46·8; (b) 80; (c) 12.
20. (i) 116, 86·6, 31·5; (ii) 0·0075; (iii) 0·0068.

3. (b) 0·9998; (c) 12; (d) 0·0002.
5. 0·16; 4·6%.
7. 0·175.
9. 0·901.
11. 91%; 10, 3·08; 5·2%.

14. 0·8%.

17. 131·4, 16·8; 0·617.
19. (i) 0·894; (ii) 0·048.

Exercise 11.7 (p. 434)

1. 29/36; 19/24; 3/5.
2.
$$\begin{pmatrix} & A_0 & A_1 & A_2 \\ A_0 & 0 & 1 & 0 \\ A_1 & \frac{1}{4} & \frac{1}{2} & \frac{1}{4} \\ A_2 & 0 & 1 & 0 \end{pmatrix} ; \frac{1}{8}$$

3. (i) (2/5 9/20 3/20); (ii) (8/29 8/29 8/29 5/29); (iii) (0 0 0 1).
4. (a) 4/9; (b) 61/324; (2/5 9/20 3/20).   **5.** 9.
6. (i) (15/47 20/47 12/47);
$$\begin{pmatrix} & 1 & 2 & 3 \\ 1 & \frac{1}{9} & \frac{5}{18} & \frac{11}{18} \\ 2 & 0 & \frac{1}{4} & \frac{3}{4} \\ 3 & 0 & 0 & 1 \end{pmatrix}$$

7. (i) $\{A, B, C\}$; (ii) $\{D\}$; 3.
8.
$$\begin{pmatrix} & A & B & C \\ A & 0 & \frac{1}{2} & \frac{1}{2} \\ B & \frac{1}{3} & 0 & \frac{2}{3} \\ C & \frac{5}{6} & \frac{1}{6} & 0 \end{pmatrix}$$
9. $\begin{pmatrix} 1 \\ -2 \\ 1 \end{pmatrix}$; 6.

10. (3/19 10/19 6/19).

Exercise 11.8 (p. 438)

1. 1·386, 0·078; $\frac{1}{3}$; 1·33, 1·143 ≤ x ≤ 1·600.
2. $\frac{1}{4}\sqrt{[(1+3e)(1+e)]}$.   **4.** 0·496; 0·6976.
6. 0·294; 3 to 1.
7. (i) $P(XW) = \frac{1}{9}, P(XB) = \frac{1}{18}, P(XR) = \frac{1}{6}, P(YW) = \frac{4}{9}, P(YB) = \frac{2}{9}$; (a) $\frac{5}{9}$, (b) $\frac{1}{5}$;
(ii) (a) 1/15, (b) 4/15.
8. (i), (ii) $3p_1 - 3p_2 + p_3$; (iii) $3(p_2 - p_3)$.
9. (i) $(p-q)/p$; (ii) $q^2/(p^2 + pq)$.   **10.** 143/250; 45/143.
11. $(5/6)^n$; $(1/6)^n$ $(r=1)$, $(r/6)^n - (r-1)^n/6^n$ $(r = 2, 3, \ldots, 6)$; $6 - \sum_{r=1}^{5} (r/6)^n$.
12. (i) $p^{n-1}(1-p)$; (ii) p^n; binomial, Np^n.
13. (a) $(6-|r-7|)/36$; 7/27; (b) $\binom{n}{r} p^r (1-p)^{n-r}$; 224/729.
14. (a) 0, 5/36, 5/36; (b) 5/648, 0·147.   **15.** (i) 1/34; (ii) 5 cm.
16. (a) 0·027, (b) 0·343; 0·72.   **18.** (a) 1/9, (b) 11/27, (c) 49/216.
19. (i) 0·265; (ii) $\frac{1}{3}(n^2-1)$.

Miscellaneous Exercise 11 (p. 441)

1. (i) (a) 8/1045, (b) 24/209; (ii) 7, 5/12.
2. 3/10, 0, 1/15, 1/20; no.
3. (i) $a/(a+b)$; (ii) $(a+3)/(a+b+5)$;
   (iii) $a(a+1)(a+2)b(b+1)/[(a+b)(a+b+1)(a+b+2)(a+b+3)(a+b+4)]$.
4. (i) 5/24; (ii) 2/3, 11/126.   5. 17/60; 31/350.
6. (i) 1/25, (ii) 9/100, (iii) 3/20; (i) 0·011, (ii) 0·011, (iii) 0·117.
7. 0·595; 0·016.   8. (a) 0·367; (b) 0·389.
9. (ii) 15/16.   10. (i) $2/[3n(3n+1)]$; $5(n+1)/[3(3n+1)]$.
11. 0·025;   12. e^{-kt_0}; (i) e^{-3kt_0}, (ii) $e^{-3kt_0}(1-e^{-kt_0})^2$.
13. (i) 181/216; (ii) 19/9.   15. (iii) 0·371.
16. (a) 0·950, (b) 0·224, (c) 0·185.   17. 0·167; 0·594.
18. $\frac{1}{2}$.$(-3, 0), \frac{1}{8}$; $(-1, 0), \frac{3}{8}$; $(1, 0), \frac{3}{8}$; $(3, 0), \frac{1}{8}$. $1, 2, 3 \ldots n$. $p = \frac{1}{2}$.
19. 1·5; 0·0045; twice.   20. (i) $1/\pi$; (ii) $\frac{1}{2}$; (iii) $\frac{1}{3}$.
21. $\lambda = 3$; $m = \frac{3}{2}$, $\sigma^2 = \frac{3}{4}$.   22. (a) 0·02, (b) 0·006.
23. $m = 80$, $\sigma = 4$; 0·055; 0·001.   24. (i) 68; (ii) 404; (iii) 471.
25. μ at 25·2 minutes past 5, $\sigma = 2·9$ min. (i) 5 h 18·8 mins; (ii) 13.
26. $P = \begin{pmatrix} 1 & \frac{3}{4} & 0 & 0 & 0 \\ 0 & 0 & \frac{3}{4} & 0 & 0 \\ 0 & \frac{1}{4} & 0 & \frac{3}{4} & 0 \\ 0 & 0 & \frac{1}{4} & 0 & 0 \\ 0 & 0 & 0 & \frac{1}{4} & 1 \end{pmatrix}$; $x^2 = \frac{1}{10}$.

Index

Area 315
Axis of revolution, force on 372–

Bounding parabola 324

Cartesian axes 307
Catenary 262–
Central force 313–
Central orbit 313–
Centre
  of oscillation 367
  of percussion 376
Compound pendulum 366
Confidence limits 422
Coordinates
  intrinsic 305–
  polar 304, 313–
Couple, as vector 231

Damped harmonic oscillations 274–

Elastic bodies, oblique impact of 328–
Enveloping parabola 325–
Equilibrium
  of forces in three dimensions 233–
  of rigid body 236–
  stable or unstable 255–

Force
  central 313–
  in three dimensions 231–
  on axis of lamina 372–

Galilean transformation 334–

Harmonic oscillations, damped 274

Hinge
  rough 240
  smooth 239

Impact, oblique 328–
Instantaneous centre of rotation 382
Intrinsic coordinates 305

Lamina
  motion in a plane 381–
  rotation about a fixed axis 358–

Markov chain 428
Mass, variable 292–
Moment of force
  about line 231
  about point 232
Motion
  general, of lamina 384–
  of connected particles 329–
  of projectile 319–
  of system of particles 336–
  on smooth curve 309–
  referred to cartesian axes 307–
  two-dimensional, in resisting medium 347–
  under central force 313–
  under variable forces 280–
  with variable mass 292–

Normal distribution 419–

Oblique impact 328–
Orbit, central 313–
Oscillations, damped harmonic 274–

Parabola, bounding, enveloping, of safety 324
Pendulum, compound 366

INDEX

Percussion, centre of 376
Poisson distribution 413–
Potential energy function 255
Probability distribution
  binomial 405
  continuous 409–
  geometric 405
  normal 419–
  Poisson 413–
  rectangular 409
Probability vector 429
Projectile 319–

Rectilinear motion
  under variable forces 280
  with variable mass 292–
Resisting medium, two-dimensional motion in 347–
Rolling, condition for 382
Rope, on rough surface 268
Rotating lamina
  about fixed axis 358
  impulse on 375
  in its plane 381–
  momentum and energy 361

Safety, parabola of 325
Sag 266
Smooth curve, motion on 309–
Stability 255–

Stable equilibrium 256
Steady state 429
Stirling's approximation 420
Straight line motion
  under variable forces 280–
  with variable mass 292–
Stretched wire 266
String
  hanging under gravity 261–
  in contact with rough surface 268

Terminal velocity 287
Transformation, Galilean 334–
Transition
  matrix 428–
  states 433

Unstable equilibrium 255

Variable mass 292–
Varignon's theorem 232
Vector
  couple as 231
  differentiation of 303
  probability 429
Virtual work 248

Work, virtual 248